A2 BIOLOGY

Michael Kent

OXFORD

UNIVERSITY PRESS

OXFORD
UNIVERSITY PRESS

Great Clarendon Street, Oxford OX2 6DP

Oxford University Press is a department of the University of Oxford.
It furthers the University's objective of excellence in research, scholarship,
and education by publishing worldwide in

Oxford   New York

Auckland   Cape Town   Dar es Salaam   Hong Kong   Karachi
Kuala Lumpur   Madrid   Melbourne   Mexico City   Nairobi
New Delhi   Shanghai   Taipei   Toronto

With offices in

Argentina   Austria   Brazil   Chile   Czech Republic   France   Greece
Guatemala   Hungary   Italy   Japan   Poland   Portugal   Singapore
South Korea   Switzerland   Thailand   Turkey   Ukraine   Vietnam

Oxford is a registered trade mark of Oxford University Press
in the UK and in certain other countries

© Michael Kent 2009

The moral rights of the author have been asserted

Database right Oxford University Press (maker)

First published 2009

Acknowledgements
'The Five Freedoms' as defined by the Farm Animal Welfare Council (page 69)
is Crown copyright material and is reproduced under the terms of the Click-Use
Licence Number C2007001380 with the permission of the Controller of Her Majesty's
Stationery Office and the Queen's Printer for Scotland.

We are grateful for permission to reproduce the following photographs;

Pages 16/17 Jeff Hunter/Gettyimages; P18 Bruce Coleman; P21 Barrie Watts/Photolibrary; P23 PD
images; P24 Martin f. Chillmaid/Photographers Direct; P27 Spectrumphotofile; P37 Crack Palinggi/
REUTERS; P48 Francesco Ridolfi/Istockphoto; P61 Radugroza/Photographers Direct; P64 orkneypics /
Alamy; P65 Nigel Cattlin/Photolibrary; P67 Keith Naylor/Shutterstock; P68(T) Jean Chung /
OnAsia/Jupiter image; P68(B) Gene Krebs/Istockphoto; P71 Ankevanwyk/Dreamstime; P75 Bruce
Coleman; P77 Freezingpictures/Dreamstime; P81 Corbis; P85(T), Mary Plage/Photolibrary; P85(MB)
blickwinkel / Alamy; P87 Hulton Archive/Stringer/Gettyimages; P95 Michael Dodd/Photographers
Direct; P98(T) Bruce Coleman; P98 (L) Bruce Coleman; P100(T) Bruce Coleman; P100(L) Bruce
Coleman; P102 Darwin Dale/Science Photo Library; P103 Scientifica/Visuals Unlimited / Alamy;
P110(T) www.offwell.info; P110(L) www.offwell.info; P114(L) Grand Tour/Corbis; P114R Robert
Read/Alamy; P115 Bruce Coleman; P118 Andrew Darrington/Alamy; P119(T) Sandra Ford/Fotolia;
P119(L) Mike Dunning/Getty Images; P122 Steve Horrell/Science Photo Library; P124(T)
xtremesailing/Istockphoto; P124(L) Femorale.com; P125 Dr Gerald Boalch/Marine Biological
Association; P126/127 Mira/Alamy; P128(T/L) Stephen Dalton/NHPA; P130 DoctorKan/Istockphoto;
P133 Science Photo Library; P145 blickwinkel/Alamy; P152 Omikron/Science Photo Library; P157(T)
Omikron/Science Photo Library; P157(L) Robert Cumming/Shutterstock; P162(L) Daniel76
/Dreamstime; P162R NASA; P170 Science Photo Library; P172; Associated press; P187 Science Photo
Library; P194 Lewis Linda/Photolibrary; P205 Michelle Del Guercio/Photolibrary; P210 Philippe
Psaila/Science Photo Library; P214 Simon Fraser/RVI, Newcastle-Upon-Tyne/ Science Photo Library;
P218 Lester Lefkowitz/CORBIS; P223 Nigel J. Dennis; Gallo Images/CORBIS; P225 Nigel Cattlin; P227
Dr. David Phillips/Gettyimages; P228 Purestock /Jupiterimages.

In a few cases we have been unable to trace the copyright holder prior to publication. If notified
the publishers will be pleased to amend the acknowledgements in any future edition.

British Library Cataloguing in Publication Data

Data available

ISBN: 978-0-19-915270-4

10 9 8 7 6 5 4 3 2 1

Printed in Singapore by KHL Printing Co Pte, Ltd.

Paper used in the production of this book is a natural,
recyclable product made from wood grown in sustainable forests.
The manufacturing process conforms to the environmental
regulations of the country of origin.

Welcome

If you are just about to start your A2 Biology studies having successfully completed your AS assessments, welcome. You will find that much of what you learnt for AS Biology will prove invaluable for your A2 studies. Your understanding of general biological principles, especially those related to cell biology and biological chemistry, will provide a good foundation for your A2 studies.

A2 Biology for AQA covers the subject content of Unit 4 and Unit 5. Unit 4 is entitled **Populations and environment** and contains the following three topics:

- how living organisms form ecosystems through which energy is transferred and chemical elements are cycled
- how human activity affects the ecological balance in a variety of ways
- how genetic variation and isolation may lead to the formation of new species

Unit 5 is called **Control in cells and organisms** and has the following three topics:

- stimulus and response – the biology of the nervous and endocrine systems
- homeostasis and the maintenance of a constant internal environment
- genes and genetic expression

Within these units, you will be covering some of the key biological issues of the twenty-first century. The aims of *A2 Biology for AQA* are to equip you with the factual knowledge to understand these issues, to help you develop the intellectual skills to use that biological knowledge wisely, and, above all, to help you enjoy and participate fully in this exciting and challenging subject.

Michael Kent
Wadebridge, Cornwall 2008

Contents

Unit 5: Control in cells and organisms

Chapter 10 NUCLEIC ACIDS, THE GENETIC CODE, AND MUTATIONS

Chapter 11 APPLYING rDNA TECHNOLOGY

Chapter 12 CONTROL IN CELLS AND ORGANISMS REVIEW

Introduction

A2 Biology for AQA has been written specifically for students taking the new specifications in GCE Advanced level Biology for the Assessment and Qualifications Alliance, the largest of the three English examination boards.

The main part of the book covers all of the subject content you will need to know for the assessments of Units 4 and 5. For easy access, the topics are covered in a series of double-page spreads that follow the order given in the Specification. Nevertheless, the spreads are self-contained and can be studied in a different order than that given in this book. However, you should take note of any prior knowledge required for an understanding of the concepts described in a particular double-page spread. Prior knowledge, where required, is listed in a box in the left hand margin of a spread.

As well as this prior knowledge, it is assumed that all students taking A2 Biology have completed the AS assessments successfully and have a good background in biology at GCSE level. It is also assumed that students have some knowledge of basic physics and chemistry.

In addition to the boxed list of prior knowledge, each main double-page spread contains:

- a list of objectives which relate directly to the AQA Specification

- a body of text that covers the subject content, with key terms identified in **bold** print and defined in the *Glossary* at the back of the book (the *Glossary* also includes terms from the AS course to help you with revision)

- check your understanding: this contain two types of questions. The first are relatively straightforward and answers for most of these can be found in the in the text or figures. The second type, presented under a dashed rule, are more difficult and are designed to 'stretch and challenge' you. Answers for all the questions can be found in the e-book inside the back cover, and in the teacher's resources.

Although the flexible design of the spreads means that you can easily follow your own route through the course, it is assumed that all of Unit 4 is studied before starting Unit 5.

How Science Works is an integral part of the new GCE AS and A level Biology Specifications; it is also an integral part of this book. As well as the spreads containing *How Science Works* synoptic questions at the end of Unit 4, most of the other spreads contain elements of *How Science Works*, and a specific example of each of the criteria detailed in the Specification is given in *How Science Works* on pages 10–11.

A2 Biology for AQA contains several double-page spreads dedicated to investigations and their interpretation. They have been designed to help students prepare for practical work, especially the investigations that form an important part of Unit 5. These are explained in more detail on page 9.

AQA A2 Biology

AQA is the largest of the three English examination boards. It sets and marks examinations, including GCSEs and A levels. AQA has a very useful website at **www.aqa.org.uk**. You can download examination papers and mark schemes from there to help you with your studies.

The Biology Specification

AQA's *AS and A Level Biology Specification* builds on its earlier biology course, and aims to encourage you to:

- study biology in a modern context
- become enthusiastic about biology
- show that you can bring different ideas together
- develop your practical skills and data analysis skills
- appreciate *how science works* and its importance in the wider world

You will learn about the biology behind contemporary issues such global warming, the use of pesticides, gene therapy, and recombinant DNA technology. You will also explore aspects of cloning, cancer, and how biologists are helping develop new techniques in the fight against inherited and non-inherited diseases.

The A2 Biology course is divided into three units. Units 4 and 5 cover biological knowledge and understanding, and Unit 6 covers practical skills. But you can also expect to be assessed on your performance in your class practicals that support the biological ideas in Units 4 and 5. Both units build on your AS studies.

Unit 4: Populations and environment

Unit 4 should be studied first. The Specification has been written so that you can complete this unit during the autumn term of Year 13. This is so you can enter for the Unit 4 examination in January. But you do not have to enter then. You can enter for the Unit 4 examination in June instead.

Unit 5: Control in cells and organisms

Unit 5 includes synoptic assessments (see next page) of the biological principles covered in all the previous units. It should therefore be studied after Unit 4.

Unit	Form of assessment	Length of paper	Type of questions	% of total A level mark
4	written paper	1 hour 30 minutes	6–9 short-answer questions plus two longer questions involving continuous prose and *How Science Works*	16.7
5	written paper	2 hours 15 minutes	8–10 short-answer questions plus two longer questions (a data-handling question and a synoptic essay)	23.3

Unit 6: Investigative and practical skills

Unit 6 assesses your practical skills and data-analysis skills. The unit addresses your ability to

- demonstrate and describe ethical, safe, and skilful practical techniques, selecting appropriate qualitative and quantitative methods
- make, record, and communicate reliable and valid observations and measurements with appropriate precision and accuracy
- analyse, interpret, explain, and evaluate the methodology, results, and impact of your own and others' experimental and investigatory activities in a variety of ways

A2 Biology for AQA does not include detailed instructions for practical work. However, it does include plenty of opportunities to develop and practise data-handling skills (for example, in spreads 12.07 and 12.08, Data handling: synoptic questions III and IV).

As in AS Biology, there are two routes for the assessment of practical skills:

- Route T with a Practical Skills Assessment (**PSA**) and an Investigative Skills Assignment (**ISA**) which is set by AQA but marked in your school or college
- Route X with a Practical Skills Verification (**PSV**) and an Externally Marked Practical Assignment (**EMPA**)

For the PSA and PSV, your teacher will assess your practical skills during several laboratory practicals.

Synoptic questions and Stretch and Challenge

Synoptic questions are a feature of AQA A2 Biology; they require you to make connections between different areas of biology. Synoptic assessment is introduced in spread 6.10, and there are synoptic questions on spreads 6.11–6.12 (focusing on How Science Works) and spreads 12.07–12.08 (focusing on data handling). Another aspect of synoptic assessment is the essay question, and spread 12.09 gives some background to tackling this type of question.

Stretch and Challenge questions are designed to stretch you and as well as those that occur throughout the book (presented under a dashed rule), there are also more Stretch and Challenge questions on spreads 6.05–6.09 and 12.02–12.06.

The examinations

The Specifications set out exactly the knowledge and understanding you need in order to be successful in your A2 Biology studies. In *A2 Biology for AQA*, each double-page spread contains learning objectives based on the Specification. Look carefully at what you need to be able to do. It might be something relatively straightforward, like recalling a fact or a definition. But often the objectives are designed to encourage you to think more deeply about the subject. You might need to understand something and be able to describe or explain it. Or you may be given information to interpret. In some cases, examination questions will be posed in an unfamiliar context. Don't be put off by this. You should be able to answer these questions by using the information provided and by applying the biological principles that you have learnt through the course.

How Science Works is a key component of the AQA A2 Biology specifications. Its inclusion emphasizes that biology is more than a body of scientific knowledge. It is also a way of looking at the world.

Listed below are aspects of *How Science Works* in which you are expected to become proficient. The list also shows where to find one example of each of the criteria in *A2 Biology for AQA*, although most spreads contain one or more aspects of *How Science Works*, with special accounts being given in *Science@work* sections. In addition, spreads 6.11 and 6.12 contain synoptic questions on *How Science Works*.

Specification summary

A Use theories, models, and ideas to develop and modify scientific explanations

The Hardy–Weinberg principle predicts that allele frequencies do not change from generation to generation (spread 5.06).

B Use knowledge and understanding to pose scientific questions, define scientific problems, present scientific arguments and scientific ideas

You will present logical scientific argument in the synoptic essay in Unit 5 (spread 12.09).

C Use appropriate methodology, including ICT, to answer scientific questions and solve scientific problems

The importance of ICT is considered in relation to the automation of DNA base sequencing (spread 11.07).

D Carry out experimental and investigative activities, including appropriate risk management, in a range of contexts

Risk assessment and management are a central part of the planning of fieldwork (spread 1.05).

E Analyse and interpret data to provide evidence, recognizing correlations and causal relationships

You will analyse and interpret data relating to the distribution of organisms, recognizing correlations and causal relationships (spread 6.04).

F Evaluate methodology, evidence, and data, and resolve conflicting evidence

Global warming gives opportunities for evaluating evidence and data and considering conflicting evidence (spread 4.01).

G Appreciate the tentative nature of scientific knowledge

Data relating to populations allow you to appreciate the tentative nature of conclusions that may be drawn (spread 1.10).

H Communicate information and ideas in appropriate ways using appropriate terminology

Writing extended prose in the essay questions gives an opportunity to demonstrate your communication skills (spread 12.09).

I Consider applications and implications of science and appreciate their associated benefits and risks

You consider ethical issues arising from the enhancement of animal and plant productivity (spread 3.06).

J Consider ethical issues in the treatment of humans, other organisms, and the environment

Ethical issues are central to the consideration of cancer preventions, treatments, and cures (spread 10.13).

K Appreciate the role of the scientific community in validating new knowledge and ensuring integrity

The Asilomar Conference and the use of rDNA technology illustrates the important role of the scientific community (spread 11.04).

L Appreciate the ways in which society uses science to inform decision making

Reclamation of heathland relies on science to inform decision making (spread 6.11).

A2 Biology for AQA covers all the theory essential for success in the AQA examinations. However, the following two tables of Internet resources have been compiled to help students delve more deeply into biological topics of particular interest to them. Table 1 lists some sites that cover a wide range of biological topics and which are relevant to both AS and A2 Biology. Table 2 gives other sites that relate to specific A2 Biology study areas. Although only relatively few sites are listed, each provides links to many other useful Internet resources.

Like the Internet resources table in *AS Biology for AQA*, these tables come with a 'health warning'. Although at the time of writing the sites were available via the web addresses given, web addresses are notoriously changeable. If you do fail to obtain access via the address, usually a simple search by using key words in a search engine such as *Google* will get you to the desired destination. For example, the website for the Institute of Biology can be obtained by writing 'Institute of Biology' in the *Google* search box and pressing 'enter'.

Topic	Organization	Web address
Most areas	Institute of Biology	http://www.iob.org/
Most areas	Access Excellence	http://www.accessexcellence.org/
Most areas	University of Central Lancashire	http://www.biology4all.com/
Most areas	Biotopics	http://www.biotopics.co.uk/index.html
Nobel Prizewinners in physiology or medicine	Nobel Prize	http://nobelprize.org/nobel_prizes/medicine/
Cell biology	Cells Alive	http://www.cellsalive.com/cells/3dcell.htm
Biochemistry	Chemistry Society	http://www.chemsoc.org/networks/learnnet/cfb/index2.htm
Health and disease	World Health Organization	http://www.who.int/en/
Plant biology	Science and Plants for Schools	http://www-saps.plantsci.cam.ac.uk/
Natural history	BBC science and nature	http://www.bbc.co.uk/sn/
Natural history	Natural History Museum	http://internt.nhm.ac.uk/eb/index.shtml

Table 1 General websites for A2 Biology

Topic	Organization	Web address
Ecology fieldwork	British Ecological Society	http://www.britishecologicalsociety.org/
Ecology fieldwork	Fieldwork Knowledge Library	http://www.fieldworklib.org/
Fieldwork (including identification and statistics)	Field Studies Council	http://www.field-studies-council.org/
Mitochondria and respiration	Mitochondria research	http://www.mitochondrial.net/
Farming and wildlife	Farmwildlife	http://www.farmwildlife.info/
Climate change	Defra	http://www.defra.gov.uk/environment/climatechange/index.htm
Climate change	Intergovernmental Panel on Climate Change	http://www.ipcc.ch/
Population statistics	National Statistics online	http://www.statistics.gov.uk/glance/#population
Nervous system	Neuroscience for kids	http://faculty.washington.edu/chudler/neurok.html
Diabetes	Diabetes UK	http://www.diabetes.org.uk/
Evolution and natural selection	Public Broadcasting Service	http://www.pbs.org/wgbh/evolution/
Evolution and natural selection	University of Berkeley	http://evolution.berkeley.edu/
Bioethics	Nuffield Council on Bioethics	http://www.nuffieldbioethics.org/
Biotechnology (includes DNA, enzymes, and genetic engineering)	National Centre for Biotechnology Education	http://www.ncbe.reading.ac.uk/ncbe/protocols/menu.html
DNA structure and function	DNA Interactive	http://www.dnai.org/
DNA structure and function	Cold Spring Harbor Laboratory	http://www.dnaftb.org/dnaftb/
Genome	The Wellcome Trust	http://genome.wellcome.ac.uk/
Cancer causes, prevention, and treatment	Cancer Research UK	http://info.cancerresearchuk.org/youthandschools/
Stem cells	US National Institute of Health	http://stemcells.nih.gov/
Cloning	University of Utah	http://learn.genetics.utah.edu/units/cloning/
Polymerase chain reaction	Dolan DNA learning centre	http://www.dnalc.org/ddnalc/resources/pcr.html
DNA fingerprinting	University of Washington	http://protist.biology.washington.edu/fingerprint/dnaintro.html

Table 2 Subject-specific websites for A2 Biology

Preparing for assessment

Units 4 and 5

The subject content of Units 4 and 5 is externally assessed through question papers that require written answers in spaces provided. The double-page spreads in *A2 Biology for AQA* cover all the content listed in the Specification. One of your main tasks during your A2 level course will be to learn and understand the required subject content, and apply your knowledge and understanding to answer questions.

Always check the current Specification for any amendments.

Unit 6

A2 Biology for AQA does not give detailed instructions for carrying out practical work; these are provided in the support material. However, it does provide opportunities within spreads for developing other investigative skills that form the part of the assessment for Unit 6. These spreads are:

Unit	Spread(s)	Topic (s)
4	1.03–1.04	Investigating populations
4	1.06	Analysing and interpreting data
4	6.02	Interpretation: photosynthesis and limiting factors
4	6.03	Investigating cellular respiration
4	6.04	Fieldwork investigations
4	6.11–6.12	How Science Works: synoptic questions I and II
5	10.04	Interpreting nucleic acid data
5	12.07–12.08	Data handling: synoptic questions III and IV

Use the *Objectives*

Use the objectives for each spread to help you organize your learning. They relate directly to the subject content in the Specification. They tell you what you should be able to do when you have finished studying the topics covered. Before the assessments in January or June, check that you can carry out all the stated objectives for the units you are being assessed in. Note that in addition to being able to describe and explain key concepts, you will be expected to apply them to new and unfamiliar situations.

Use the *Check your understanding*

Most of these questions are designed to help you check that you can recall and comprehend key concepts covered in each spread. However, the last question is often more demanding and is designed to stretch and challenge you. These questions are separated from the others by a dashed rule. They may be posed in an unfamiliar context, or the answer may require you to use knowledge and understanding about biological principles that are covered in earlier parts of the Specification.

Immediately after you have finished a spread, read the questions and write down your answers. Then check through the text to make sure that your answers are correct. If you cannot find the answer, or if you do not understand the question, make a note of it and discuss it with your lecturer or teacher at the earliest opportunity.

Answers to all the questions are provided on the CD copy of this book (found inside the back cover) and in the *Planning and Resource Pack* for teachers/lecturers.

Use the *Glossary*

One of the challenges of studying biology is getting to grips with the wealth of biological terms. It can be a bit daunting when first confronted with a lot of new words. From the start of your studies, you are advised to build up your biological vocabulary as you go along. For each topic, you might like to compile your own list of key terms and use the glossary or a biology dictionary to define them. In assessments, questions will include terms mentioned in the Specification; it will be assumed you understand the meaning of these terms. And in your written answers you will be expected to make appropriate use of biological vocabulary.

The glossary of biological terms at the back of this book includes terms introduced at both AS and A2 levels. You can use it for reference and also to structure your revision. In addition to a glossary of biological terms, *A2 Biology for AQA* also has a glossary of instructional terms used in the questions posed in the formal assessments.

Use past examination papers

The AQA website provides links to specimen question papers and mark schemes for all the units. This is a valuable resource and you are strongly recommended to make full use of it. After revising each unit, print out the relevant specimen questions, attempt to answer them, and go through the mark scheme to check your answers.

Note that for Units 4 and 5 of A2 Biology, in addition to the quality of your biological knowledge and understanding, the quality of your written work will also be assessed in the longer questions that require continuous prose. You will be expected to

- produce text that is legible with grammar, spelling, and punctuation sufficiently accurate to make its meaning clear
- adopt a writing style that is appropriate for a scientific subject, and
- organize your material clearly and coherently, using biological vocabulary when appropriate

Discuss your work

If you have the opportunity, talk to other students about the biology you are studying. If you are a member of a class, you might perhaps want to form a small group (at least three, but no more than five) that meets to discuss major topics. Discussions can be very useful. In order to express yourselves in your own words, you will need to clarify your thoughts. It's important that members of the group are encouraged to question anything that is not clear. In this way, you will develop a better understanding of biology.

Unit 4: Living organisms form structured communities within dynamic but essentially stable ecosystems through which energy is transferred and chemical elements are cycled. Humans are part of the ecological balance and their activities affect it both directly and indirectly. Consideration of these effects underpins the content of this unit and should lead to an understanding that sustainability of resources depends on effective management of the conflict between human needs and conservation.

It is expected that candidates will carry out fieldwork involving the collection of quantitative data from at least one habitat and will apply elementary statistical analysis to the results.

AQA Approved Specification (July 2007)

Populations and environment

Coral reef ecosystems have an amazingly high biodiversity, but they are very fragile. Global warming and pollution threaten them. Their continued existence will depend on effective management to resolve the conflicts between human needs and biological conservation.

OBJECTIVES

By the end of this spread you should be able to

- define and identify the main components of an ecosystem
- recall that a population refers to all the organisms of one species in a habitat
- recall that populations of different species form a community

Prior knowledge

- the species concept (*AS Biology for AQA* spread 17.01)

Savannah grassland, one of about 40 biomes in the biosphere

Where organisms live

The Earth is the only part of the known Universe to contain living organisms. Even here, most active organisms are confined to a thin surface layer no more than 28 km deep. This living part of the Earth is called the **biosphere**. It consists of streams, rivers, lakes and seas, land down to a soil depth of a few metres, and the atmosphere up to an altitude of a few kilometres.

Biologists refer to the largest subunits of the biosphere as **biomes**. Each biome is defined in terms of its living organisms and their interactions with the environment. There are about 40 biomes. Some are aquatic (occur in water) whereas others are terrestrial (occur on land). The locations of the major terrestrial biomes, such as tropical rainforest, deserts, and tundra, correspond broadly with climatic regions (temperature and precipitation being the most important climatic factors).

Within each biome, there are different ecosystems. These are generally regarded as the basic functional unit of ecology. An **ecosystem** contains a dynamic but essentially self-sustaining and stable community of organisms interacting with each other and with their non-living surroundings. A good example of an ecosystem is a pond.

The place within an ecosystem in which an individual organism lives is called its **habitat** or, if it is very small, its **microhabitat**. Most ecosystems contain several habitats.

Defining a population and a community

A **population** is a group of individuals of the same species that live in a particular area at any one time. A **community** consists of a group of populations of different interacting species living in an area at one time.

Studying interactions between organisms

Ecology or **environmental biology** is the scientific study of how the interactions of individual organisms with their physical and biological environments determine the distribution and abundance of populations (and species). Those who study the interactions are called ecologists.

Ecology is a relatively new science. The term comes from two Greek words (*oikos*, meaning 'house' and *logos*, meaning 'discourse'). It was first used by the German biologist Ernst Haeckel in 1866. To many non-scientists, the word 'ecology' has a different meaning. It conjures up images of activists campaigning to preserve environmental quality and promoting ways of living in harmony with nature. Although the scientific use of the term ecology does not imply environmental concern, the activities of environmental conservationists draw heavily on the knowledge of ecologists.

An organism's total surroundings, including living, dead, and non-living components, make up its **environment**. Environmental factors that result from the activities of living organisms are called **biotic factors**. These include feeding relationships and disease.

A typical community has

- **autotrophs** ('self feeders'; also known as **primary producers** or simply producers, for example green plants) that manufacture their own food
- **herbivores (primary consumers)** that eat autotrophs
- **secondary consumers** that eat herbivores
- **tertiary consumers** that eat secondary consumers
- **decomposers**, mainly bacteria and fungi, that obtain nutrients by breaking down the remains of dead organisms

Environmental factors resulting from the non-living part of an ecosystem are called **abiotic factors**. These include

- **climatic factors** such as light, temperature, water availability, and wind
- **edaphic factors** associated with the soil, such as texture, pH, temperature, and organic content
- **topographic factors** such as the angle and aspect of a slope

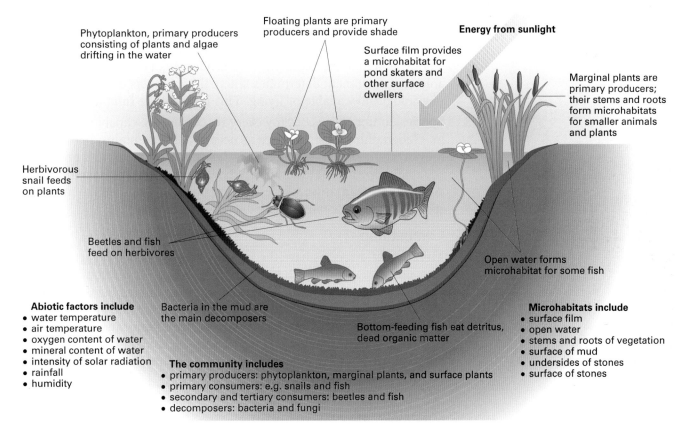

The main components of a pond ecosystem (with acknowledgement to Chris Clegg)

Check your understanding

1 Define an ecosystem.
2 Distinguish between a community and a population.
3 Explain why it is more accurate to define the biosphere as the global ecosystem rather than the global community.

OBJECTIVES

By the end of this spread you should be able to

- discuss the concept of the ecological niche
- explain why the niche a species occupies within a habitat is governed by adaptation to both biotic and abiotic conditions

Prior knowledge

- populations and communities (1.01)

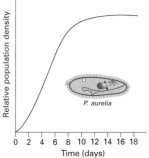

(a) *P. aurelia* grown alone

Relative population density

P. aurelia

0 2 4 6 8 10 12 14 16 18
Time (days)

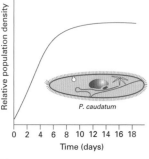

(b) *P. caudatum* grown alone

Relative population density

P. caudatum

0 2 4 6 8 10 12 14 16 18
Time (days)

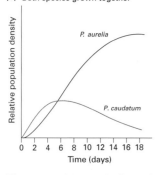

(c) Both species grown together

Relative population density

P. aurelia

P. caudatum

0 2 4 6 8 10 12 14 16 18
Time (days)

The competitive exclusion principle: Gause's experiment

Ecological niche defined

The **ecological niche** of a species is the sum of all the physical, chemical, and biological factors it needs to survive and reproduce. This includes its temperature requirements, all the resources it exploits (such as food and water), and the space it occupies. Time also forms part of the ecological niche – members of two similar species may appear to have the same niche because they live in the same habitat and use the same resources, but because they are active at different times of the day their niches are actually separate.

The ecological niche of a species depends not only on where its members live but also on what they do. The analogy is sometimes made that if a species' habitat is its 'address', the niche is that habitat plus the species' 'profession'. In other words, a species' niche is its ecological role or how it fits into an ecosystem.

Fundamental niches and realized niches

The ecological niche of a species reflects the complete set of conditions, abiotic as well as biotic, to which it is adapted. The physical, abiotic environment determines where the members of a species are physiologically capable of living. This is called the **fundamental niche**.

Species do not always occupy every place they are capable of living in. Geographical barriers, such as oceans and mountains, may have prevented them from moving from their place of origin to a potentially suitable environment, or they may have been excluded from it by competition with other species. The part of the fundamental niche that a species *actually* occupies is called its **realized niche**.

Competition

Members of a species have the same ecological niche. Those living in the same area, therefore, have to share resources. If a resource is limited, this results in competition. Competition between members of the same species is called **intraspecific competition**. It limits the growth of a population and has led to the evolution of special forms of behaviour such as ritualized fighting and territoriality. Competition between two different species for limited resources is called **interspecific competition**.

In 1934 G. F. Gause, a Russian ecologist, studied competition between two species of single-celled organisms, *Paramecium aurelia* and *P. caudatum*. He cultured the two species separately under controlled conditions, supplying a constant amount of food each day. Both populations grew rapidly until they reached a maximum density after which the density leveled off. Gause then cultured the two species together. The population of *P. aurelia* grew more slowly than when on its own, but the population of *P. caudatum* was eliminated.

Gause concluded that the two species were competing for the food and that *P. aurelia* had the competitive edge, feeding more efficiently and reproducing more rapidly than *P. caudatum*. Gause formulated the **competitive exclusion principle**. This states that species with the same requirements cannot live permanently together in the same habitat because one is bound to be a stronger competitor than the other.

Investigating barnacle niches

Marine biologist Joseph H. Connell tested the competitive exclusion principle on two species of acorn barnacle, *Semibalanus balanoides* and *Chthalamus stellatus*, growing on an exposed rocky shore on the Isle of Cumbrae in Scotland.

Populations of adult *S. balanoides* are most densely concentrated from the mid shore down to the low water spring tide mark. Adult *C. stellatus* are confined to the upper shore but both species have planktonic larvae (that is, they drift in the moving waters of the sea). These larvae settle on hard surfaces from the mid shore up to the mean high water neap tide mark.

Connell investigated the effect of *S. balanoides* on *C. stellatus*, by removing all the young *S. balanoides* that were touching or surrounding *C. stellatus*. He did this after the period of larval settlement in experimental plots at several different tidal levels. At the same time, he marked out adjacent control (untouched) plots. Over the next year, he found that populations of *C. stellatus* were able to survive much better in the experimental plots than in the control plots. Direct observations showed that *S. balanoides* was faster growing and smothered, crushed, or undercut *C. stellatus*. Connell concluded that the realized niche of *C. stellatus* was much smaller than its fundamental niche, and that this was due to competition with *S. balanoides*.

In a separate experiment, Connell moved rocks encrusted with *S. balanoides* from the mid tide level to the high water mark. He discovered that *S. balanoides* failed to survive because it is unable to tolerate the more extreme physical conditions at high tide levels. He concluded that its fundamental niche and realized niche are similar.

Acorn barnacles are crustaceans. They have been likened to modified shrimps lying on their backs and enclosed within a series of calcareous plates into which they can be withdrawn for protection. They have six pairs of two-branched legs. By kicking these they filter feed by catching plankton.

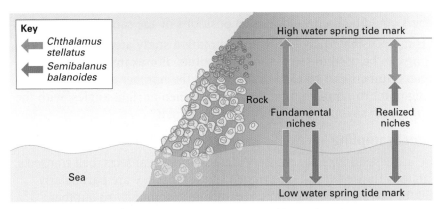

Key
- Chthalamus stellatus
- Semibalanus balanoides

High water spring tide mark

Rock

Fundamental niches

Realized niches

Sea

Low water spring tide mark

The realized and fundamental niches of two species of barnacle

Check your understanding

1 Define ecological niche.

2 Distinguish between the fundamental niche and realized niche of a species.

3 In Connell's test of the competitive exclusion principle, name the resource the barnacles competed for.

4 Suggest why Connell's field experiment shows that competition can occur, but it does not show that it necessarily *does* occur in normal situations.

OBJECTIVES

By the end of this spread you should be able to

- critically appreciate some of the ways in which the numbers and distributions of organisms may be investigated

- use counting along transects and random sampling with quadrats to obtain quantitative data

Prior knowledge

- populations and communities (1.01)

Representative sampling

Most ecological investigations involve collecting data about the abundance of organisms and their pattern of distribution. Unless the study area is small and it is possible to count all the individuals within a given population, it is necessary to take representative samples. A truly representative sample should be

- taken at random to ensure that every member of the population has an equal chance of being selected
- unbiased by the sampling equipment or procedure used
- large enough to provide a sufficiently precise estimate of the population

Random sampling

Although there are many ways to sample randomly, they are all designed to ensure that individual organisms at each point within the study area have an equal chance of being sampled. For example, if a study site is a conveniently flat and square area measuring 10 m by 10 m, the area could be marked by lengths of string laid out along two edges at right angles to each other. The string could be marked off at 1 m intervals and these intervals numbered 1 to 10. The numbered marks along the vertical axis and the horizontal axis of the marked out square could then be used to define coordinates of any sampling point within the study area. Pairs of numbers between 1 and 10 drawn at random from a table of random numbers could be used to locate the positions of the sampling points.

If it is not possible to lay out strings to mark a study area, random walking can be used to select sampling points. For example, a pair of random numbers can be used to determine the number of paces from starting point to sample point, first forward then at right angles, with the spin of a coin saying whether to turn left or right.

Transect sampling

There are several types of transect including line transects, belt transects, continuous transects, and interrupted transects. They are particularly useful for investigating the transition of one community to another, particularly when the communities occur in a linear sequence,

Transect sampling

(a) a line transect

(b) a continuous belt transect

(c) an interrupted belt transect

In the line transect, all individuals touching the line are recorded.

In the belt transect, all individuals occurring within the quadrats placed in the numbered locations are recorded.

for example from the low tide mark up to the high tide mark on a rocky seashore.

Typically when a **line transect** is used, a tape or string is laid along the ground in a straight line between two poles to indicate the limits of the transect. Sampling is confined strictly to recording all individual organisms actually touching the line. A single line is unlikely to be representative and it is usually necessary to take a series of lines.

A **belt transect** is a strip of chosen width through the habitat. It is made by laying two parallel line transects, usually 0.5 m or 1 m apart, between which all individuals are recorded.

A **continuous transect** is a line transect or belt transect in which the whole line or belt is sampled. In an **interrupted transect**, samples are taken at points along the line or belt transect. The points are usually at regularly spaced horizontal or vertical intervals. Vertical intervals are usually more appropriate when studying a distribution pattern that is likely to be related to height, for example up a rocky shore or mountain.

Frame quadrat

Once a sampling point has been selected at random, you can use a frame quadrat to take samples and estimate the **population density** of plants and sedentary or sessile animals (that is, the number of individuals per unit area). **Sedentary animals**, such as sea anemones and snails, move only slowly or infrequently. **Sessile animals**, such as adult barnacles, are immobile. A **frame quadrat** is usually a metal, plastic, or wooden frame that forms a known area, such as 0.25 m^2 or 1 m^2.

Frame quadrats can be of any regular shape. The most common are square, but circular and even hexagonal frames can be used. Whatever the shape, the frame quadrat must be of a known area.

To estimate population density, all individuals present within the chosen quadrats are counted and their numbers recorded. It is important to be consistent, for example, by recording all individuals completely or partially visible. By convention, when using a square frame quadrat you count only organisms that are completely within the quadrat or touching its top or left-hand side.

The density of each species can be expressed as an average per square metre, or the number counted can be extrapolated to give the total number for the whole study area.

To obtain an accurate estimate of population size, a series of quadrat samples are usually required. You can determine the number of samples to be taken using a **running mean**. A mean is calculated after each new quadrat sample. The mean values will fluctuate each time, but will gradually settle down until a point is reached where taking another sample has only a very small effect on the mean. When this happens, the sample size is adequate. Taking more quadrat samples would be unrewarding and a waste of time.

Sampling with frame quadrats along a transect line

Check your understanding

1 Distinguish between a line transect and a belt transect.

2 What is a frame quadrat?

3 How should quadrats be positioned within a sampling area?

- - - - - - - - - - - - -

4 When using a square frame quadrat, the edge effects are minimized by counting only those organisms clearly within the perimeter and those touching the top and left-hand sides (including the top left-hand corner). Suggest a sampling protocol for reducing the edge effects when using a circular quadrat.

By the end of this spread you should be able to

- use percentage cover and frequency as measures of abundance
- use the mark and recapture technique to estimate the population size of mobile species

Prior knowledge

- random sampling using quadrats (1.03)

A point quadrat can be used to measure percentage cover.

Percentage cover and frequency of occurrence

Percentage cover, the proportion of an area covered by the growth form of a plant or by a sedentary animal such as a barnacle or mussel, can be estimated using a frame quadrat divided into a 10 × 10 grid with string.

For each species being investigated, you record the number of squares fully occupied, partly occupied, and unoccupied. Each fully occupied square represents a 10% cover. You then need to estimate how many full squares all the partly occupied squares represent so that you can calculate the total percentage cover.

A frame quadrat divided into a 10 × 10 grid can also be used to estimate the **frequency of occurrence** of a species. You record 'present' or 'absent' for each species being investigated for each of the 100 squares. The 'present' score for each species can be used as an index of frequency or in an estimate of abundance. If the 'present' score is expressed as a percentage of the total number of squares, the result will be a **percentage frequency**. As with other forms of quadrat sampling, individuals overlapping the edges of squares must be treated consistently. One common method, as described in spread 1.03, is to include an individual if it touches the top or left-hand sides of a square and to exclude it if it touches the bottom or right-hand sides.

A **point quadrat** (or **pin frame**) is designed for sampling stationary organisms, usually plants. It consists of a frame of adjustable height and a cross bar bearing a number of holes (usually 10) through which a 'pin' such as a knitting needle can be passed. The point quadrat is used to record the presence or absence of organisms at a number of points. You pass the needle through each hole in turn and record each species touched by the tip of the needle as it is lowered to the ground.

Point quadrats are usually used to measure percentage cover. With a 10-needle point quadrat, each 'hit' with a needle represents a 10% cover.

$$\text{Percentage cover} = \frac{\text{hits} \times 100}{\text{hits} + \text{misses}}$$

The recording can be confined to organisms that form the canopy and shade other organisms, or it can be made for all organisms touched by each needle. In this case, the percentage cover may well exceed 100% because several organisms will overlap and shade the same patch of ground.

Abundance scales

When it's not possible or not necessary to measure abundance in absolute, quantitative terms, frame quadrats and point quadrats can be used to estimate abundance in relative, qualitative terms using one of several **abundance scales**. In the **ACFOR scale** each species present at a sample site is assigned to one of the following five categories:

A = abundant   O = occasional

C = common   R = rare

F = frequent

When comparing two or more sites, it might be necessary to include an additional category 'N' to represent 'none present'.

ACFOR scales use arbitrary values for the five categories. As a result they can be highly subjective – two investigators may use different criteria. For example, one investigator may regard a species as 'abundant' if its frequency exceeds 90% whereas another may regard the species as being abundant if it has a frequency over 80%. Also, the relative abundances of conspicuous organisms tend to be overestimated and those of inconspicuous ones underestimated. However, ACFOR scales can be made less subjective if the criteria for assigning species to each category are standardized.

The mark-release-recapture technique

The **mark-release-recapture technique** is used to estimate the population size of mobile organisms such as fish. It involves capturing the organisms, marking them in a way that causes no harm, and returning them to their natural environment. After allowing sufficient time for the marked individuals to mix thoroughly with other members of the population, a second sample is taken and the number of marked and unmarked individuals recorded. The population size is estimated using the following relationship:

Estimate of population size (N) equals the product of the number of organisms caught and marked in the first sample (S_1) and the total number of organisms in second sample (S_2) divided by the number of marked individuals in the second sample (recaptures, R), that is,

$$N = \frac{S_1 S_2}{R}$$

This population estimate is known as the **Lincoln index**. It is based on the assumption that, for example, if a second sample contains 10% of the original marked individuals then these represent 10% of the total population. Reliance on this assumption depends on certain conditions being met:

1 Organisms are captured randomly; there is no bias towards a particular group.

2 The mark is not lost between release and recapture.

3 Marking does not hinder the movement of organisms or harm them in any way.

4 For the second sample, marked and unmarked individuals have an equal chance of being caught.

5 Marked organisms mix randomly with unmarked individuals.

6 Organisms disperse evenly within the study area.

7 During the study period, changes in population size due to immigration, emigration, death, and birth are negligible.

Check your understanding

1 Define percentage cover.

2 If 15 point quadrats, each with 10 needles, were used to sample an area, and the number of hits for one species was 20, estimate the percentage cover.

3 Use the Lincoln index to estimate the population size of three-spined sticklebacks in a pond. After 55 individuals were marked and released, 68 individuals were captured and of these 15 were marked recaptured.

4 The use of the Lincoln index as an estimate of population size assumes marked and unmarked individuals have an equal chance of being caught. Suggest what might cause this assumption to be wrong.

OBJECTIVES

By the end of this spread you should be able to

- appreciate the need for appropriate risk management before carrying out fieldwork
- consider the ethical issues of carrying out fieldwork

What is risk management?

In the context of fieldwork, a **risk** can be defined as the likelihood that carrying out a specific piece of work might be harmful. Before carrying out *any* fieldwork, a risk assessment should be made in which individual hazards are identified. A **hazard** refers to the ability of a procedure, a piece of equipment, a substance, or a biological agent to cause harm. For example, one of the hazards associated with water when carrying out fieldwork on aquatic habitats is drowning. However, the risk of drowning will depend on the location and nature of the water, and on the fieldwork procedures used. When dipping nets in a school pond, the risk of drowning will be very low compared with the risk of drowning when studying rock pools on a seashore.

Risk management involves using safe working practices so that risks associated with specific activities can be reduced to acceptable levels. Although a zero risk is probably impossible to achieve, health and safety legislation requires schools and colleges to provide environments that are safe and without undue risk to health. Before each specific piece of fieldwork, information on how to carry it out safely must be provided and, where appropriate, training given. It is the responsibility of staff *and* students to take reasonable care to ensure the health and safety of themselves and others – risk management is everyone's responsibility.

Fieldwork procedures

Each institution will have its own particular **risk assessment procedures** for fieldwork, but generally they will aim to establish:

- the hazards (biological, chemical, and physical)
- the risks involved
- the people at risk
- the procedures that will be adopted to prevent or minimize the risk

The outcome of the risk assessment must be recorded and appropriate safety information given to all those at risk. When you are involved in fieldwork you should make sure that you have read all the appropriate health and safety information in advance. You should also pay close attention to the person in charge before and during the fieldwork as particular hazards and risks may be emphasized. Most schools and colleges also have a safety handbook giving general information about safe working practices. It is important that everyone is aware of these.

If you are going to carry out fieldwork as part of an individual project, you will need to be involved in the risk assessment along with your supervisor. Your particular fieldwork will require its own specific risk assessment and risk management procedures, but there are some basic rules you should consider.

- Make sure that you and your supervisor are clear about the objectives of the fieldwork and that you have identified the potential hazards and appropriate risk management procedures.
- Include in your risk assessment travelling to and from the sites.
- Never work alone without the permission of your supervisor.

Students carrying out fieldwork on a rocky shore.

- Design your fieldwork carefully to take into account your experience and the sites you will visit.
- Include in your risk assessment any physical disabilities and bring them to the attention of your supervisor.
- Make sure that you have suitable clothing and equipment for all the weather conditions you might encounter.
- Carry and be able to use a first aid kit.
- Check the weather forecast before starting and be prepared to abandon the fieldwork if conditions become unsuitable.
- Arrange a responsible contact person and give them the full details of your intended fieldwork locations, routes, and times. Keep to these arrangements or inform your contact of any change – you should also inform them of your safe return.

Ethics and fieldwork

Each piece of fieldwork will have its own ethical issues which must be considered in advance. General considerations include

- potential damage to the physical environment
- potential harm to the health and safety of organisms in the habitat being studied

The physical environment, such as a fragile sand dune complex, may be damaged by trampling. Plants and small animals may also be damaged directly by being trampled on or their microhabitat made more dangerous. For example, when studying a rocky shore, boulders might be turned over so that the marine organisms living under them can be found. Unless the boulder is returned to its former position, right side up, the organisms may die from exposure to the elements or to predators. Collecting or disturbing any organisms should be minimized. Any live organisms removed from the habitat for identification should be handled with due regard for their health and safety, and they should be returned to their habitat as soon as possible.

To minimize damage to rural environments and to its inhabitants, those carrying out fieldwork should observe the Country Code and pay particular attention regarding instructions on access to and conduct whilst on private land.

Proposals for project work should include ethical considerations and you should not begin your project until your proposals have been officially approved and signed off.

Check your understanding

1 Distinguish between a hazard and a risk.
2 Explain why after turning over a boulder to search for organisms you should return the boulder to its original position.

By the end of this spread you should be able to

- analyse and interpret data relating to the distribution of organisms, recognizing correlations and causal relationships

- appreciate the tentative nature of conclusions that may be drawn from such data

Prior knowledge

- the need for random sampling (1.05 and *AS Biology for AQA* spread 10.01)

- the importance of chance in contributing to differences between samples (*AS Biology for AQA* spread 10.01)

- the mean and standard deviation, normal distribution about a mean (*AS Biology for AQA* spread 10.01)

Formula for Spearman's rank correlation coefficient

$$r = 1 - \frac{6\Sigma d^2}{n^3 - n}$$

Where
d = the difference in rank of the values of each matched pair
n = the number of pairs
Σ = the sum of

Fieldwork investigations usually generate data that needs to be analysed before you can interpret it. There is a wide range of statistical analyses that can be used, but the following are those specified by AQA for A2 Biology.

Standard error

The **standard error** (SE) is a measure of how reliable is the sample mean as an estimate of the true mean of the whole population. It is calculated using the following equation:

$$SE = \frac{s}{\sqrt{n}}$$

where s = standard deviation and n = sample size

The equation reflects the fact that SE is directly affected by sample dispersion (that is, the spread of the values from the mean, or the standard deviation around the mean). SE is also inversely related to sample size. The sample mean calculated from a small sample with a large dispersion is likely to be very different from the true mean.

If the mean number of snails within 21 quadrats of area 1 m² was 5.0, and the standard deviation was 1.28, the standard error would be 0.28. The results could then be summarized as

Mean number of snails per m² = 5.0 ± 0.280 ($n = 21$)

95% confidence limits

The standard error is used to calculate the 95% confidence limits for the sample mean.

For data with a **normal distribution**, the **95% confidence limits** for the true mean (that is, the mean of the whole population from which the samples have been taken) is defined as the range of values between the sample mean plus two standard errors and the sample mean minus two standard errors. Therefore, if the mean number of snails per 21 quadrats is 5.0 and the standard error equals 0.28, the 95% confidence limits would be 5.0 ± 0.56, that is, from 4.44 to 5.56.

The 95% confidence limit shows us that we can be 95% certain that the true mean density of snails lies within the range of 4.44 per m² to 5.56 per m². Conversely, on average, 5 times out of 100, the 95% confidence limits will *not* contain the true mean of the population.

Spearman's rank correlation coefficient

Where a correlation is suspected between samples of two variables, **Spearman's rank correlation coefficient** is used to test the strength and direction of the correlation. It involves ranking the two sets of data separately, giving the highest value the rank of 1 and so on. For example, in a study of shore crabs found in 1 m² quadrats on rocky shores that have different exposures to wave action, the data collected for the crabs and the data for exposure to wave action would be ranked separately.

Spearman's rank correlation coefficient (r) is calculated using the formula shown in the box. The Spearman's rank correlation coefficients can range from −1.0, indicating a perfect negative correlation, to +1.0, indicating a perfect positive correlation. The nearer the correlation is to +1 or −1, the stronger is the relationship between the variables. A correlation coefficient of 0.0 indicates no correlation. Because results can occur by chance, it is necessary calculate the significance of the correlation coefficient at a chosen confidence level (usually 95%).

A strong correlation does not imply a causal relationship between two variables. For example, even if the density of shore crabs and exposure to wave action had an *r* of −1.0 and it was more than 95% significant, you cannot assume that increasing wave exposure *causes* a reduction in the density of shore crabs.

Chi-squared test

The **Chi-squared test** assesses the difference between observed (*O*) and expected (*E*) values. In fieldwork, it can be used to compare two or more sets of data to find out if any differences between them are significant. For example, the table below shows the analysis of results of catching crayfish in traps at six different locations (A–F) in a river, using the formula:

$$\text{Chi squared} = \sum \frac{(O - E)^2}{E}$$

The expected numbers are calculated based on the assumption that there is no difference between the traps.

Positive correlation

Negative correlation

No correlation

Scattergraphs showing three different types of correlation

Crayfish trap location	A	B	C	D	E	F	Total
Observed numbers (*O*)	16	25	7	17	35	20	120
Expected numbers (*E*)	20	20	20	20	20	20	120
$O - E$	−4	+5	−13	−3	+15	0	
$(O - E)^2$	16	25	169	9	225	0	
$(O - E)^2/E$	0.8	1.25	8.45	0.45	11.25	0	22.2

As with most statistical tests, a **null hypothesis** is adopted and tested. This assumes there is no difference between the numbers of crayfish trapped in each location. The value of chi squared gives the likelihood or the probability that the null hypothesis is correct. In our example, the value of chi squared is 22.2.

A chi-squared table gives the probability of obtaining the value for different degrees of freedom. A **degree of freedom** is a statistical term that relates to the number of free variables in the system. In this case it is the number of samples minus 1, that is, 5 (6 − 1). Using a chi-squared table shows that the probability (*p*) of obtaining a value of 22.2 at 5 degrees of freedom is less than 0.001: there is a less than 0.1% probability of the null hypothesis being correct. Ecologists usually say that the 5% level ($p = 0.05$) is significant. Our value of *p* is much less than 0.05, so the null hypothesis is rejected. We can conclude that there is a significant difference in the number of crayfish trapped in each location and that it would be worthwhile carrying out further investigations to find out why.

Check your understanding

1 Copy and complete the following:

Standard error is directly proportional to _____ and inversely proportional to _____.

2 Define standard error.

3 Distinguish between a correlation and a causal relationship.

OBJECTIVES

By the end of this spread you should be able to

- understand why population size may vary as a result of abiotic and biotic factors

- explain why intraspecific competition can limit population growth

- interpret sigmoid (S-shaped) population growth curves and J-shaped population growth curves

- distinguish between density dependent factors and density independent factors

Prior knowledge

- populations and communities (1.01)

- ecological niche (1.02)

Populations are dynamic, constantly changing components of ecosystems. They are commonly described using the following terms:

- **population size** – the number of individuals in a population

- **population density** – the number of individuals per unit area

- **population growth** – a change in the number of individuals; this is referred to as positive growth when the numbers increase and negative growth when they decrease

- **population growth rate** – the change in number of individuals per unit time; again, this may be positive or negative

S-shaped population growth curves

The **population growth curve** of a population colonizing a new habitat in which conditions are initially ideal (there is an abundant food supply; no competitors, predators, or disease; optimal temperature; and so on) has a sigmoid or S-shape, as shown here. Populations of bacteria growing on agar, or yeast growing in broth, typically have this type of population growth curve.

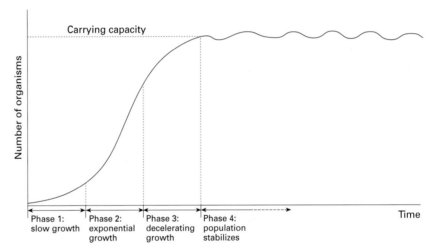

A sigmoid or S-shaped population growth curve

Initially, the population growth curve rises gently, showing that the population growth rate is slow (phase 1, sometimes called the **lag phase**). Growth is slow because there are few reproducing individuals; those that reproduce sexually may not be able to find a mate when the population density is low.

In phase 2 of the population growth curve, the population grows at its **biotic potential**, that is, its maximum rate. The birth rate exceeds the death rate and the population size doubles at regular intervals. This phase is called the **exponential growth phase** (or the **log phase**, because the curve is linear when plotted on a logarithmic scale).

Exponential growth cannot go on forever. Eventually the population is prevented from increasing further by various factors in the environment, such as lack of food or space, increased intraspecific competition for resources, a build-up of toxic chemicals produced by the organisms, or an increase in the incidence of parasitic infections and disease.

The population growth rate slows down (phase 3, during which the population growth curve becomes less steep), and stops (phase 4, during which the population growth curve levels off).

Collectively, the factors that limit the size of a population are referred to as the **environmental resistance**.

Boom and bust

The maximum population size that can be sustained over a relatively long period of time by a particular environment is called the **carrying capacity**. In some circumstances, a population increases so rapidly during the exponential growth phase that it overshoots the carrying capacity. Because the environment cannot support the population, a population crash usually follows. Populations that show this type of J-shaped population growth curve are sometimes referred to as **'boom-and-bust' populations**. Overpopulation and intense intraspecific competition for limited resources can damage the environment, leading to a new, lower carrying capacity. After the crash, the population fluctuates around the new carrying capacity.

Environmental factors that limit population size

Exponential population growth is not often seen except under controlled laboratory conditions or when a species colonizes a new habitat. In natural populations a complex of environmental factors interact, some tending to decrease the population size, others tending to increase it.

These factors may be classified as follows:

- **Density-dependent factors** – as the population density (the number of organisms in a given area) increases, density-dependent factors affect a larger proportion of the population. They are usually biotic and include intraspecific competition for space and other resources, availability of food, disease, and predation.

- **Density-independent factors** – these factors affect the same proportion of the population no matter how great its density. They are usually abiotic. The most important are weather (short-term changes in atmospheric conditions, such as rainfall and temperature) and climate (long-term changes in atmospheric conditions). A sudden freeze can have a dramatic effect on populations, killing a fixed percentage of organisms irrespective of the population density.

Environmental factors determine the size of a population by affecting

- the **birth rate** (number of births/number of adults in the population)
- the **death rate** (number of deaths/number of adults in the population)
- the amount of movement into the population (**immigration**)
- the amount of movement out of the population (**emigration**).

In favourable conditions, the population will grow if

(Number of births + number of immigrants) is greater than
(number of deaths + number of emigrants)

In unfavourable conditions, the population will decrease if

(Number of births + number of immigrants) is less than
(number of deaths + number of emigrants)

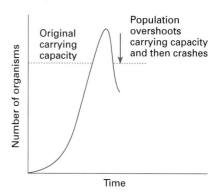

A 'boom-and-bust' population growth curve, for example, of a population of Daphnia (water fleas) in a culture. The population increases so much it overshoots the carrying capacity.

Check your understanding

1 Name two abiotic factors that affect population size.

2 Why does intraspecific competition limit population growth of yeast cultured in the laboratory?

3 Name the type of population growth curve that shows a population explosion followed by a dramatic decline.

4 Why is temperature a density-independent factor?

5 Suggest why laboratory cultures of yeast often have an S-shaped population growth curve followed by a decline phase.

OBJECTIVES

By the end of this spread you should be able to

- explain how population size may vary as a result of interspecific competition and predation

Prior knowledge

- variation in population size I (1.07)

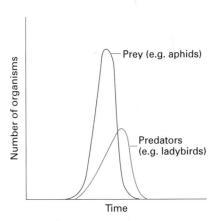

Changes in the population densities of ladybirds (predators) and aphids (prey)

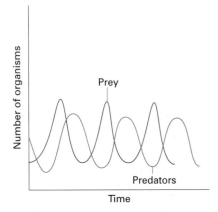

Graph based on the mathematical prediction of the relationship between an efficient predator and its prey in a confined space

Individual organisms do not live in isolation in a community. They interact continually with each other. The interactions may be between members of the same species (**intraspecific interactions**) or between members of different species (**interspecific interactions**).

Interspecific interactions

Interspecific interactions are as many and varied as the organisms themselves. One way of classifying them is according to whether the effects of interacting species on each other are positive (+) or negative (−). In a + − relationship, one species benefits at the expense of another (such as a predator–prey relationship). In a − − relationship, both species are harmed (for example, interspecific competition). In a + + relationship both species benefit (these relationships are called mutualisms).

Types of predator–prey relationships

Theoretically, there are two main types of predator–prey relationship. One applies to small predators and prey which have seasonal populations that peak in the summer. As conditions become favourable (for example, food plants become more abundant and the temperature reaches an optimal range), increases in the prey population are followed by increases in the predator population. At the end of the summer, the sizes of both populations fall as they enter an overwintering stage in their life cycles. In nature, ladybirds and aphids tend to have this sort of relationship, as shown in the top graph opposite.

The other type of theoretical predator–prey relationship applies to a population of efficient predators when they meet a population of prey in a confined space. The predator population is limited by its food supply. In turn, the prey population is determined by the number of individuals killed by predators. The result would be wave-like oscillations in the population densities of both predator and prey. The oscillations tend to be out of phase with one another, with the predator oscillations lagging behind those of the prey, as shown opposite.

Canada lynx and snowshoe hares

One of the most studied predator–prey relationship is that of the Canada lynx (*Lynx canadensis*) and the snowshoe hare (*Lepus americanus*). Charles Elton, one of the pioneers of animal ecology, analysed the records of lynx and hare furs traded with the Hudson's Bay Company in Canada over a 200 year period. He showed that peaks and troughs in the population density of hare were followed by corresponding changes in lynx population density, and that this pattern was shown for as long as records have been kept. The graphs have been interpreted as showing a stable relationship in which prey and predator regulate each other in a cyclical manner. The interpretation usually goes something like this:

> When predators are scarce but prey are numerous, the predator population grows quickly. Inevitably, as predation increases, the size of the prey population falls. When the prey population falls, competition between predators for food is increased and predator numbers fall as some die of starvation. With fewer predators, the prey population starts to rise again, thus completing the cycle.

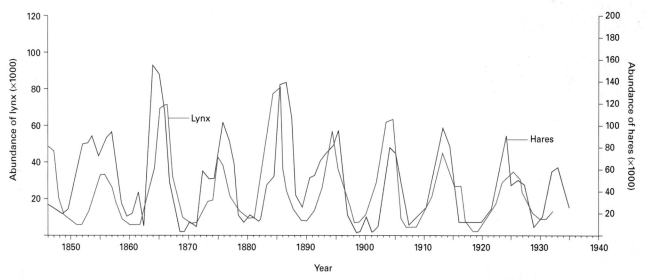

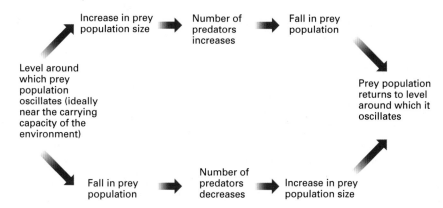

Changes in the abundance of lynx and snowshoe hares based on number of fur skins lodged with the Hudson's Bay Company in Canada

Predator and prey populations are therefore regulated by a **negative feedback** mechanism that keeps the populations balanced at levels that the environment can support.

Hypothetical negative feedback mechanism controlling the size of predator and prey populations

The description just given, however, is an oversimplification. On islands where lynx are absent, snowshoe hare populations show similar cycles to those on the mainland.

An alternative explanation for the oscillations is that high hare densities affect the grasses they eat. When heavily grazed, the grasses have a lower nutrient content, and they produce shoots with high levels of distasteful chemicals. This leads to increased hare mortality. On the mainland, therefore, although the lynx population probably oscillates in response to the hare population, the hare population is probably regulated by cyclic changes in the plants the hares eat.

It is likely that most cyclic patterns of population growth are caused by the interaction of many factors. So far, no predator–prey oscillation has been proven to be driven solely by interactions between a single predator and its prey.

Canada lynx and snowshoe hare

Check your understanding

1 How did Elton estimate the population size of snowshoe hares and lynx in Canada?

2 Which piece of evidence shows that oscillations of snowshoe hare populations are not regulated entirely by their predator population?

- - - - - - - - - - - - - -

3 In a test-tube study of predator–prey relationships between single-celled organisms, the introduction of the predator always resulted in the annihilation of the prey, followed by death by starvation of the predator. In the real world, predators rarely wipe out prey completely. Suggest differences between laboratory situations and real life which might account for the different outcomes.

OBJECTIVES

By the end of this spread you should be able to

- discuss the significance of growth in human populations
- interpret population growth curves, survival curves, and age pyramids

Prior knowledge

- variation in population size (1.07–1.08)

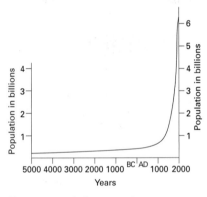

Human population growth curve

Population size and growth

More than 6 billion people inhabit the Earth; double the population of the 1950s. When humans were primitive hunter-gatherers, the world population was probably less than a quarter of a million. The development of agriculture about 10 000 years ago brought an upswing in population growth rate so that the human population reached about 100 million around 5000 BC. It continued to grow steadily until the start of the Industrial Revolution.

Since the Industrial Revolution the human population has been going through a phase of exponential growth. Social and economic changes including increased mechanization and advances in technology have stimulated an increase in the birth rate. Advances in medicine and hygiene have reduced the death rate, particularly of newborn babies. The world population has therefore increased dramatically as countries have become industrialized.

In 1650, the world population was about 500 million; by 1850 it had doubled to 1 billion; 80 years later it had again doubled. The doubling time is now about 40 years. This means that if the population continues to rise unabated it will reach 12 billion in 40 years, 24 billion in 80 years, and so on. It is difficult to predict future trends, but it is self evident that exponential growth cannot be sustained indefinitely.

Carrying capacity

The effect of human population growth will depend on the carrying capacity of the Earth, that is, the maximum number of people that the world can sustain for a long period of time. Unfortunately the knowledge about carrying capacities gained from animal studies do not apply to human populations. No one can say precisely what our population limit is. New discoveries and inventions can change the carrying capacity of the land dramatically. This is what happened during the Industrial Revolution of the eighteenth century, and continued with the green revolution in the twentieth century when farming techniques improved agricultural outputs enormously. Today, world food production is growing at a faster rate than the world's population, and developments in genetic engineering are likely to increase food productivity even more.

Pessimists believe that the world population already exceeds the carrying capacity of the Earth, and that unless we do something very soon to curb our population growth and (just as important) to stop overexploiting natural resources, the result will be war, famine, and disease on a global scale. If this happens, our population growth curve would probably turn out to be J-shaped and of the boom-and-bust variety. Optimists, on the other hand, believe that our population growth curve will eventually become sigmoid-shaped: the growth rate will slow down and the world population will stabilize below the carrying capacity. United Nations analysts are among the optimists: they predict that the world population will stabilize at about 12 billion in 120 years' time.

Survival curves

Survival curves keep track of the fate of any given **birth cohort** (a birth cohort refers to individuals born at about the same time). The curves show the percentage of individuals still living at a given age.

Survival curves (also called survivorship curves) of populations are compiled by plotting survival of a cohort against age. The survival may be expressed as a percentage or as a log of the number of survivors from each 1000 births. Age may be expressed in actual years when comparing human populations, or per cent lifespan when comparing different species. There are three basic types of survival curve.

- Type I applies to species that have a high survival rate of the young, live out most of their expected lifespan, and die in old age.

- Type II applies to species that have a relatively constant death rate throughout their lifespan. Death is often from predation or disease.

- Type III is found in species that have many young, most of which die very early in their life.

Most human populations have a type I survival curve. But in populations where infant mortality is very high, survival curves approach type III.

Population pyramids

Demography is the statistical study of the size and structure of populations (for example, the age or sex distribution in the populations) and of the changes within them. Demographic trends can be shown in special diagrams called **population pyramids**, graphs that show the distribution of a population by age and sex.

The population pyramids of more developed countries are quite different from those of less developed countries. In many less developed countries, a large percentage of the population is younger than 15 years. The number of women reaching child-bearing age (generally taken to be between 15 and 49 years) will continue to rise steadily, so that the population size will continue to soar for many years.

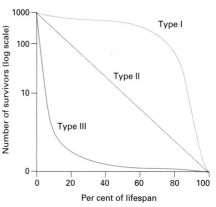

Three types of survival (survivorship) curve

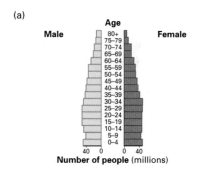

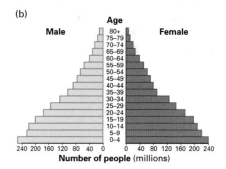

*Population pyramids for (**a**) more developed countries in North America or Europe and (**b**) less developed countries in Africa, Asia, Central or South America*

Check your understanding

1 Define carrying capacity.

2 Coral has a relatively constant rate of mortality throughout its lifespan. What type of survival curve does it have?

- -

3 Study the graph below.

   a Compare the different mortality of infants in Great Britain and Niger at the end of the twentieth century, and England in the seventeenth century.

   b Suggest reasons for the differences.

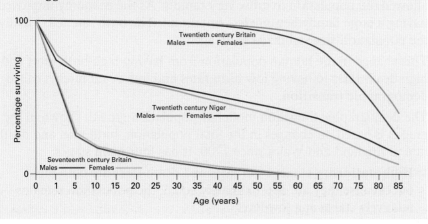

Survivorship curves for males and females in Great Britain in 1999, Niger in 1999, and in England in the seventeenth century

35

OBJECTIVES

By the end of this spread you should be able to

- calculate population growth rates from birth rate and death rate data

- relate changes in size and structure of human populations to different stages in demographic transition

Prior knowledge

- human populations I (1.09)

Relative growth rate (% per year)	Doubling time (years)
0.5	139
1.0	70
2.0	35
5.0	14
8	9

Doubling times for different relative growth rates

Calculating population growth rates

The population growth rate refers to the change in size of a particular population per unit time. The particular population might be people in a country, plants in a given area of a field, or microorganisms in a litre of culture medium.

Assuming there is no net emigration or immigration,

$$\text{Population growth rate} = \text{birth rate} - \text{death rate}$$

When the death rate exceeds the birth rate, the growth rate becomes negative.

In populations in which emigration and immigration do not balance,

$$\frac{\text{Population growth}}{\text{rate}} = (\text{birth rate} - \text{death rate}) + \frac{(\text{immigration rate}}{- \text{emigration rate})}$$

Absolute growth rate is the change in numbers per unit time. Relative growth rate is the percentage by which numbers increase per unit time. These are quite different. For example, if in one year the size of a population increased from 50 million to 51 million, it would have an absolute growth rate of 1 million people and a relative growth rate of 2%. But if in the following year the population again increased by 1 million from 51 million to 52 million, it would have the same absolute growth rate (that is, 1 million) but its relative growth rate would be 1.96%.

Relative growth rates can be used to predict the doubling times of a population, that is, the time it takes for the number of individuals to double. The table opposite shows the doubling times for five different relative growth rates, assuming these growth rates remain constant.

Demographic transition: a change in the population

Globally, most of the growth in human populations in recent years has taken place in the less developed countries. Some people believe that these countries will go through the same stages of population growth as the more developed countries of Europe and North America.

Before the Industrial Revolution, the population of Britain, for example, was relatively low and stable, with high mortality and high birth rates. The Industrial Revolution brought improved agriculture, better nutrition, and better medical knowledge. Birth rates became high and death rates low (similar to the present situation in the less developed countries). This allowed the population to grow very rapidly. As the economy improved, having a large family became less important. The birth rate fell. Finally, the population size stabilized at all ages.

This change in the human population from having high birth rates and high death rates to having low birth rates and low death rates is called **demographic transition**.

The demographic transition model of population growth shown in the graphs is based on changes in the total population, birth rate, and death rate in England and Wales since 1700. It has four main stages:

1 **High stationary stage:** birth rates and death rates are both high but because they are approximately equal, the population remains relatively stable at a low level.

2 **Early expanding stage:** the population starts to expand rapidly, mainly because of a decrease in death rate.

3 **Late expanding stage:** the birth rate declines but the population continues to expand as the birth rate still exceeds the death rate.

4 **Low stationary stage:** birth rates and death rates are approximately equal.

Although demographic transitions have already happened in more developed countries and are already happening in some less developed countries, it does not mean that *all* less developed countries will go through similar demographic transitions as they become more economically developed.

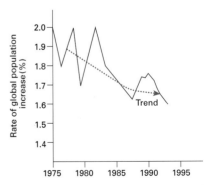

Improved medical care is an indication of demographic transition in a less developed country.

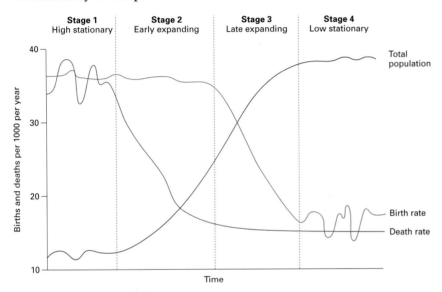

Demographic transition model of population growth. Curves showing changes in the total population, birth rate, and death rate in England and Wales since 1700 approximate to this model.

Check your understanding

1 Distinguish between the absolute growth rate and the relative growth rate.

2 What trends characterize a country in demographic transition?

- -

3 a Describe and suggest an explanation for the trend in the graph opposite.

b Comment on the reliability of the graph.

c Compare the relative growth rate in 1975 with that in 1994 and the doubling times associated with them.

Estimates of the rates of global population increase between 1975 and 1994. The figures are based on birth rates as a percentage of the population. They are estimates because some countries do not have reliable data.

OBJECTIVES

By the end of this spread you should be able to

- describe the structure of ATP

- explain how ATP is synthesized from ADP

- recall that ATP provides the immediate source of energy for biological processes

Prior knowledge

- biological processes and energy (*AS Biology for AQA* spread 3.01)

- active transport (*AS Biology for AQA* spread 5.06)

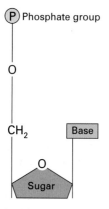

The basic structure of a nucleotide

ATP structure

Adenosine triphosphate (ATP) is a nucleotide. **Nucleotides** are nitrogen-containing organic substances which play a vital role in every aspect of an organism's life. Each nucleotide is made of three parts:

- a nitrogen-containing organic **base**

- a five-carbon sugar (a **pentose**)

- one or more **phosphate** groups

Nucleotide molecules occur singly (**mononucleotides**) or combined in numbers from two to many thousands (**polynucleotides**). DNA and RNA are polynucleotides. ATP is a mononucleotide in which the base is **adenine** and the sugar is **ribose**. Adenine and ribose together make up **adenosine**. Attached to the ribose of the adenosine are three phosphate groups, as shown below.

The ATP cycle

The covalent bond linking the second and third phosphate groups in ATP is unstable. It is easily broken by **hydrolysis**. When this happens a phosphate group (P) is removed, and ATP becomes **ADP (adenosine diphosphate)**. At least 30 kJ of energy are released during this reaction. It is called an **exergonic reaction** because energy is released.

ATP can be resynthesized from ADP and inorganic phosphates by a condensation reaction. The energy needed to drive this reaction comes from respiration. The **ATP cycle** shows the relationship between ATP, ADP, and respiration. An enzyme called **ATP synthase** catalyses the formation of ATP from ADP and a phosphate group.

ATP: cellular energy to drive biological processes

Biochemists often refer to ATP as the 'energy currency of cells' because all cells use ATP as the source of energy to drive their metabolic machinery. Organisms may have many different fuel molecules. Proteins, lipids, and carbohydrates can all supply energy. However, these cannot drive biological processes in living cells until they are converted to ATP.

In cells, the exergonic breakdown of ATP is coupled to energy-consuming (**endergonic**) reactions. Phosphate groups liberated from

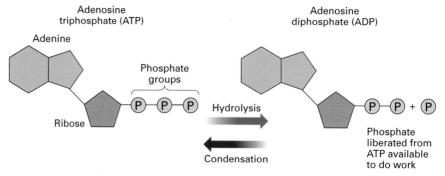

The breakdown and formation of adenosine triphosphate

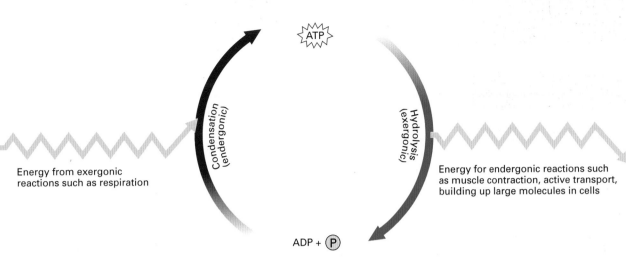

The ATP cycle

ATP attach to other molecules (a process called phosphorylation). This energizes the molecules, enabling them to do work. For example, a contracting muscle cell requires about 2 million ATP molecules per second to drive its biochemical machinery. A runner uses the equivalent of about 75 kg of ATP during a marathon race. However, the human body has only a small store of ATP (approximately 100 g in an average person). This store would be exhausted after only one second of the race if the runner could not continuously regenerate ATP from ADP and inorganic phosphate by respiration of nutrients, mainly carbohydrate and fat.

The need to regenerate ATP in the ATP cycle is not confined to muscle cells. Every living cell in all organisms, from the smallest bacterium to the largest to plant and animal, requires a continual source of ATP regenerated in the ATP cycle.

A collaborative understanding of ATP synthesis and function

Science, especially modern science, is not an individual pursuit. It usually involves the collaboration of several teams of scientists working in different parts of the world. Our understanding of the synthesis and functioning of ATP has resulted from this kind of international collaboration. In 1997 three biochemists shared the Nobel Prize in Chemistry – Paul D. Boyer and John E. Walker 'for their elucidation of the enzymatic mechanism underlying the synthesis of adenosine triphosphate (ATP)' and Jens C. Skou 'for the first discovery of an ion-transporting enzyme, Na$^+$, K$^+$-ATPase'.

Boyer and his co-workers at the University of California Los Angeles, USA, proposed a mechanism by which the enzyme ATP synthase catalyses the formation of ATP from ADP and inorganic phosphate. Walker and his co-workers at the Medical Research Council's Laboratory of Molecular Biology Cambridge, UK, established the structure of the enzyme and verified the mechanism proposed by Boyer. Skou, working at Aarhus University in Denmark, discovered another enzyme called **sodium, potassium-stimulated adenosine triphosphatase** (Na$^+$, K$^+$-ATPase). Both enzymes are membrane bound.

(More information about the laureates and their work can be found on the Nobel Prize website, http://nobelprize.org.)

Check your understanding

1 Describe the chemical structure of an ATP molecule.

2 Name the enzyme that catalyses the synthesis of ATP from ADP and a phosphate group.

3 Why does ATP have to be continuously regenerated by an active cell – why can't it simply be stored and regenerated periodically?

4 In what active transport mechanism does sodium, potassium-stimulated adenosine triphosphatase play an essential role? What is the basic function of this active transport mechanism?

2.02 Photosynthesis: an overview

OBJECTIVES

By the end of this spread you should be able to

- summarize the main features of photosynthesis
- distinguish between the light-dependent and the light-independent reactions in a typical C3 plant

Prior knowledge

- ATP (2.01)
- plant cell structure and ultrastructure (*AS Biology for AQA* spreads 13.04–13.05)

Most plants have no structures for ingesting and digesting food. They have no mouth and no alimentary canal, yet plant material is rich in carbohydrates, proteins, and fats. Instead of obtaining their food from other organisms, plants make it for themselves using simple ingredients. They are **autotrophs** ('self feeders').

What is photosynthesis?

A typical plant takes in carbon dioxide (from the air) and water (from the soil) and uses these to make sugars and other complex substances. Oxygen is released as a waste product. The energy in the chemical bonds of the raw materials (carbon dioxide and water) is less than the energy in the chemical bonds of the products. Therefore the reaction is endergonic and requires an external source of free energy. This energy is supplied by sunlight that falls on the plant. A green substance, **chlorophyll**, enables the plant to trap light energy and use it to make sugars. The process of using sunlight to build up complex substances from simpler ones is called **photosynthesis**.

Photosynthesis is a complex process which takes place in a series of small steps. However, it can be summarized by the simple equation:

$$6CO_2 + 12H_2O \xrightarrow{\text{light and chlorophyll}} C_6H_{12}O_6 + 6O_2 + 6H_2O$$

$$\underbrace{\text{Carbon dioxide} + \text{water}}_{\text{Raw materials}} \longrightarrow \underbrace{\text{glucose} + \text{oxygen} + \text{water}}_{\text{products}}$$

The water molecules on both sides of this equation reflect the fact that the oxygen produced by photosynthesis is derived from water. Two water molecules are required to form one oxygen molecule. The hydrogen in glucose is also derived from water.

Experiments using radioactive isotopes and isolated chloroplasts, the sites of photosynthesis, suggest that there are two main sets of reactions in photosynthesis:

- **light-dependent reactions**, in which water is broken down into hydrogen and oxygen using light energy
- **light-independent reactions**, in which the hydrogen reacts with carbon dioxide to form a carbohydrate. Water is re-formed in this reaction.

The light-dependent reactions take place only if sufficient light energy is available. The light-independent reactions take place in light or darkness.

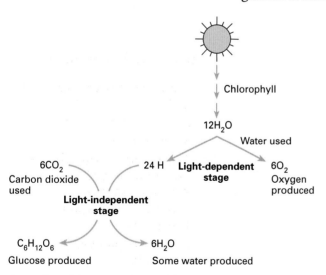

An outline of the two main sets of reactions that make up photosynthesis

Converting glucose to other substances

The glucose formed by photosynthesis is used as the raw material for other chemical reactions. It is the main substrate used in respiration. Some of the glucose is converted to other carbohydrates: cellulose to form cell walls, sucrose to be transported to other parts of the plant, and starch for storage. Some of the glucose is combined with minerals from the soil to make proteins and other complex organic substances. Although light is needed to make glucose, it is not needed to turn the glucose into these other substances.

Photosynthesis: the basis of life

Green life has been steadily pumping out oxygen as a waste product of photosynthesis for millions of years. Some of the oxygen is used as a raw material for respiration, but most of it has accumulated in the atmosphere. So the very existence of our oxygen-rich atmosphere depends on the photosynthesizing activities of green life.

Animals cannot make their own food. They obtain complex organic substances by eating other organisms. These organisms ultimately depend on the ability of plants to harvest energy from sunlight to make food from carbon dioxide and water. Life on Earth is almost entirely solar powered.

The site of photosynthesis

Although leaves are the main sites of photosynthesis in most plants, it can take place in any part that is green. These green parts have **chloroplasts**, which contain all the biochemical machinery necessary for photosynthesis. The light-dependent stage takes place in the grana of the chloroplast, while the light-independent stage occurs in the stroma.

Outline of a chloroplast

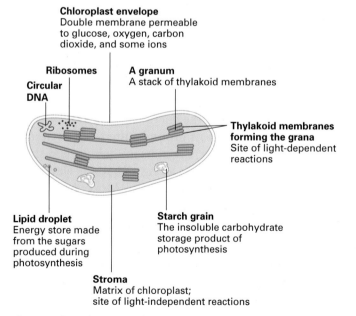

Chloroplast envelope
Double membrane permeable to glucose, oxygen, carbon dioxide, and some ions

Ribosomes

Circular DNA

A granum
A stack of thylakoid membranes

Thylakoid membranes forming the grana
Site of light-dependent reactions

Lipid droplet
Energy store made from the sugars produced during photosynthesis

Starch grain
The insoluble carbohydrate storage product of photosynthesis

Stroma
Matrix of chloroplast; site of light-independent reactions

*Each chloroplast consists of two membranes enclosing a gelatinous matrix called the **stroma** which contains ribosomes, circular DNA, and enzymes used in photosynthesis. Suspended in the stroma are **thylakoids**. These are disc-like membrane sacs, stacked in a group to form a **granum** (plural grana). The space inside each thylakoid is connected with the other thylakoids in the stack, forming a continuous fluid-filled compartment called the thylakoid space. The thylakoid membranes contain photosynthetic pigments, including chlorophyll.*

Check your understanding

1 During photosynthesis, what gas is

  a a raw material

  b a product?

2 Give the precise location in the chloroplast of a typical terrestrial plant of

  a the light-dependent reactions

  b light-independent reactions

3 Suggest why the main carbohydrate stored in chloroplasts is starch rather than glucose or sucrose.

41

OBJECTIVES

By the end of this spread you should be able to

- explain what happens when light energy excites electrons in chlorophyll

- explain the role of electron transport chains in generating ATP and reduced NADP

- define photolysis and explain its significance

Prior knowledge

- chloroplast ultrastructure (*AS Biology for AQA* spread 13.05)

- ATP (2.01)

- photosynthesis: an overview (2.02)

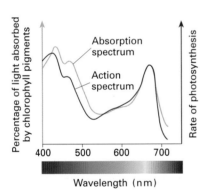

Comparison of the action spectrum for photosynthesis in a typical green plant and the absorption spectrum for its pigments

The energy that drives the light-dependent reactions of photosynthesis comes from sunlight. Visible sunlight consists of a range of different wavelengths that combine to give white light, but which when seen separately appear as different colours. As sunlight falls on a plant, some wavelengths are absorbed, while others are reflected or transmitted. The pattern of absorption and reflection depends on coloured substances called pigments. The most important are chlorophylls.

Chlorophyll and other photosynthetic pigments

Chlorophyll contains a **porphyrin** ring with magnesium at the centre linked to a long hydrocarbon chain. Chlorophyll is a green pigment which absorbs light mainly in the red-orange and blue-violet parts of the visible light spectrum. It is the most important pigment in plants because it is the only one that takes a direct part in photosynthesis.

As well as chlorophyll, there are other photosynthetic pigments in chloroplasts called **accessory pigments**. These do not participate directly in the light-dependent reactions, but they absorb light of other wavelengths outside the range of chlorophyll and convey the energy to chlorophyll which then uses it in photosynthesis.

Absorption spectrum and action spectrum

The **absorption spectrum** refers to the wavelengths of light that a particular photosynthetic pigment absorbs. The **action spectrum** refers to the wavelengths of light that bring about photosynthesis. The absorption spectrum of the combined photosynthetic pigments in a plant coincides very closely with the action spectrum of photosynthesis for that plant, supporting the notion that these pigments harvest light for photosynthesis.

The production of ATP and reduced NADP

The light-dependent reactions occur in chlorophyll bound to the thylakoid membranes in a chloroplast. Light absorbed by chlorophyll provides energy to convert ADP and inorganic phosphate (P) to ATP, a process called **photophosphorylation**.

When chlorophyll absorbs light, the energy level of one of its electrons is raised: the electron goes from its **ground state** to an **excited state**. During the light-dependent stage, electrons in their excited state leave the chlorophyll molecule and are passed along a chain of **electron carrier** molecules called an **electron transport chain**. Most of the electron carriers in the series are proteins. Each has a stronger affinity for electrons than the previous molecule. This keeps electrons moving down the chain.

A **redox reaction** takes place during the transfer of an electron from one carrier molecule to the next. One carrier molecule donates an electron (is oxidized), while the other molecule accepts it (is reduced). **Nicotinamide adenine dinucleotide phosphate (NADP)** is the final electron acceptor in the electron transport chain. When it gains an electron, NADP becomes reduced.

As electrons pass down the electron transport chain in the series of redox reactions, they release energy which is used to make ATP. This method of ATP production is called photophosphorylation because light energy is used to add a phosphate group to ADP.

Photolysis

Light energy trapped by chlorophyll is also used to split water molecules and release electrons; this process is known as **photolysis**. Photolysis occurs when water molecules are split with the aid of light energy.

When light-activated chlorophyll loses an electron it becomes positively charged. In this oxidized state the chlorophyll becomes 'electron hungry'; it needs to replace the electron before it can react again. Photolysis involves the enzymatic extraction of electrons from water to replace each electron lost from chlorophyll. This results in water molecules being split. In addition to electrons, protons and oxygen are also produced:

$$2H_2O \rightarrow 4H^+ + 4e^- + O_2$$

The electrons replace those lost from chlorophyll, the protons (H^+) reduce NADP, and the oxygen is a waste gas or used in respiration.

In summary, the light-dependent stage of photosynthesis uses energy from sunlight to make ATP, and it splits water to release protons and electrons which reduce NADP to NADPH. The ATP and NADPH are used in light-independent reactions to synthesize carbohydrates.

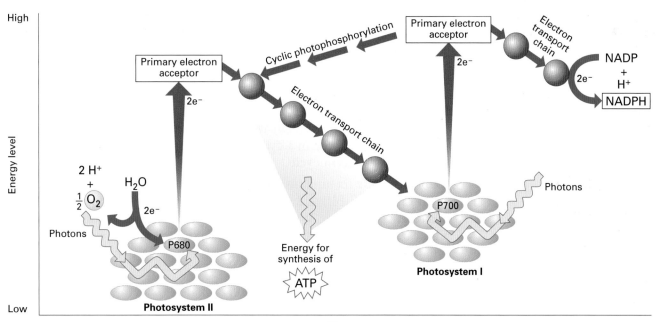

The Z scheme: a diagrammatic representation of electron flow during the light-dependent stage. Photophosphorylation and photolysis are intimately linked in a series of reactions involving two photosystems. NADP, the final acceptor of electrons and protons, is reduced to NADPH. It stores the electrons and protons until they can be transferred to carbon dioxide in the light-independent stage. Every molecule of NADPH formed requires two electrons from photosystem I. These electrons are replaced by those from photosystem II which regains its electrons from water. The energy from the excited electrons passing down the electron transport chain is used to make ATP.

Check your understanding

1 Name the three main products of the light-dependent stage of photosynthesis.

2 What is the name of the final acceptor in the electron transport chain?

3 Which ions are produced when water molecules dissociate?

- -

4 Suggest why pure chlorophyll fluoresces when exposed to light.

By the end of this spread you should be able to

- understand that light-independent reactions in a typical C3 plant depend on carbon dioxide being accepted by ribulose bisphosphate (RuBP) to form two molecules of glycerate 3-phosphate (GP)

- explain why ATP and reduced NADP are required for the reduction of GP to triose phosphate

- recall that RuBP is regenerated in the Calvin cycle

- recall that triose phosphate is converted to useful organic substances

Prior knowledge

- light-dependent reactions (2.03)

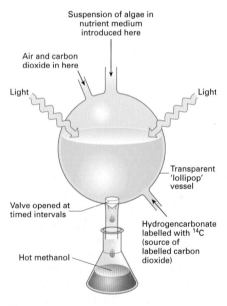

The 'lollipop' apparatus. Algae were cultured in the 'lollipop' into which hydrogencarbonate (the source of carbon dioxide) labelled with radioactive ^{14}C was injected. The metabolic pathway of the light-independent reactions was traced by analysing the radioactive compounds formed at different times after injection of the hydrogencarbonate.

Calvin's 'lollipop apparatus'

The light-independent reactions of photosynthesis result in carbon dioxide being fixed. **Carbon dioxide fixation** occurs when the carbon dioxide is assimilated into the plant by being converted into biochemical products. The overall reaction is the reduction of carbon dioxide to glucose.

The light-independent reactions take place in the stroma of chloroplasts. Enzymes convert carbon dioxide to glucose in a series of reactions discovered by Melvin Calvin using his 'lollipop apparatus'. The reactions use ATP and NADPH, the products of the light-dependent stage.

In Calvin's experiments, the 'lollipop' was a thin transparent vessel containing a suspension of *Chlorella*, a single-celled alga. Carbon dioxide labelled with radioactive carbon (^{14}C) was added to the suspension and light was shone on it. A sample from the suspension was collected in a tube containing boiling methanol at timed intervals from a few seconds to several minutes. The boiling methanol killed the algae quickly, inactivated their enzymes, and stopped the reactions instantaneously. The radioactive compounds that had been formed were extracted from the samples using paper **chromatography** and analysed by **autoradiography**.

The Calvin cycle

In this way, Calvin was able to trace the biochemical path taken by carbon in photosynthesis. His investigations revealed that carbon dioxide is converted to glucose in a cyclical chain of reactions which is now known as the **Calvin cycle**.

Carbon dioxide fixation involves a five-carbon compound called ribulose bisphosphate (RuBP) and an enzyme called ribulose bisphosphate carboxylase (rubisco). With the catalytic help of rubisco, RuBP combines with carbon dioxide to form an unstable six-carbon compound which splits immediately into two molecules of a three-carbon compound, glycerate 3-phosphate (GP):

$$RuBP + carbon\ dioxide \rightarrow 2GP$$
$$5C \qquad 1C \qquad 2 \times 3C$$

(where C represents a carbon atom)

Plants that fix carbon dioxide into three-carbon GP are called **C3 plants**.

GP is then reduced to a triose phosphate (a phosphorylated three-carbon sugar) called glyceraldehyde 3-phosphate (GALP). This reduction requires hydrogen and energy from ATP: both are supplied by the light-dependent reactions.

About one-sixth of the total triose phosphate is then used to make glucose. This can be converted into other carbohydrates, for example, sucrose, starch, and cellulose; and also into amino acids, fatty acids, or glycerol. The remaining five-sixths is converted back to RuBP. The regeneration of RuBP requires phosphate from ATP.

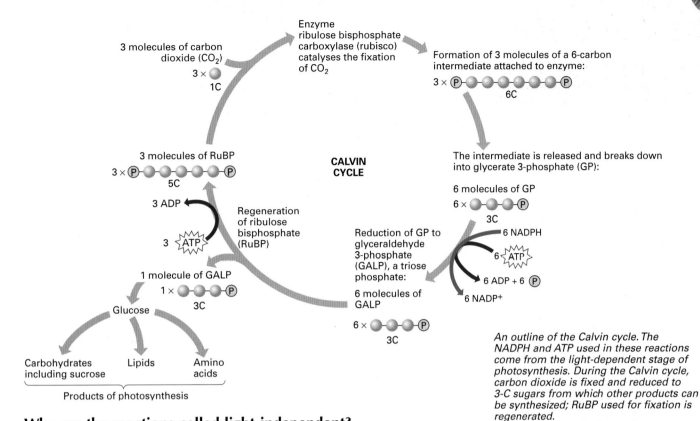

An outline of the Calvin cycle. The NADPH and ATP used in these reactions come from the light-dependent stage of photosynthesis. During the Calvin cycle, carbon dioxide is fixed and reduced to 3-C sugars from which other products can be synthesized; RuBP used for fixation is regenerated.

Why are the reactions called light-independent?

The light-independent reactions can take place in the dark if sufficient ATP and NADPH are available (for example, under experimental conditions using isolated chloroplasts). However, under most circumstances, sufficient ATP and NADPH are available only when ATP and NADPH are being generated by the light-dependent reactions that take place in sunlight. Although the reactions of the light-independent stage do not depend directly on light, they usually take place simultaneously with the reactions of the light-dependent stage.

Check your understanding

1 Why are C3 plants so named?
2 What is the role of NADPH in the Calvin cycle?
3 Describe two functions of ATP in the Calvin cycle.
4 What happens to the triose phosphate formed in the Calvin cycle?

- -

5 In bright sunlight where the temperature is high, carbon dioxide levels are low, and oxygen levels are high, oxygen and carbon dioxide compete for the same active site on rubisco (ribulose bisphosphate carboxylase, the enzyme involved in the fixation of carbon dioxide). Consequently in addition to catalysing the reaction between RuBP and carbon dioxide, rubisco also catalyses the oxidation of RuBP in a process called **photorespiration**. When oxidized, RuBP forms only one molecule of GP and some by-products, including glycolate (a two-carbon compound). GP enters the Calvin cycle, but the by-products are oxidized in a complex series of reactions which salvages some of the carbon dioxide. Photorespiration is regarded as wasteful because, unlike normal respiration or photosynthesis, the by-products become oxidized without producing any ATP or NADPH. It has been shown that under some conditions, photorespiration may reduce the photosynthesizing efficiency of plants by as much as 50 per cent. Some plants that live in conditions which favour photorespiration, such as sugar cane, have evolved a biochemical mechanism which overcomes this problem.

a Define photorespiration.
b Why is photorespiration regarded as wasteful?

By the end of this spread you should be able to

- describe how temperature, carbon dioxide concentration, and light intensity affect photosynthesis
- explain compensation point
- explain the principle of limiting factors as applied to photosynthesis

Prior knowledge

- photosynthesis (2.01–2.04)
- temperature and enzyme-catalysed reactions (*AS Biology for AQA* spread 3.03)

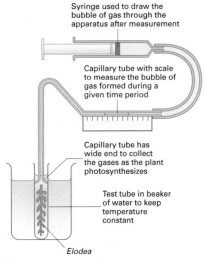

Syringe used to draw the bubble of gas through the apparatus after measurement

Capillary tube with scale to measure the bubble of gas formed during a given time period

Capillary tube has wide end to collect the gases as the plant photosynthesizes

Test tube in beaker of water to keep temperature constant

Elodea

Measuring the rate of photosynthesis in Canadian pondweed

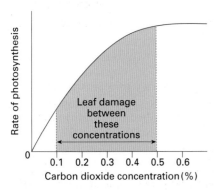

The effect of carbon dioxide level on the rate of photosynthesis

Measuring the rate of photosynthesis

The rate of photosynthesis can be estimated from

- the volume of oxygen released by a plant
- the volume of carbon dioxide taken in by a plant per unit time
- the amount of carbohydrate produced per unit time

In school and college investigations on photosynthesis an aquatic plant is commonly used, as shown here. The volume of gas or the rate at which bubbles of gas are released from the plant are taken as estimates of the photosynthesis rate. However, this does not give an accurate measure of photosynthesis. Assuming that all the gas is oxygen, some generated by photosynthesis is used by the plant for respiration. Respiration goes on all the time, even when photosynthesis is at its height.

Using oxygen liberation underestimates the true rate of photosynthesis. What is actually being measured is the rate of photosynthesis above a point called the **compensation point**. This is defined as the point at which the rate of photosynthesis in a plant is in exact balance with the rate of respiration, so there is no net exchange of carbon dioxide or oxygen.

Factors that affect the rate of photosynthesis

Photosynthesis is affected by many factors, both external (in the environment) and internal (in the plant). External limiting factors include light intensity, the wavelength of light, carbon dioxide levels, temperature, wind velocity, and water and mineral supplies. Internal factors include type and concentration of photosynthetic pigments, enzyme and water content, and leaf structure and position. Temperature, carbon dioxide concentration, and light intensity are three external factors that are relatively easy to manipulate.

Temperature

Changes in temperature have little effect on the reactions of the light-dependent stage because these are driven by light, not heat. However, the reactions of the Calvin cycle are catalysed by enzymes which, like all enzymes, are sensitive to temperature. The effect of temperature on these reactions is similar to its effects on other enzymes. The optimum temperature varies for each species, but for many temperate plants it is between 25°C and 30°C.

Carbon dioxide levels

The average carbon dioxide content of the atmosphere is about 0.04 per cent. An increase in carbon dioxide concentration up to 0.5 per cent usually results in an increase in the rate of photosynthesis. However, concentrations above 0.1 per cent can damage leaves. Therefore the optimum concentration of carbon dioxide is probably just under 0.1 per cent, as the graph shows. In dense, warm, and well lit vegetation, low levels of carbon dioxide often limit the rate of photosynthesis.

Light intensity

The rate of photosynthesis is directly proportional to light intensity up to a maximum point after which the rate levels off because the photosynthetic pigments have become saturated with light. Some other factor (for example, availability of carbon dioxide, or amount of chlorophyll) stops the reaction from going any faster. Very high light intensities may actually damage some plants, reducing their ability to photosynthesize.

The **light compensation point** (the light intensity at which the rate of photosynthesis is exactly balanced by the rate of respiration) varies for different plants. Two major groups have been identified: **sun plants** and **shade plants**. Sun plants include most temperate trees, such as oak. They photosynthesize best at high light intensities. Shade plants include those of the shrub layer, such as ferns. Their light compensation point is relatively low, but they cannot photosynthesize very efficiently at high light intensities. Consequently sun plants out-compete shade plants at high light intensities.

The principle of limiting factors

The **principle of limiting factors** states that when a physiological process such as photosynthesis depends on more than one essential factor being favourable, its rate at any given moment is limited by the factor at its least favourable value and by that factor alone. This factor is called the **limiting factor**.

When other factors are kept constant, an improvement in the value of the limiting factor leads to an increase in the rate of the process. Conversely when the rate of the process does not increase in response to an improvement in an important factor, some other factor is limiting the process. For a process to go at its maximum rate, all factors must be at their optimum level.

Under natural conditions plants are subjected to many factors simultaneously. In all cases, the rate of photosynthesis is determined by the limiting factor (the least optimal factor, the factor that is furthest from its optimum value). The graph opposite shows how light intensity, temperature, and carbon dioxide concentration all interact to limit the rate of photosynthesis.

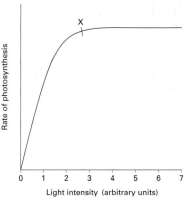

The effect of light intensity on the rate of photosynthesis

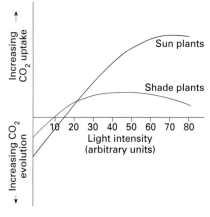

The response of sun plants and shade plants to changes in light intensity

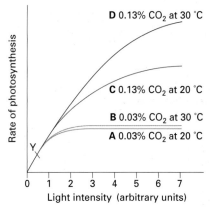

The interaction of different factors and their effect on the rate of photosynthesis. Above a light intensity of about 3 units, the factor limiting the rate of photosynthesis in A and B is carbon dioxide concentration (indicated by the significantly higher rate of photosynthesis in C); temperature is limiting the rate of photosynthesis in C; and a factor other than light intensity is beginning to limit the rate of photosynthesis in D.

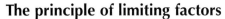

Check your understanding

1 Why is the effect of temperature on photosynthesis similar to its effect on the breakdown of hydrogen peroxide using catalase?

2 How does the light compensation point of a shade plant differ from that of a sun plant?

3 In the graph opposite, which factor is limiting the rate at point Y?

- -

4 Water is essential for photosynthesis. In an investigation, one batch of *Chlorella* (a green alga) was placed in water in which the oxygen atoms were replaced by ^{18}O, a heavy isotope. Then a second, separate batch of *Chlorella* was placed in unlabelled water and given a supply of carbon dioxide labelled with ^{18}O. Analysis by mass spectrometry showed that only the first batch of *Chlorella* gave off ^{18}O. What does this tell you about the source of the oxygen generated by photosynthesis?

OBJECTIVES

By the end of this spread you should be able to

- explain how growers apply a knowledge of limiting factors such as temperature, carbon dioxide concentration, and light intensity in commercial glasshouses

- evaluate such applications using appropriate data

Prior knowledge

- stomata and gas exchange in plants (*AS Biology for AQA* spread 15.04)

- factors limiting the rate of photosynthesis (2.05)

Tomatoes growing in a commercial glasshouse

Enhancement of the environment in commercial glasshouses is just one practical application of biology that can increase food production dramatically. Farmers cultivating wheat and other cereal crops outdoors have little control over the external climate, usually the most important determinant of yields. Plants grown in commercial glasshouses can be maintained in optimal conditions to maximize growth rate and quality.

Water and minerals

Because of the high costs of glasshouse construction and the control of its internal environment, only plants with a sufficiently high monetary value are suitable for commercial glasshouse cultivation. One such plant is the tomato.

All plants require water to survive, and tomatoes are no exception. Like other C3 plants, tomatoes usually respond to water deficiency by closing their stomata. This reduces their ability to take in carbon dioxide and leads to a significant reduction in photosynthesis. Consequently, water deficiencies lead to lower yields.

Having too much water is not good either. Soil flooded with water becomes anaerobic. Lack of oxygen reduces the ability of the roots to take up minerals. This has a detrimental effect on the growth and quality of tomatoes. Commercial glasshouses usually have an automated irrigation system to ensure that soil water levels do not limit tomato growth.

In the natural environment, mineral deficiencies commonly limit plant growth. Several minerals (including phosphorus, copper, manganese, zinc, magnesium, and iron) are involved in energy transfers that directly affect photosynthesis. Potassium is involved in the opening and closing of stomata; too little potassium would therefore impact on photosynthesis efficiency.

In commercial glasshouses, where supplementation is easily achieved, minerals are rarely limiting.

As well as water and minerals, the three environmental variables which are most commonly controlled in commercial glasshouses are temperature, carbon dioxide concentration, and light intensity.

Temperature

Temperature management is very important for the successful cultivation of tomatoes. Poorly controlled temperatures, in addition to reducing photosynthesis, can increase disease problems and lead to changes in fruit colour and loss of quality (for example, high temperatures can lead to the fruit being orange instead of deep red).

Although varieties differ in their temperature optima, most tomatoes produce the largest yields of highest quality fruits when day temperatures are in the range of 19–24°C and when night temperatures remain above 12°C but below 17°C.

As well as varying between day and night, optimal temperatures also vary for different stages of plant development. In one study, the optimum range for germination was 16–29°C whereas that for vegetative growth was 21–24°C.

The temperature of many commercial glasshouses is controlled by thermostats linked to various systems including heating furnaces, ventilation fans, evaporative cooling pads, and shades.

Carbon dioxide

Atmospheric carbon dioxide levels are about 0.035%. Those in glasshouses can easily go down to 0.02%. Like most other C3 plants, tomatoes are highly responsive to changes in the level of carbon dioxide around their leaves. Increasing carbon dioxide levels up to 1% significantly increases growth and fruit yields. Increases above 1%, however, have a harmful effect, causing plants to become thickened and twisted and fruits to turn purple.

Carbon dioxide levels are usually enhanced by burning fossil fuels such as butane or methane. Effective use of this technology requires that glasshouses be closed for long periods each day. On hot days, this can quickly result in the temperature rising above its optimum. In temperate regions, therefore, carbon dioxide enhancement is usually confined to the winter months. In Florida, the frequent need for ventilation of the greenhouse, even in winter, restricts carbon dioxide enhancement; it cannot be maintained for more than an hour or so on most days before the cooling system has to be turned on.

Light intensity

Two major limiting factors for tomato production in northern latitudes is low light intensity and short day lengths. The response of tomatoes to changes in light intensity is typical for a C3 plant whose first product of photosynthesis is a three-carbon sugar phosphate: the photosynthetic rate increases with increased light intensity up to a maximum, after which the rate levels off. Very high light intensities are detrimental causing fruit to suffer cracking and sun scalding. For glasshouses in very sunny locations, shading is essential to maintain the production of high quality fruit.

Commercial tomato growers use artificial lighting to control light intensity and the length of exposure to light. However, the light has to have the correct balance of wavelengths because tomatoes respond to the quality of light as well as its intensity. When artificial light has excess blue light with very little red light, tomato plants become shorter and their fruits become harder and darker in colour. When the light has an excess of red over blue, the growth becomes soft with long internodes, resulting in lanky plants.

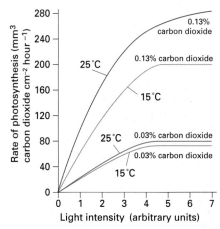

Graph showing the rate of photosynthesis of tomato plants under different environmental conditions

---

Check your understanding

1 Wheat grown in glasshouses gives far greater yields than grown outdoors. Suggest why wheat is not grown commercially in glasshouses.

2 Why is it not possible to give *the* optimum temperature for cultivation of tomatoes?

3 Why do commercial glasshouse tomato growers not usually enhance carbon dioxide in Florida?

4 What type of response do tomato plants have to light intensities up to the optimum?

- -

5 Use the graph to help explain the environmental conditions under which it might be profitable to increase the carbon dioxide content of the air in the glasshouse.

OBJECTIVES

By the end of this spread you should be able to

- describe the main features of aerobic respiration

- show that glycolysis takes place in the cytoplasm and involves the oxidation of glucose to pyruvate with a net gain of ATP and reduced NAD

Prior knowledge

- ultrastructure of mitochondria (*AS Biology for AQA* spread 4.06)

- ATP (2.01)

What is respiration?

To most people respiration means breathing. But to biologists, it is the process by which energy in organic molecules is made available for an organism to do biological work. To avoid confusion, this process is sometimes called **cellular respiration**.

Cellular respiration occurs in every living cell. All cells need energy in the form of ATP for activities such as movement, growth, reproduction, and repair. Organisms obtain ATP by breaking down energy-rich organic molecules either made by themselves (for example, by photosynthesis) or obtained from other organisms (such as by predation). There are two types of cellular respiration:

- aerobic, which requires oxygen and generates relatively large amounts of ATP

- anaerobic, which does not require oxygen but generates relatively little ATP

Aerobic respiration

Aerobic respiration refers to cellular respiration that depends on oxygen. It is often summarized by the equation

$$C_6H_{12}O_6 + 6O_2 \longrightarrow 6CO_2 + 6H_2O$$

glucose   oxygen   carbon dioxide   water

The equation implies that glucose is broken down in a single step, but this is a gross simplification. If glucose is oxidized in this way outside the cell, it burns and creates a flame. Cells have to oxidize glucose in a much more controlled manner so that the heat generated does not destroy them.

Aerobic respiration is a complex process in which energy-rich molecules are broken down in a series of steps. During the breakdown, energy is released which is used to synthesize ATP from ADP and inorganic phosphate. The heat produced during respiration does not usually cause organisms to burn because its release is spread over many biochemical reactions. However, fires associated with bacteria in decomposing haystacks remind us of the dramatic consequences of not allowing the heat from respiration to escape.

There are four major stages in aerobic respiration:

1 glycolysis

2 the link reaction

3 the Krebs cycle

4 the electron transport chain

Glycolysis

Glycolysis involves a series of reactions that takes place with or without oxygen in the cytosol (the fluid part of the cytoplasm) of **eukaryotic cells**. Glycolysis starts with one molecule of glucose and ends with two molecules of **pyruvate** (pyruvic acid), a three-carbon compound. During the process, a nucleotide called **nicotinamide adenine dinucleotide** (NAD) is reduced (it has hydrogen added to it) and there is a net production of two molecules of ATP. Although this is a relatively small energy yield, glycolysis takes little time to complete and does not require oxygen. It can therefore provide energy immediately.

Before glycolysis can start it has to be given some **activation energy**. This is provided by two molecules of ATP which break down by **hydrolysis** to ADP and inorganic phosphate. The glucose is energized by two phosphorylations (the addition of phosphates) to form fructose 1,6-diphosphate and then two molecules of glyceraldehyde 3-phosphate (GALP). To activate glucose, the cell invests two molecules of ATP, with no return.

The reduction of NAD and the synthesis of ATP

Each GALP molecule is converted by a series of reactions to pyruvate. The conversion involves the removal of hydrogen (oxidation). This reaction is coupled with the production of ATP and the reduction of NAD. Each **redox reaction** generates enough energy to synthesize two molecules of ATP. Therefore, for each molecule of glucose broken down during glycolysis, four molecules of ATP are produced. However, because two molecules of ATP are used as activation energy, glycolysis produces a net gain of two ATP molecules for each glucose molecule.

What is NAD?

NAD is **nicotinamide adenine dinucleotide**, an organic molecule formed from two **nucleotides**. It is synthesized in our bodies from vitamin B_3. NAD is regarded as a **coenzyme** because it assists enzymes called dehydrogenases which catalyse the removal of hydrogen from a substrate.

During glycolysis, NAD plays a pivotal role. Its reduction by the addition of hydrogen to form NADH is coupled with the phosphorylation of ADP to ATP. This type of phosphorylation in which the phosphate group is obtained from a molecule other than ATP, ADP, or AMP (adenosine monophosphate) is called **substrate-level phosphorylation**.

To ensure continuous supplies of NAD for cellular respiration, the NADH produced during glycolysis must be oxidized back to NAD. If this did not happen NAD would soon run out and the production of ATP would halt. In aerobic respiration, NAD is regenerated when NADH releases hydrogen into mitochondria. The hydrogen enters the electron transport chain to generate more molecules of ATP. However, in anaerobic respiration when oxygen is unavailable, NAD is regenerated by fermentation, a process in which no more ATP molecules are generated.

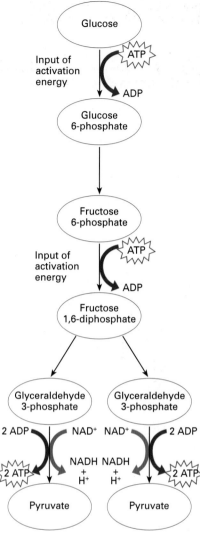

An outline of the main stages of glycolysis, in which one molecule of glucose is broken down to 2 molecules of pyruvate. In addition to pyruvate, glycolysis produces 4 molecules of ATP (with a net production of 2 molecules; 2 are used as activation energy) and 2 molecules of NADH.

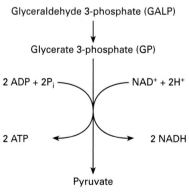

The reduction of NAD is coupled with substrate-level phosphorylation during glycolysis.

Check your understanding

1 Define cellular respiration in terms of the type of energy it generates.
2 For glycolysis, distinguish between the gross production of ATP and the net production of ATP.

2.08 Cellular respiration II

OBJECTIVES

By the end of this spread you should be able to

- understand the significance of the link reaction which involves pyruvate combining with coenzyme A to produce acetyl coenzyme A

- recall that acetyl coenzyme A has a two-carbon fragment that combines with a four-carbon molecule to produce a six-carbon molecule which enters the Krebs cycle

- understand that the Krebs cycle involves a series of oxidation–reduction reactions which generate reduced coenzymes and ATP by substrate-level phosphorylation and which lose carbon dioxide

Prior knowledge

- cellular respiration I (2.07)

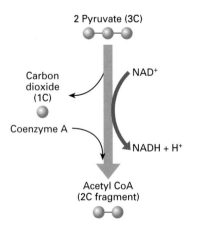

The link reaction: conversion of pyruvate to acetyl coenzyme A by oxidative decarboxylation

The link reaction

The **link reaction** links glycolysis to the Krebs cycle. It occurs only if oxygen is available. In the link reaction, shown below left, pyruvate molecules combine with **coenzyme A** to form **acetyl coenzyme A** (acetyl CoA). The conversion is called **oxidative decarboxylation** because it involves the removal of carbon dioxide (decarboxylation) and the loss of hydrogen (oxidation) from the pyruvate.

Acetyl CoA diffuses from the cytosol into mitochondria, the site of the Krebs cycle.

In the mitochondria, the two-carbon acetyl fragment of acetyl CoA combines with oxaloacetate (a four-carbon compound) to form citrate (a six-carbon compound). The coenzyme A component detaches and is recycled. Two acetyl fragments enter the Krebs cycle for each molecule of glucose that undergoes glycolysis.

The Krebs cycle

The **Krebs cycle** is a key phase of aerobic respiration of glucose. The cycle results in the six-carbon citrate, formed from the combination of the two-carbon acetyl fragments with a four-carbon compound, oxaloacetate, being progressively broken down by a series of reactions and then reformed. The diagram opposite shows the detail of this process. Like the link reaction, the Krebs cycle involves **oxidative decarboxylation**. Two molecules of carbon dioxide are produced in the cycle for every acetyl fragment that enters. The carbon dioxide is excreted as a waste gas.

During the Krebs cycle, hydrogen is stripped off the organic molecules and taken up by coenzymes, either NAD or **flavine adenine dinucleotide** (**FAD**). These redox reactions are exergonic. They release enough energy to synthesize one molecule of ATP for each turn of the cycle. This synthesis of ATP is called **substrate-level phosphorylation** because the addition of a phosphate group to ADP is coupled with the exergonic breakdown of a high-energy substrate molecule.

Each molecule of glucose that enters glycolysis generates two molecules of ATP in the Krebs cycle because one molecule of glucose gives two molecules of acetyl CoA. This immediately doubles the ATP production of glycolysis. However, more importantly, the release of hydrogen during the Krebs cycle provides the reducing power to generate many more ATP molecules in the next stage of aerobic respiration: the electron transport chain.

Check your understanding

1 What is the significance of pyruvate combining with coenzyme A to produce acetyl coenzyme A?

2 In the Krebs cycle, how is citrate formed?

3 Name the coenzymes that are reduced during the Krebs cycle.

- -

4 Acetyl coenzyme A is a complex organic compound that has many more than two carbon atoms. Suggest why the AQA specifications state that 'acetyl coenzyme A is effectively a two-carbon molecule...'.

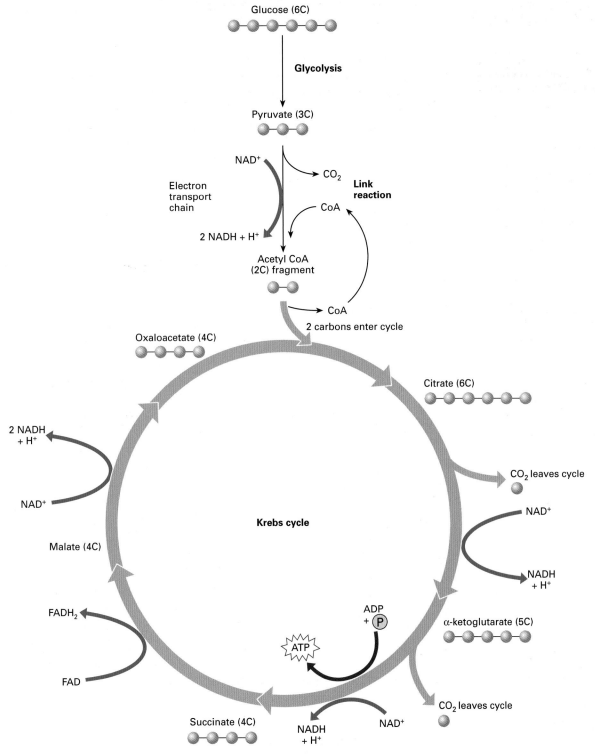

An outline of the Krebs cycle in which two molecules of ATP are generated for each glucose molecule that undergoes glycolysis. Glycolysis takes place in the cytoplasm, while the Krebs cycle along with oxidative phosphorylation (see spread 2.09) occur in the mitochondria.

2.09 Cellular respiration III

OBJECTIVES

By the end of this spread you should be able to

- explain how the synthesis of ATP is associated with the transfer of electrons down the electron transport chain and the passage of protons across mitochondrial membranes

Prior knowledge

- cellular respiration I and II (2.07 and 2.08)

The electron transport chain

The hydrogen atoms carried by reduced NAD and FAD molecules generated in the Krebs cycle are the source of the electrons that are passed along a series of electron carriers which form the **electron transport chain**. For each turn of the Krebs cycle, three molecules of NADH and two molecules of $FADH_2$ are generated. Because the Krebs cycle turns twice for each molecule of glucose respired, six molecules of NADH and four molecules of $FADH_2$ are produced for each molecule of glucose.

A chain of redox reactions

NADH and $FADH_2$ pass on electrons when they donate hydrogen to the next carrier in the system so that a redox reaction takes place (for example, NADH becomes oxidized by losing hydrogen and its electron, whereas the next carrier becomes reduced by gaining hydrogen and the electron). At one point in the electron transport chain, hydrogen atoms split to form protons and electrons. The chain concludes with the reduction of oxygen to form water, a reaction catalysed by an iron-containing enzyme called **cytochrome oxidase** that imparts a brownish colour to mitochondria. An outline of the electron transport chain is shown below. If oxygen, the final acceptor of electrons and protons, becomes absent, or if cytochrome oxidase is inactivated by a metabolic poison such as cyanide, the reactions in the electron transport chain stop.

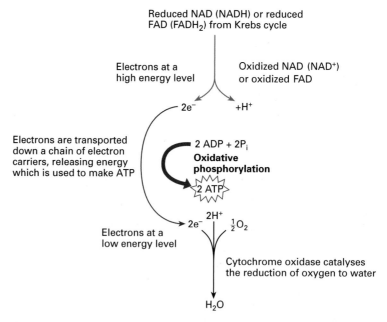

An outline of the electron transport chain. This liberates energy which is used ultimately to synthesize ATP by oxidative phosphorylation. Note that oxygen is the final acceptor of electrons and protons. Without oxygen this type of respiration cannot take place.

The site of the electron transport chains and ATP production

The redox reactions take place on the inner membranes of mitochondria in which the electron carriers are embedded. The inner membrane is highly folded into structures called **cristae** which increase the surface area on which the reactions take place.

According to the **chemiosmotic theory** shown below, each individual redox reaction releases energy, but not enough to make ATP. Instead, the energy is used to pump protons (H^+) from the mitochondrial matrix into a compartment between the inner and outer mitochondrial membranes. The protons accumulate so that a steep concentration gradient develops between the compartment and the mitochondrial matrix. The inner membrane is generally impermeable to protons except at special structures called **stalked particles** which function as **chemiosmotic channels**. Therefore, the protons can only diffuse back from the compartment into the matrix through the stalked particles. The energy associated with the flood of protons through the chemiosmotic channels of stalked particles drives the synthesis of ATP. The phosphorylation of ADP is catalysed by an enzyme called **ATPase** located in the bulbous ends of the stalked particles:

$$ADP + P_i \xrightarrow{\text{ATPase}} ATP$$

Theoretically, 32 molecules of ATP are generated by the electron transport chain for each glucose molecule that undergoes glycolysis. The oxygen-dependent synthesis of ATP within mitochondria using energy released from redox reactions is called **oxidative phosphorylation**.

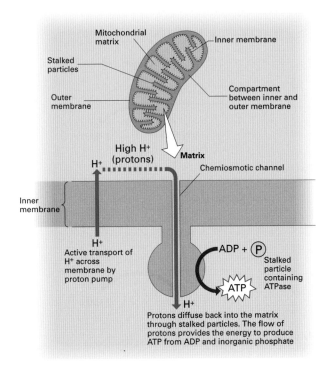

The chemiosmotic theory of how ATP is synthesized from the energy released by the redox reactions in the electron transport chain

Check your understanding

1 Precisely where in the cell of a eukaryote is the electron transport chain located?

2 In relation to oxidative phosphorylation, where is the ATPase located and what is its function?

- - - - - - - - - - - - - - - - - -

3 In the USA in the 1940s, some doctors prescribed low doses of a drug called dinitrophenol (DNP) to help patients lose weight. After the death of a few patients, this method of weight loss was deemed unsafe and was abandoned. Subsequently, it was discovered that DNP makes the lipid layer of inner mitochondrial membrane leaky to protons. Explain how this leads to weight loss.

OBJECTIVES

By the end of this spread you should be able to

- understand that anaerobic respiration involves glycolysis followed by the production of ethanol or lactate and the regeneration of NAD

- compare anaerobic respiration with aerobic respiration

Prior knowledge

- cellular respiration I–III (2.07–2.09)

Anaerobic respiration

Anaerobic respiration takes place in the absence of oxygen. As with aerobic respiration, it involves **glycolysis** and generates ATP and NADH in the same way. However, in anaerobic respiration the NADH is oxidized by **fermentation**, a process in which no more ATP molecules are generated.

Fermentation

In the absence of oxygen, the pyruvate generated by glycolysis can follow one of two metabolic pathways: alcoholic fermentation to produce ethanol, or lactate (lactic acid) fermentation to produce lactate.

Alcoholic fermentation occurs in plants and yeast that are respiring anaerobically. Pyruvate is first converted to ethanal by **decarboxylation** (the removal of carbon dioxide). The carbon dioxide is released as a gas. This feature of fermentation is taken advantage of in bread-making. The decarboxylated pyruvate is then reduced to ethanol using hydrogen from NADH, as shown in diagram (a) below. Ethanol is another product of yeast fermentation that has been exploited for thousands of years.

Lactate fermentation occurs in animals that are respiring anaerobically. The pyruvate from glycolysis is converted in a single step to lactate (lactic acid). The reaction is catalysed by lactate dehydrogenase and the process requires hydrogen from NADH, as shown in diagram (b).

Alcoholic and lactate fermentation enable NAD to be regenerated so that the glycolysis of more glucose molecules can take place.

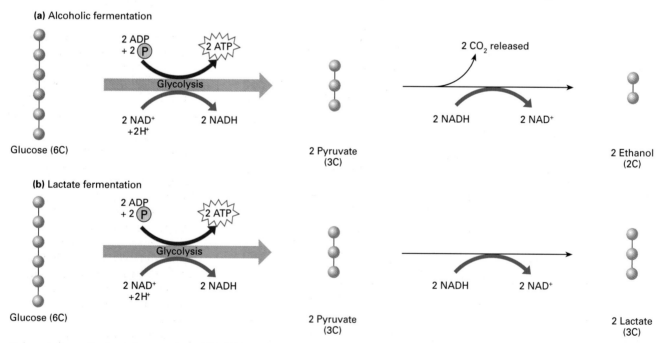

(a) Alcoholic fermentation

Glucose (6C) → Glycolysis (2 ADP + 2 P → 2 ATP; 2 NAD⁺ + 2H⁺ → 2 NADH) → 2 Pyruvate (3C) → 2 CO₂ released; 2 NADH → 2 NAD⁺ → 2 Ethanol (2C)

(b) Lactate fermentation

Glucose (6C) → Glycolysis (2 ADP + 2 P → 2 ATP; 2 NAD⁺ + 2H⁺ → 2 NADH) → 2 Pyruvate (3C) → 2 NADH → 2 NAD⁺ → 2 Lactate (3C)

Fermentation – the fate of pyruvate and NADH in anaerobic conditions: (a) alcholic fermentation in plants and yeast (b) lactate fermentation in animals

Glycolysis transfers only a small proportion of the energy in glucose to ATP. Nevertheless, anaerobic organisms have to satisfy all their energy needs by glycolysis, despite its low level of ATP production. When oxygen is available, aerobic organisms can release a much greater proportion of the energy in glucose to make many more ATP molecules via the Krebs cycle and ultimately via the electron transport chain.

Anaerobic respiration also occurs in aerobes, such as ourselves, when there is insufficient oxygen available to meet the energy demands of cells. An adult human usually produces about 150 g of lactate per day. But during heavy exercise, over 30 g of lactate can be produced in minutes. The production of large amounts of lactate is associated with muscle fatigue, often experienced as cramps.

The efficiency of ATP production

Theoretically, the complete aerobic respiration of each molecule of glucose results in a net production of about 36 ATP molecules. Estimates vary according to the tissue being studied, and there is also some uncertainty about the number of ATP molecules produced from each molecule of reduced NAD. Nevertheless, most experts agree that when oxygen is available, a cell can harvest as much as 40 per cent of the potential energy within one glucose molecule. Most of this production depends on the activities of the electron transport chain.

A much smaller percentage yield is provided by anaerobic respiration. If oxygen is unavailable, the electron transport chain cannot work. This in turn stops the Krebs cycle, because the reduced NAD and FAD cannot be reoxidized. Consequently, anaerobic respiration results in a gross production of only 4 molecules of ATP for each molecule of glucose, with a net production of only 2 molecules of ATP.

Anaerobic respiration	Aerobic respiration
does not require oxygen	requires oxygen
involves glycolysis	involves glycolysis, the link reaction, Krebs cycle, and electron transport chain
generates small amounts of ATP very quickly	generates large amounts of ATP slowly
takes place in the cytoplasm	takes place in the cytoplasm and mitochondria
requires activation energy provided by 2 molecules of ATP	requires activation energy provided by 2 molecules of ATP
has a gross production of 4 ATP molecules per glucose molecule	has a maximum gross production of about 38 ATP molecules per glucose molecule
has a net production of 2 ATP molecules per glucose molecule	has a maximum net production of 36 ATP molecules per glucose molecule
generates reduced NAD (NADH)	generates reduced NAD (NADH)
regenerates NAD by alcoholic fermentation or lactate fermentation	regenerates NAD in the Krebs cycle

Anaerobic and aerobic respiration of glucose compared

Check your understanding

1 Is carbon dioxide released in alcoholic fermentation or in lactate fermentation?

2 Which type of respiration generates ATP more quickly, anaerobic or aerobic?

3 In terms of ATP production per molecule of glucose, which type of respiration is more efficient, anaerobic or aerobic?

- - - - - - - - - - - - - - - -

4 In the respiration of fats, each fatty acid is converted to acetyl CoA by a process called beta oxidation. Essentially, this involves as many two-carbon acetyl fragments being split off the hydrocarbon chain as possible and then combining each of these with coenzyme A. Each acetyl CoA enters the Krebs cycle in the usual manner.

Suggest why the aerobic respiration of one molecule of fatty acid can generate many more molecules of ATP than is generated from one molecule of glucose.

OBJECTIVES

By the end of this spread you should be able to

- describe how energy is transferred through an ecosystem

- recall that photosynthesis is the main route by which energy enters an ecosystem

- describe how energy is transferred through the trophic levels in food chains and how some is always dissipated as heat

- distinguish between pyramids of numbers and pyramids of biomass

Prior knowledge

- biological processes and energy (*AS Biology for AQA* spread 3.01)

- populations and communities (1.01)

- photosynthesis (2.02)

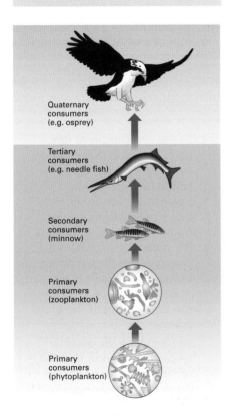

Most food chains start with a producer and end with a top consumer.

Life cannot exist without energy. Energy from the Sun sustains most terrestrial and aquatic ecosystems. The energy flows in one direction through the ecosystem. Green plants and other photoautotrophs harness light energy from the Sun for photosynthesis. Animals and other consumers eat the autotrophs and each other. This one-way flow of energy is a consequence of the laws of thermodynamics: energy can be neither created nor destroyed, and when energy is transformed from one type to another, some is dissipated as heat.

The transfer of food energy in an ecosystem can be shown in a number of ways including food chains, food webs, and pyramids.

Food chains and food webs

A **food chain**, such as that in the diagram below, shows a sequence of organisms in which each organism is the food of the next one in the chain. Each organism within the chain represents a **trophic level**. The arrows show the direction of flow of energy, with the arrow head always pointing away from the organism being eaten.

The number of links in a food chain is usually limited to four or five, because at each trophic level, a high percentage of the useful energy consumed as food is dissipated as heat, and lost from the food chain.

Food chains help to show how substances, for example insecticides, can be transferred from one species to another in an ecosystem. However, they do not reflect the complex feeding relationships that exist in most ecosystems, and they give the false impression that each organism feeds on only one other type of organism.

Most organisms have a wide range of food items, and belong to more than one trophic level. The food chains in an ecosystem are usually interconnected, and this can be shown as a **food web**. An example is shown on the next page.

Even food webs have their limitations. In large ecosystems, the relationships are often so complex that related species with similar diets are grouped together on the food web. Only a part of the ecosystem is usually shown on a food web (most, for example, omit decomposers). A food web showing all the feeding relationships of every species would have so many lines that it would be almost incomprehensible. Also, food webs give no information about how *much* energy flows from one part of an ecosystem to another.

Pyramids of numbers and biomass

A **pyramid of numbers** shows the quantitative relationships between the number of organisms at each trophic level. Each trophic level is represented by a rectangle, the length of which is proportional to the number of organisms at that level. Starting with the rectangle for the first trophic level (primary producers) as the base, the next rectangle is drawn on the centre of the first, and so on. Simple ecosystems, with many more organisms at the lower trophic levels than at higher ones, have a pyramid of numbers with the expected shape, as in (a) on the page opposite. However, pyramids of numbers do not always have this shape. A single large tree, for example as in (b), can form an ecosystem which supports many thousands of insects, and when parasites are included in a pyramid

A pond food web

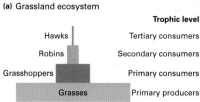

(a) Grassland ecosystem

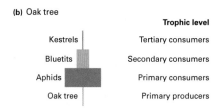

(b) Oak tree

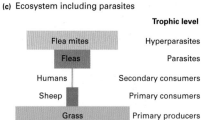

(c) Ecosystem including parasites

Pyramids of numbers for: (a) a grassland ecosystem (b) an oak tree ecosystem (c) an ecosystem that includes parasites

of numbers the upper level can be much larger than the one below it (c). These are called inverted pyramids. They emphasize the fact that numbers of organisms do not always reflect the amount of energy: one large tree, for example, contains much more energy than the organisms that feed on it.

Pyramids of biomass like the one shown below take account of this difference in size between organisms. In such pyramids, the horizontal axis is the total dry mass (**biomass**) of organisms at a particular trophic level, not their number.

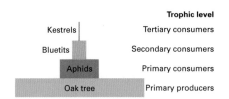

Pyramid of biomass for an oak tree ecosystem

Check your understanding

1 In a food chain, what do the arrows indicate?
2 What is the main route by which energy enters an ecosystem?
3 Why are there rarely more than five links in a food chain?
4 What is biomass?

- -

5 A lipid-soluble organic chemical DDT (dichlorodiphenyltrichloroethane) was the first major synthetic pesticide. Introduced in the mid-1940s to control insect-transmitted diseases in American troops, DDT has been remarkably effective in killing insect pests such as malarial mosquitoes and head lice which transmit typhus. DDT is cheap and effective. However, ecologists found that animals higher up a food chain have more DDT at higher concentrations than those lower in the food chain. For example, the osprey is the top predator in one North American aquatic food chain. The bird was found to contain over 8 million times more DDT than the water taken in by primary producers at the bottom of the food chain. These high concentrations were not enough to kill the birds, but they caused their egg shells to be weakened so that few offspring survived. Suggest why DDT had a higher concentration in osprey than in the organisms lower in the food chain.

OBJECTIVES

By the end of this spread you should be able to

- distinguish between pyramids of biomass and pyramids of energy
- understand how energy flow diagrams can show quantitative aspects of energy transfer between trophic levels

Prior knowledge

- energy transfer I (3.01)

Comparing pyramids of biomass and energy

Pyramids of biomass do not give a meaningful comparison between masses of different trophic levels in the same ecosystem, or between biomasses of the same trophic level in different ecosystems. This is because the biomass usually refers to the **standing crop**, the amount of dry mass measured at any one time. Whereas the standing crop of trees, for example, may result from the accumulation of biomass over 100 years or so, that of algae may represent only a few days' growth. Moreover, the biomass of populations of small organisms may fluctuate considerably throughout the year. The shape of the 'pyramid' therefore depends on the time of the year the sampling takes place. In aquatic ecosystems, this can result in an inverted pyramid of biomass.

If the total biomass accumulated by each trophic level over the whole year was measured, we would usually obtain a pyramid of biomass that has a pyramidal shape. But this still may not be an accurate reflection of energy relationships in an ecosystem because two organisms of the same mass do not necessarily have the same energy content.

The type of pyramid that best reflects energy relationships in an ecosystem is a **pyramid of energy**. In this pyramid the horizontal axis shows the energy entering each trophic level in an ecosystem over a given period of time (often a whole year).

A pyramid of biomass and a pyramid of energy for the same marine ecosystem may look quite different, as shown below. The pyramid of biomass shows that at any one particular time, the biomass of small zooplankton (small animals drifting in the sea) is much greater than the mass of phytoplankton (microscopic primary producers that also drift in the sea). This inverted pyramid results from the fact that most members of the phytoplankton have a life expectancy of a few days whereas zooplankton might live for several weeks, and fish might live for several years. It does not take into account the rapid turnover rate of phytoplankton (that is, their rapid replacement by reproduction). The pyramid of energy for the same marine ecosytem shows that the rate of energy flow decreases at each higher trophic level.

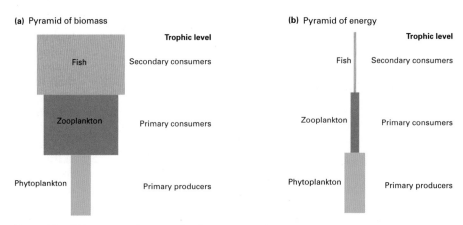

Pyramids of biomass and energy for the same marine ecosystem

Energy flow diagrams

Pyramids of energy give a clear visual image of energy relationships within an ecosystem. However, most are incomplete because they do not show **detritivores** and **decomposers**. These organisms play a vital role in the energetics of ecosystems by eating and breaking down dead organic matter. Up to 80 per cent of all energy incorporated into primary producers may not be eaten by consumers, but is transferred directly to detritivores or decomposers.

Pyramids of energy also do not show how much energy passes to each part of the ecosystem. An alternative way of illustrating the transfer of energy through an ecosystem is by an **energy flow diagram**, as shown below. This type of diagram shows the amounts of energy (usually expressed as kJ m^{-2} y^{-1}) entering different parts of an ecosystem, and what happens to that energy. Using diagrams like this, it is possible to compare the efficiencies of energy transfer at different trophic levels or in different ecosystems.

*Detritivores are organisms that rely on **detritus**, decomposing particles of organic matter, for food. Prawns (Palaemon spp.) are partial detritivores which feed on decaying seaweeds trapped in the pools of rocky shores.*

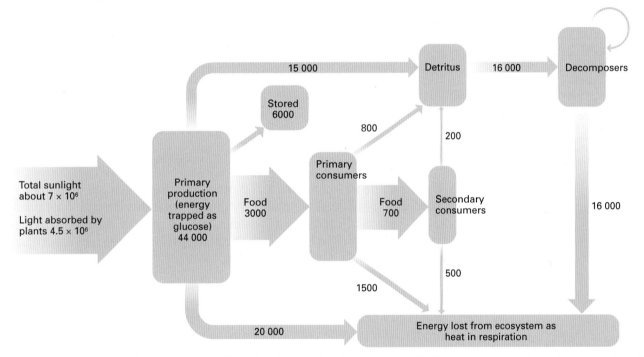

Energy flow diagram for a forest ecosystem. Note that the transfer of energy obeys the first law of thermodynamics; energy is neither created nor destroyed as it passes from one trophic level to the next. The figure shows energy in kilojoules per square metre per year (kJ m^{-2} y^{-1}).

Check your understanding

1 Which is the best way of comparing two ecosystems, by a pyramid of biomass or a pyramid of energy? Explain your answer.

2 In a rock pool, seaweeds had a biomass of 30 g m^{-2} and the animals had a total biomass of 1000 g m^{-2}.

  a What would be the shape of the pyramid of biomass?

  b Account for this shape.

- -

3 Energy enters an ecosystem in the form of light. In what form does it eventually return to the abiotic environment?

3.03

OBJECTIVES

By the end of this spread you should be able to

- compare natural ecosystems and those based on modern intensive farming in terms of energy input and productivity

- recall that net productivity is defined by the expression

  Net = gross − respiratory
  productivity productivity loss

Prior knowledge

- energy transfer I and II (3.01–3.02)

The fate of solar energy falling on a leaf. Most of the solar energy is not absorbed by the leaf, but is either reflected, transmitted, or used to evaporate water. When solar energy of a suitable wavelength is absorbed by a chloroplast, only about 20% is converted into organic molecules; the remainder is converted to heat.

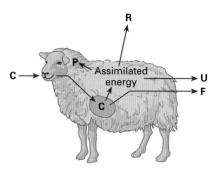

The flow of energy through a consumer.

$$C = P + R + U + F$$

where

C = energy consumption (food ingested)

P = secondary productivity (energy stored in new body tissues or used to make reproductive products)

R = respiration (energy transferred to the environment as heat as a result of the body's metabolic activities)

U = energy in urine

F = energy in faeces

Energy and food production: efficiency of primary productivity

One square metre of the Earth's surface can receive up to 5 kJ of solar energy per minute. However, only a small percentage of incident solar energy (generally about 2% but usually no more than 8%) actually ends up as new plant tissue, as shown here.

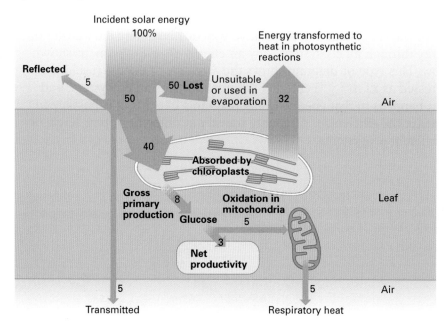

The total rate at which plants and other autotrophs synthesize organic material (such as carbohydrate) is called **gross productivity**. It includes organic material used in respiration as well as organic material stored as new plant tissue. The rate at which autotrophs store organic material as new tissue is called **net productivity**:

Net productivity = gross productivity − respiratory loss

Only the energy that is stored in plant tissues is available to consumers. Net productivity values, therefore, are often used to compare the productivity of different ecosystems.

Efficiency of secondary productivity

Usually only a small proportion of the energy stored in plant tissue is transferred to the next trophic level. Most plants die and decompose without being used by animals.

The rate at which energy or organic material is built up (**assimilated**) into the tissues of a consumer (primary, secondary, tertiary, or quaternary) is called **secondary productivity**. Energy flow diagrams can be compiled to show what happens to the food consumed by an animal. For example, when a herbivore such as a sheep consumes plant material, some remains undigested and is egested in the faeces. Digested plant material is absorbed by the herbivore, and some energy in this material is used in respiration and transferred to the environment as heat; some will be transferred to the environment in urine and other secretions; and some will be used for growth of new tissue and reproduction as shown in the energy budget diagram opposite.

Energy budgets

The flow of energy through an organism can be summarized in an energy budget expressed by the following equation:

$$C = P + R + U + F$$

According to the first law of thermodynamics, the energy input must equal the energy output. Therefore the equation can be used to estimate unknown variables. For example, assuming the other variables are known, P can be calculated from the equation

$$P = C - (F + R + U)$$

The equation can also be used to calculate the assimilated energy, that is, the energy absorbed by the gut that is subsequently used to produce new tissues, or is lost in respiration and the urine. Therefore,

$$\text{Assimilated energy} = P + R + U$$

and

$$\text{Assimilation efficiency} = [(P + R + U)/C] \times 100\%$$

Trophic efficiency

It is generally stated that about 10% of the energy in one trophic level ends up in the next trophic level. However, this figure (called **trophic efficiency**) is only a rough average. Some herbivores eating large amounts of indigestible plant material have a trophic efficiency of less than 1%; zooplankton feeding on phytoplankton, on the other hand, may have a trophic efficiency of up to 40%. Nevertheless, a diminishing amount of energy is passed from one trophic level to the next. This is why the number of trophic levels in an ecosystem is limited to four or five.

Intensive farming

In agriculture, trophic efficiencies can be increased enormously by artificially breeding new varieties of plants and animals that have higher trophic efficiencies than their natural counterparts, and by growing organisms intensively indoors in a confined space under controlled environmental conditions. Glasshouses, for example, are designed to maximize the conversion of light energy to the plant product desired, be it flowers or fruit. In factory farms, the animals are usually given a highly digestible and easily assimilated diet supplemented with growth stimulants, and their movements are restricted to ensure that food consumed by an animal is converted efficiently into useful products such as eggs, meat, or milk.

• •

Check your understanding

1 Why is net productivity used to compare the efficiencies of different ecosystems?

2 If an area of 8000 m^2 of grassland is needed to keep a cow, the productivity of grass is $21\,000$ kJ m^{-2} y^{-1}, and the production of new cow tissue is 5×10^6 kJ m^{-2} y^{-1}, calculate the percentage of the energy in the grass that is used in the production of new tissue. Show your working.

3 Suggest two reasons why keeping farm animals indoors would reduce respiratory losses.

OBJECTIVES

By the end of this spread you should be able to

- discuss the ways in which productivity is affected by farming practices

- compare the use of natural and artificial fertilizers

Prior knowledge

- energy transfer (3.01–3.03)

A farmer muck spreading. The delivery of this valuable material is poorly controlled compared with the application of pellets of inorganic fertilizer. The heavy machinery needed to spread the natural fertilizer also means that it cannot be applied to growing crops such as wheat, or on wet soils which might become badly compacted, damaging soil structure.

The recent increase in agricultural production has been brought about not only by glasshouse growing and factory farming (spread 3.03) but also by many other methods. These include improved soil management.

Soil management

Soil, the upper weathered layer of the Earth's crust, is a vastly underrated resource. Many people think of it as dirt, but soil, though rarely more than 2 m deep, provides the nutrients and water for the plants which, directly or indirectly, provide most of the food we eat. The environmental factors that operate in soil are called **edaphic factors**. The four most important are outlined below.

- Physical structure: this depends mainly on the relative proportions of different sized particles and the way they interact with the organic content of the soil. The proportions of clay, sand, and silt determine soil texture. The distribution, shape, and stability of aggregates of soil particles make up soil structure.

- Water and air content: the amount of water in a soil depends on its water-holding capacity and the extent to which water can move down through the soil. Both of these properties are affected by soil texture and structure. Soil air is present in spaces between soil particles that are not filled with water. The amount of air therefore depends upon the texture and structure of the soil and also on the amount of water it contains. Waterlogged soils usually lack oxygen. This inhibits root activity, slows down decomposition of organic matter, and leads to mineral deficiencies.

- Chemical composition: this includes the nutrient content and the amount of humus in the soil. Humus is organic matter that has been broken down by fungi and bacteria. Soils with a high humus content retain water well without preventing drainage. They therefore tend to be damp but not waterlogged. Humus also has a negative electrical charge which tends to bind positively charged mineral ions to the soil, preventing the minerals from being washed out of the soil by rain (a process called **leaching**).

- pH: the hydrogen ion concentration of soils affects the solubility and availability of minerals in the soil, as well as affecting plant enzymes.

Farming techniques used to improve soils include

- ploughing: by mechanically breaking apart large clods of soil to remove weeds and obtain a good tilth (that is, a crumbly porous soil with a stable granular structure)

- liming: the addition of agricultural lime to increase soil pH which neutralizes acidic soils and promotes clumping of soil particles

- drainage and irrigation: to optimize water and air content

- the application of fertilizers

Improving soil fertility

Soil fertility and crop production can be improved by using **natural** or **artificial** fertilizers. The addition of natural fertilizers to a soil has the added advantage of promoting humus formation and improving soil texture. However, compared with artificial fertilizers, natural fertilizers such as manure and sewage sludge are often more difficult to handle and take a longer time to have an effect. They also have a more variable composition and their action is less predictable.

Artificial **inorganic fertilizers** come in the form of pellets, granules, or liquids that contain one or more inorganic nutrients (most commonly a mix of nitrogen, phosphorus, and potassium). They are expensive to make, especially in terms of energy used, but they are relatively easy to apply, and are quick acting and reliable. Used appropriately, inorganic fertilizers promote plant growth, but heavy use may damage plants by 'burning' or 'scorching' them, and may increase soil acidity. Also, artificial fertilizers are more easily leached into aquatic environments than are natural fertilizers. Leaching of nitrates and phosphates can lead to **eutrophication**, an over-enrichment of nutrients in the water that encourages the rapid growth of algae and plants which, when they die and decompose, deoxygenates the water.

Without fertilizers, heavy harvesting of most commercial crops would deplete the soil of minerals. According to the **principle of limiting factors**, a deficiency of any one mineral limits growth. The most common limiting factors are nitrogen, phosphorus, and potassium. Used in the correct amounts, fertilizers can increase crop yields greatly, but their application obeys the law of diminishing returns. Above a certain level, the increase in crop yield for each unit of fertilizer gets less. Even worse, above a critical level, adding fertilizers actually reduces crop yields and damages the environment.

The most appropriate fertilizer must be chosen for the crop and the soil. Crop yield will be boosted only if a fertilizer contains a nutrient that has been acting as a limiting factor.

A bag of combined NPK fertilizer (N = nitrogen, P = phosphorus, and K = potassium). The fertilizer may take the form of granules that can be applied in precise amounts by light machinery to a young cereal crop.

Check your understanding

1 What effect does liming have on an acidic soil?

2 Which type of fertilizer releases nutrients more slowly, a natural fertilizer or an artificial fertilizer?

- -

3 When applying a fertilizer, any financial benefit of increased yields should exceed the cost of the fertilizer and its application, including labour, machinery, and fuel. Study the table. Assume a value of £100 per tonne of crop and a cost of 50p per kg of inorganic fertilizer.

  a For each level of fertilizer application, calculate the cost–benefit ratio, which is given by the value of increased yield per cost of fertilizer. Show your working.

  b Describe how the cost–benefit ratio changes as the rate of fertilizer application increases.

Fertilizer applied (kg hectare^{-1})	Crop yield (tonnes hectare^{-1})
0	2.7
50	3.4
100	4.3
150	4.8
200	4.9

Yields of a cereal crop at different levels of inorganic fertilizer application

OBJECTIVES

By the end of this spread you should be able to

- describe what a pest is
- consider how intensive cultivation of crops and rearing of domestic livestock can affect the incidence of pests
- compare the use of chemical pesticides, biological agents, and integrated systems in controlling pests on agricultural crops

Prior knowledge

- populations and ecosystems (1.01)
- food chains and food webs (3.01)

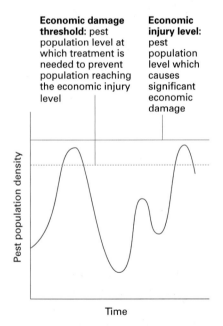

Economic damage threshold: pest population level at which treatment is needed to prevent population reaching the economic injury level

Economic injury level: pest population level which causes significant economic damage

The economic injury level and economic damage threshold of pest populations

Intensive farming and pests

A **pest** is any organism that has an undesirable effect. In farming, it may cause harm economically, or affect the health of crops or livestock.

Pests can have a devastating effect on crops, especially in **monocultures** in which only one crop is grown. Monoculture systems are simpler than natural ecosystems and usually lack pest predators. Also, growing the same crop on the same land every year allows pest populations to build up.

Pests can have an equally damaging effect on domestic livestock reared in crowded conditions. Parasitic worms, for example, have evolved a very high reproductive output because in uncrowded natural populations there is only a small chance of finding a new host. In crowded conditions, the high reproductive output of the parasite can lead to its rapid transmission from one host to another.

Whereas in a natural environment, a pest may affect only one or a few plants and animals, on an intensive farm they can destroy whole crops of cereal, whole flocks of sheep, or whole herds of cattle. For this reason, effective pest control is an essential part of modern farming.

Most pests cause significant economic harm only when their population reaches a certain level, called the **economic injury level**. To prevent fast-growing pest populations from reaching this level, control measures have to be started at a lower pest population level, called the **economic damage threshold**. Pests may be controlled chemically, biologically, by cultural methods, or by a combination of methods.

Cultural methods of pest control

Cultural methods such as weeding, tillage, and crop rotation are among the most common methods of pest control. Weeding and tillage help to remove weeds; overturning the soil may expose insect pests to predatory birds.

Crop rotation often prevents the build-up of pests that occurs in monocultures, but it is only effective when a pest cannot attack successive crops. In Europe crop rotations that include potato, oilseed rape, and wheat are popular because they discourage the build-up of several potentially damaging pest species.

Crop damage can also be minimized by growing the crop at a particular time in the life cycle of the pest. The crop is sowed or harvested at times when the pest can do least damage, for example, before insect pests lay eggs, or before adults emerge from their dormant state. Other cultural methods of pest control include

- removing the remains of crops and badly damaged plants which might harbour pests
- creating physical barriers (for example, apple trees are protected from codling moth caterpillars by putting sticky bands on their trunks)
- covering the soil with organic material (mulching) which prevents light from reaching weeds
- **intercropping** – planting two different crops in the same field; for example, undersowing cereal crops with rye grass provides suitable conditions for ladybirds which control aphids on the cereals

Chemical control

Toxic chemicals that control pest populations are called **pesticides**. There are hundreds of different types, usually classified according to the pest organism they treat. These include **fungicides**, **herbicides**, and **insecticides**. **Contact pesticides** kill pests without being eaten (for example, by penetrating the cuticles of insects); **systemic pesticides** are taken into a plant and translocated within the plant, and enter the pest when it eats the plant or sap.

Modern pesticides have played an essential part in feeding an increasing world population, but most have one or more undesirable features. **Broad-spectrum pesticides**, designed to affect a wide range of pests, may also kill harmless organisms or beneficial ones such as the predators of the pest. Other pesticides, such as DDT, are very stable. They persist in the environment and, even if used in very small concentrations, may build up in the food chain. Most pesticides lose their effectiveness as the pest evolves resistance to them. There is a continual search for alternative methods of pest control.

Biological control

Biological control is the control of pests and weeds by other living organisms or by biological agents. A predator or parasite is usually used to control the pest. The aim is not to eliminate the pest or weed (that would also wipe out the control organism), but to use the control organism to keep the pest or weed population below the economic injury level. Care must be taken to ensure that the control organism does not become a nuisance itself. Cane toads, for example, introduced into Australia to control beetle infestations of sugar cane, have become a serious pest because they kill non-target organisms such as pygmy possums as well as the beetles.

Biological agents include pheromones and genetically engineered insecticides. **Pheromones** are chemicals released, for example, by a female insect to attract a mate. Synthetic pheromones have been used to lure pests into traps laced with an insecticide.

Biological control agent: ladybirds have a voracious appetite for aphids.

Integrated pest management

In practice, no one type of pest control is ideal. The Food and Agriculture Organization of the United Nations recommends considering fully the environmental context and biology of the pest, and then using all suitable methods to maintain the pest population at a level below the economic injury level. This form of pest control is called **integrated pest management**. Pest population densities are monitored, and pesticides used only when the pest reaches the economic damage threshold.

Check your understanding

1 Define a pest.

2 Which feature of the life cycle of a parasitic worm makes it such a potentially dangerous pest where animals are reared in crowded conditions?

3 In which type of pest control, chemical or biological, might it be a disadvantage to destroy all the pests?

4 The controlled environment in glasshouses often favours pests such as whitefly and aphids. A parasitic wasp, *Encarsia formosa*, is commonly introduced to control whitefly. The temperature requirements of the wasps are more critical than those of the whitefly:

- wasp populations do not survive temperatures below 10°C
- the wasp population growth matches that of the whitefly population above about 18°C
- wasps reproduce much more rapidly than the whitefly at temperatures above 26°C

Suggest the implications of these observations on the use of the parasitic wasp to control whitefly populations in greenhouses.

A factory farm for chickens

Low intensity farming: free-range chickens

Pest control

The greatest concerns relating to chemical pesticides are their toxicity to non-target species (especially humans). Other concerns relate to their cost, effectiveness on the target species, and persistence in the environment. Some people object to the use of *any* pesticides and believe that no organism should be intentionally killed by direct human intervention.

Assuming that a farmer accepts the use of chemicals to control pests, the ideal chemical pesticide is

- cheap
- effective
- specific (it only affects the target species)
- broken down quickly to form harmless substances

Biological control agents also raise concerns as is demonstrated by the biological control of rabbit populations.

Rabbits are serious pests in parts of Britain and Australia. Their legendary reproductive output can result in rapid population growth. Overgrazing by large populations can lead to soil erosion with both economic and environmental repercussions. In Australia, all existing rabbits are thought to have originated from 24 individuals introduced in 1859. With plenty of food, unlimited space, and few natural predators, the population boomed. In a desperate attempt to control the rabbit population, in 1950 a viral disease called **myxomatosis** was introduced deliberately into Australia to control rabbit populations. The result was dramatic. Rabbit populations plummeted by more than 90%. Grazing land improved and livestock production increased.

However, within a few years rabbits began to evolve a resistance to the disease. The numbers began to rise, not as much as before, but enough to cause concern. In the 1990s another viral disease (haemorrhagic viral disease) was introduced. This has, once again, had a devastating effect on the rabbit population, but has caused considerable controversy.

Many Australians have objected to the introduction of viral diseases because of the way they disable rabbits before they die. There has also been some concern about the possibility of the virus attacking non-target animals such as domestic rabbits.

Intensive farming

The demand for cheap, plentiful food has led to the development of very intensive systems such as factory farming. By keeping animals indoors in controlled environments in which movement is restricted, and by feeding them specific diets, often with added growth supplements and antibiotics, factory farmed animals gain weight quickly and are ready for slaughter at a much earlier age than animals that are reared outdoors. For example, factory farmed cattle can average weight gains of more than one kilogram a day, making them ready for slaughter in under a year. (Antibiotics are given to farm animals mainly to control disease, but they also boost growth rates.) Factory farming is an emotive topic that can provoke strong arguments.

Some of the points made for factory farming

- It is the only way to supply cheap food in the high quantities demanded by the public.
- The conditions are controlled to maximize the health as well as growth of the animals; the animals' welfare is looked after. If it were not, animal growth and production would not be maintained.
- Less intensive forms of livestock rearing would require more countryside to be used as farmland and would harm wildlife.

Some of the points made against factory farming

- It does not allow animals to have all the Five Freedoms, and the wellbeing of animals should not be sacrificed for financial gain.
- Antibiotics, pesticides, and growth-promoting supplements may harm human health and the environment.
- Factory farming, like intensive cultivation of crops, is not cost effective because it relies on the heavy use of fossil fuels.

The Five Freedoms

The following information is reproduced from the Farm Animal Welfare Council (FAWC) which is an independent advisory body established by the Government in 1979.

The welfare of an animal includes its physical and mental state and we consider that good animal welfare implies both fitness and a sense of well-being. Any animal kept by man, must at least, be protected from unnecessary suffering.

We believe that an animal's welfare, whether on farm, in transit, at market or at a place of slaughter should be considered in terms of 'five freedoms'. These freedoms define ideal states rather than standards for acceptable welfare. They form a logical and comprehensive framework for analysis of welfare within any system together with the steps and compromises necessary to safeguard and improve welfare within the proper constraints of an effective livestock industry.

1 **Freedom from Hunger and Thirst** – by ready access to fresh water and a diet to maintain full health and vigour.

2 **Freedom from Discomfort** – by providing an appropriate environment including shelter and a comfortable resting area.

3 **Freedom from Pain, Injury or Disease** – by prevention or rapid diagnosis and treatment.

4 **Freedom to Express Normal Behaviour** – by providing sufficient space, proper facilities and company of the animal's own kind.

5 **Freedom from Fear and Distress** – by ensuring conditions and treatment which avoid mental suffering.

Check your understanding

1 How does the addition of antibiotics directly affect the production of meat in cattle?

2 Give **a** an environmental and **b** an economic reason for the intensive breeding of livestock.

3 Give two reasons why some Australians objected to the use of viruses to control rabbit populations in Australia.

- -

4 Which of the Five Freedoms are particularly hard to achieve in highly intensive livestock breeding systems?

OBJECTIVES

By the end of this spread you should be able to

- recall that chemical elements are recycled in ecosystems whereas energy flows through ecosystems

- appreciate that microorganisms play a key role in recycling chemical elements

- understand the general features of nutrient cycles

Prior knowledge

- populations and communities (1.01)

- food chains and food webs (3.01)

Recycling the building blocks of life

The Earth is a closed system for matter except for small amounts that enter the atmosphere from space. This means that all the chemical elements needed for the structure and function of living organisms come from those present in the Earth's crust when it was formed billions of years ago. These chemical elements, the building blocks of life, continually cycle through the Earth's environmental systems over timescales that vary from a few days to millions of years. The environmental systems are the **atmosphere**, **hydrosphere**, **biosphere**, and **lithosphere**. The cycles are called **biogeochemical cycles** because they include a variety of biological, geological, and chemical processes.

Chemical elements move through the environmental systems in quite a different way to energy. When an organism uses energy to carry out its activities, some of the energy is dissipated as heat in the environment and is not available to be used again by another organism. Hence energy flows in one direction through an **ecosystem**. In contrast, when an organism absorbs a chemical element from the environment and assimilates it into its body, it does so only temporarily. When the organism dies and decomposes, the element is returned to the environment and can be used again.

Nutrient cycles

Biogeochemical cycles involving chemical elements essential to life are called **nutrient cycles**. They include the carbon cycle and the nitrogen cycle. Both of these operate on a global scale. At any one time, an element such as carbon or nitrogen may be in the biotic phase or abiotic phase of a nutrient's cycle.

The **biotic phase** consists of the **biosphere** in which elements are incorporated into the compounds of organisms. An element may be passed from primary producers to consumers in the food chain, but it ultimately ends up in decomposers. These release the element back into the abiotic phase.

In the **abiotic phase**, chemical elements usually take the form of inorganic ions or are part of inorganic compounds. The elements occur either in solution in soil water or aquatic environments (the hydrosphere), or in the atmosphere as a component of gases, or in rocks or sediments (the lithosphere). Weathering of rocks and sediments releases elements into the soil, water, or atmosphere. This process may take a very long time. Once in solution or released as gases in the atmosphere, the elements can be taken up by primary producers; this might require the assistance of other organisms. In doing so, the elements move from the abiotic phase to the biotic phase, completing the cycle.

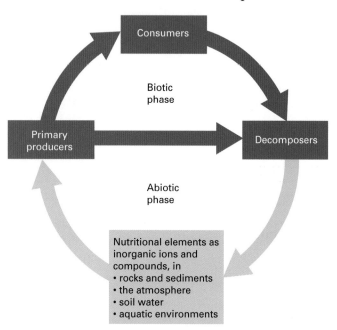

The abiotic and biotic phases of a nutrient cycle. The abiotic phase is shown in blue and the biotic phase in green.

Decomposition is a key feature of nutrient cycles. It enables elements locked in an organism's dead body to become available to other organisms. **Decomposition** results in the breakdown of organic matter into simpler compounds and eventually into inorganic materials. It involves **decomposers**: microorganisms, mainly bacteria and fungi, that obtain their nutrition from dead or decaying organic matter.

Storing elements

In addition to having abiotic and biotic phases, nutrient cycles have reservoirs and exchange pools. **Reservoirs** are those parts of the cycle where the element is held in large quantities for prolonged periods of time. **Exchange pools**, on the other hand, are where elements are held for only a short time. The length of time an element is retained in a reservoir or an exchange pool is called its **residence time**. Whereas oceans are reservoirs in which water may reside for thousands of years, a cloud is an exchange pool in which water is held for a few days or less. The biosphere acts as an exchange pool and also serves to move elements from one sphere of a cycle to another. For example, trees bring water up from woodland soil to be evaporated into the atmosphere. Cockles, oysters, and mussels extract carbon from sea water and use it to make their calcareous shells which eventually turn into limestone rock.

Dung beetles are important decomposers that occur on all continents except Antarctica. Most species feed exclusively on faeces. They play a particularly important role in farmland ecosystems, improving nutrient cycling and soil structure by consuming or burying dung. On cattle ranches, they protect livestock by removing the dung which, if left, could provide a habitat for pests such as flies. Several countries, including Australia, have introduced the creature for the benefit of their agriculture.

Check your understanding

1 How does the movement of chemical elements in an ecosystem differ from the movement of energy?

2 What role do decomposers have in a nutrient cycle?

3 Distinguish between a reservoir and an exchange pool.

- -

4 Suggest how the movement of a chemical element through an ecosystem could be monitored.

By the end of this spread you should be able to

- describe the main features of the carbon cycle

- discuss the importance of respiration, photosynthesis, and human activity in the carbon cycle

Prior knowledge

- photosynthesis (2.02)

- carbon dioxide fixation (2.04)

- respiration (2.07)

Carbon is central to life on Earth. It forms a part of all major biological molecules (carbohydrates, lipids, proteins, and nucleic acids). The **carbon cycle** below shows the pathway by which carbon atoms are passed from one organism to another, and between organisms and their environment.

Carbon in the atmosphere and the oceans

Although carbon dioxide gas makes up only a small proportion of the atmosphere (less than 0.04 per cent), it is the main source of carbon for organisms living on land. For aquatic organisms, the main source is hydrogencarbonate ions (HCO_3^-) formed from dissolved carbon dioxide (and carbonate rock). The store of carbon in the oceans is 50 times that in the atmosphere. Links in the carbon cycle between the atmosphere and the oceans mean that as carbon dioxide concentration increases in the atmosphere, more becomes dissolved in water. This carbon dioxide is then converted to hydrogencarbonate ions. The reverse happens when atmospheric carbon dioxide concentrations are low.

Fixing carbon dioxide

The carbon in the atmosphere and oceans is made available to organisms by **photosynthesis**. Green plants and other photoautotrophs fix carbon dioxide and convert it to carbohydrates and other organic molecules. The carbon is then passed along the **food chain** as an animal eats a plant, or as one animal eats another. Carbon is returned to the atmosphere or oceans mainly as carbon dioxide, by the **respiration** carried out by all organisms. Unlike the nitrogen cycle, the assimilation of carbon into autotrophs does not depend on the action of **decomposers**. Nevertheless, decomposers play an important part in the cycle by releasing carbon

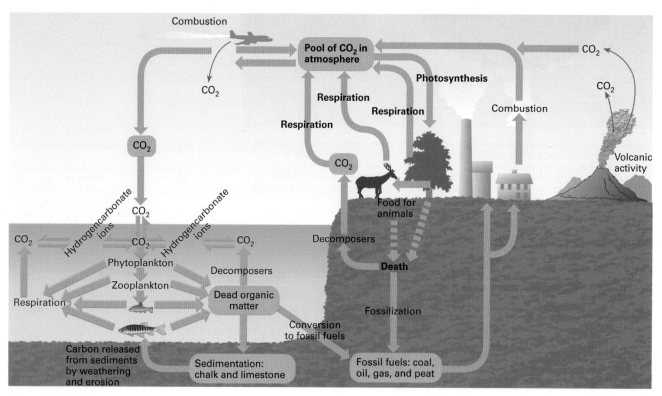

The carbon cycle

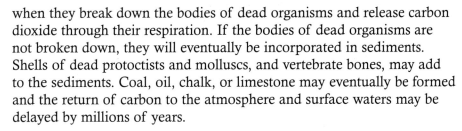

when they break down the bodies of dead organisms and release carbon dioxide through their respiration. If the bodies of dead organisms are not broken down, they will eventually be incorporated in sediments. Shells of dead protoctists and molluscs, and vertebrate bones, may add to the sediments. Coal, oil, chalk, or limestone may eventually be formed and the return of carbon to the atmosphere and surface waters may be delayed by millions of years.

Releasing carbon dioxide into the atmosphere

Carbon in the **lithosphere** (the Earth's crust, including the mineral fraction of soil) may be made available to organisms by volcanic activity, or when sedimentary rocks are broken down by weathering. However, in the short term, these processes usually play only a minor role in the carbon cycle.

In recent years, human activities have had a much more dramatic impact in releasing carbon trapped in sediments. It is estimated that more than six billion tonnes of fossil fuels (coal, oil, gas, and peat) are burned each year. A significant but smaller amount of carbon dioxide is also released into the atmosphere during deforestation and the production of cement from limestone.

Although some of the carbon dioxide produced by human activities dissolves in surface waters or is assimilated by primary producers, a significant amount is being added to the atmosphere. It has been estimated that carbon dioxide levels have increased steadily in the atmosphere from about 280 parts per million (ppm) in 1750 to over 350 ppm in the 1990s.

Atmospheric carbon dioxide and temperature changes

Scientists working in Greenland have shown that changes in carbon dioxide are not a new phenomenon. Measuring carbon dioxide concentrations in air bubbles trapped in ice cores, they found that carbon dioxide levels there have fluctuated from about 180 to 300 ppm (parts per million) over the last 250 000 years. Comparing these long-term changes in atmospheric carbon dioxide with estimates of temperature changes during the same period revealed a very strong correlation between the two variables, as the graphs here show. However, such a correlation does not *prove* that changes in carbon dioxide levels cause temperature changes. From this data, no one can be sure if a high carbon dioxide level results from a warmer atmosphere or is the cause of it.

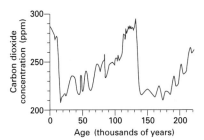

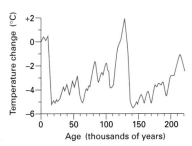

Long-term changes in atmospheric carbon dioxide concentration and temperature

Check your understanding

1 By what process does carbon in the atmosphere become available to organisms in a community?

2 What is the role of decomposers in the carbon cycle?

3 State three ways by which human activities add carbon to the atmosphere.

- -

4 Between 1960 and 1990, some of the most convincing evidence for rising atmospheric carbon dioxide levels came from measurements of air samples taken at an observatory at Mauna Loa in Hawaii. The observatory is in an ideal position as a sampling point for atmospheric gases because it is in the middle of the Pacific Ocean, remote from the local effects of continental industries. Describe and explain the trends shown in the graph.

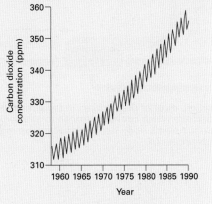

Carbon dioxide concentrations at Mauna Loa observatory, Hawaii

OBJECTIVES

By the end of this spread you should be able to

- describe the role of microorganisms in the nitrogen cycle in sufficient detail to illustrate the processes of saprobiotic nutrition, ammonification, nitrification, nitrogen fixation, and denitrification

Prior knowledge

- nutrient cycling (3.07)

Nitrogen is a vital component of every living organism. It forms part of all proteins, nucleic acids, and their products. The nitrogen cycle shown below ensures that nitrogen bound up within these organic molecules in the biosphere is constantly recycled so that it may be reused. The cycle involves saprobiotic nutrition, ammonification, nitrification, nitrogen fixation, and denitrification.

Saprobiotic nutrition: breaking down dead organisms

Saprobiotic nutrition is carried out by organisms (mainly bacteria and fungi) which feed on dead matter. The organisms are called **decomposers** because they break down solid organic matter into soluble substances, the first step in the decay of dead organisms. For example, they secrete enzymes onto solid muscle which break down proteins to soluble amino acids.

Ammonification: ammonia from urea and amino acids

Decomposers and other microorganisms are also involved in the conversion of organic nitrogen-containing compounds in urea and amino acids into ammonia. The process is called **ammonification**. Ammonia can be lost to the atmosphere, taken up by plants as ammonium ions, or converted to other forms of nitrogen.

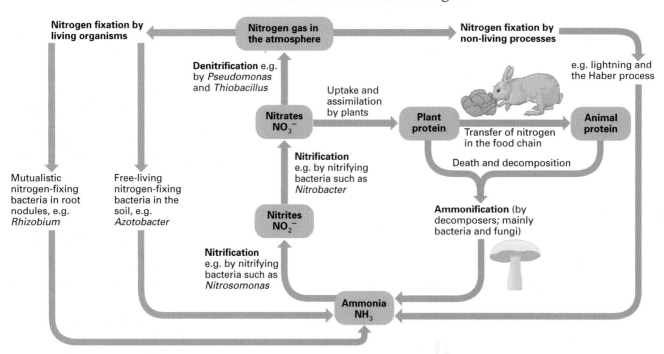

The nitrogen cycle. Note that without nitrogen fixation, most of the nitrogen in the atmosphere would not be available to living organisms. Most nitrogen fixation occurs in nitrogen-fixing bacteria which convert nitrogen to ammonia. The ammonia is then converted to nitrites and nitrates, either within root nodules of leguminous plants or by nitrifying bacteria in the soil. The plants use the nitrogen from nitrates to synthesize amino acids from other organic acids produced.

Nitrification: ammonium ions to nitrites and nitrates

Ammonium ions in the soil or water are oxidized to nitrites and nitrates by **nitrifying bacteria**. These bacteria are **chemoautotrophs**, obtaining energy from the redox (reduction–oxidation) reactions involved in nitrification. Nitrification requires oxygen, so it happens most rapidly in well aerated soils or well oxygenated bodies of water. The nitrate ions produced by nitrification can be taken up by plants and used to make proteins. Consumers obtain their nitrogen in the form of proteins when they eat plants or other animals.

Nitrogen fixation: nitrogen to nitrogen compounds

Nitrogen occurs in huge quantities in the air (the gas makes up 79% of the atmosphere). Most organisms cannot use atmospheric nitrogen, or nitrogen dissolved in water, because nitrogen gas is chemically unreactive (each molecule consists of two nitrogen atoms linked by a triple bond). Before plants and animals can use nitrogen, it must first be converted to absorbable nitrogen compounds. This conversion is called **nitrogen fixation**.

In nature, a little nitrogen fixation occurs during thunderstorms. Lightning provides the energy to oxidize nitrogen to nitrogen oxides. These gases dissolve in rain droplets to form dilute nitric acid. However, more than 95% of nitrogen fixation takes place not by lightning but biologically. The fixation is carried out by **nitrogen-fixing bacteria**. Some live free in the soil or water (the latter as cyanobacteria or blue-green algae). Other nitrogen-fixing bacteria live mutualistically in root nodules on leguminous plants such as clover, beans, gorse, and sweet peas.

Nitrogen-fixing bacteria possess **nitrogenase**, an enzyme that enables them to reduce nitrogen to ammonia or ammonium compounds:

$$\text{Nitrogen} + \text{hydrogen} \xrightarrow{\text{nitrogenase}} \text{ammonia}$$
$$N_2 + 3H_2 \longrightarrow 2NH_3$$

Nitrogen fixation is also carried out artificially by industry in the production of nitrogen fertilizers. Almost as much nitrogen is fixed in the manufacture of nitrogen fertilizer as is fixed biologically. One major industrial process, the **Haber process**, synthesizes ammonia from atmospheric nitrogen and hydrogen by passing the gases at high temperature and pressure through an inorganic catalyst. The process requires large amounts of energy, usually from burning fossil fuels.

Denitrification: nitrates back to nitrogen

The nitrogen cycle is completed by **denitrifying bacteria**. These bacteria live in conditions of low oxygen and reverse the nitrifying process, converting nitrates to nitrites, and nitrites to nitrogen gas. The process, called **denitrification**, leads to the loss of nitrogen from the biotic component of an ecosystem to the atmosphere.

Root nodules on a leguminous plant. The nodules contain nitrogen-fixing bacteria. The relationship between the bacteria and plants is an example of mutualism: the plants obtain a source of nitrogen while the bacteria obtain shelter and a constant supply of sugars from the plant.

Check your understanding

1 By what process do microorganisms convert

   a amino acids into ammonia

   b gaseous nitrogen into ammonia

   c ammonia into nitrites

   d nitrates into gaseous nitrogen?

2 Suggest why leguminous plants produce a red pigment called leghaemoglobin in their root nodules.

4.01 Global warming

OBJECTIVES

By the end of this spread you should

- be able to explain the roles of carbon dioxide and methane in enhancing the greenhouse effect and bringing about global warming

- have sufficient knowledge about global warming to evaluate data relating to global warming and its effects on the yield of crops, life cycles and numbers of insect pests, and the distribution and numbers of wild animals and plants

Prior knowledge

- the carbon cycle (3.08)

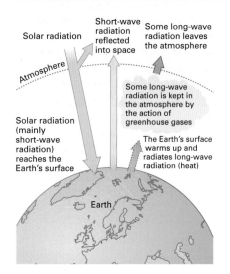

The greenhouse effect

Global warming refers to the increase in the average temperature of the Earth's near-surface air and oceans in recent decades, and its projected continuation. According to the Intergovernmental Panel on Climate Change, global warming during the 100 years ending in 2005 has amounted to a rise of $0.74 \pm 0.18°C$. This rise has been linked to increases in the levels of greenhouse gases in the atmosphere.

Greenhouse gases prevent heat from escaping from the atmosphere. This effect, called the **greenhouse effect**, is a natural phenomenon that has been essential for the evolution of life on Earth. It allows surface temperatures to be much higher than they would otherwise be.

Solar energy reaches the Earth as short-wave radiation. When this radiation strikes a surface, much of its energy is converted into heat, a form of radiation that has a long wavelength. Greenhouse gases absorb and retain long-wave radiation or reflect it back toward the surface of the Earth, as shown in the diagram. These gases therefore act like panes of glass in a greenhouse, letting visible light in, but retaining some of the heat before it escapes into space.

Two important greenhouse gases are carbon dioxide and methane. Carbon dioxide contributes up to 25% of the greenhouse effect. Most of the carbon dioxide in the atmosphere is produced naturally. However, recent increases have been attributed to human activities, particularly the burning of fossil fuels, deforestation, and cement-making.

Methane is produced by anaerobic bacteria living in the mud of boggy habitats, landfill sites, rice paddies, and in the gut of certain insects (such as termites) and ruminants (for example, cattle, sheep, and camels). Human activities such as cattle ranching have increased the release of methane into the atmosphere (a single cow can, for example, produce $200 \ dm^3$ of methane per day).

Other greenhouse gases include water vapour, chlorofluorocarbons (CFCs, used as refrigerants), and ozone (triatomic oxygen, O_3).

Most scientists agree that increases in greenhouse gases, particularly carbon dioxide and methane, are linked to global warming, but they are unsure about the effects this will have on crop yields, insect pests, and wild animal and plant populations.

Crop yields

If temperature were the only factor that changed with global warming, the effects should be to

- increase rates of photosynthesis and boost crop yields where temperature is a limiting factor to plant growth

- reduce the yield of plants that were growing in optimal temperature conditions before global warming

However, temperature does not act in isolation. Other factors, such as water availability, soil structure and texture, and incidence of pests are also likely to change. Any of these factors could limit crop yields.

Insect pests

Global warming is expected to extend the territory for insect pests carrying infectious diseases, such as the *Anopheles* mosquito which transmits malaria, ticks which carry encephalitis and Lyme disease, and sandflies which carry visceral leishmaniasis. For countries that cannot afford remedial action such as draining swamps, vaccination, and increased use of pesticides, global warming is likely to lead to a significant increase in the incidence of these diseases. For richer countries that can afford remedial action, the consequences are likely to be more economic than medical.

Wild animals and plants

Global warming is already having an effect on wild animals and plants. Some species are being forced out of their habitats, while others are flourishing. Species ranging from molluscs to mammals and from grasses to trees are changing their patterns of distribution and the timing of their reproductive cycles.

As to the future, much will depend on whether the present trends in temperature rises continue. It will also depend on the secondary effects of global warming, such as reduced ice cover, rising sea levels, and weather changes. In southern Europe, long, hot summers could lead to an increased risk of forest fires. Alterations to ocean currents, caused by increased freshwater inputs from glacier melt, would affect the distribution of marine organisms. This could have a dramatic impact on existing fisheries.

Species that appear at most immediate risk are those in the polar regions, such as polar bears in the Arctic and emperor penguins in the Antarctic. Some ecologists warn that the variable effects of global warming on different species could easily disrupt the links within a community and damage fragile ecosystems to the point of extinction.

Evaluating the effects of climate change

One of the responsibilities of scientists is to advise politicians about complex scientific issues so that they can make informed decisions about future policies. In 1988, a scientific body called the **Intergovernmental Panel on Climate Change** (IPCC) was established by the United Nations to give authoritative and reliable evaluations on the risk of climate change caused by human activity.

In 2001, it reported that humanity had 'likely' played a role in global warming. On 2 February 2007, it declared that the evidence of a warming trend is 'unequivocal', and that human activity has 'very likely' been the driving force in that change over the last 50 years.

In Oslo on 10 December 2007, the IPCC, along with Al Gore, was awarded the Nobel Peace Prize 'for their efforts to build up and disseminate greater knowledge about man-made climate change, and to lay the foundations for the measures that are needed to counteract such change'.

Polar species are at most immediate risk from global warming as their habitats disappear.

Check your understanding

1 Name two important greenhouse gases.
2 What types of species are at most immediate risk from global warming?
3 Suggest how an increase in the acreage of tropical rainforest converted to cattle ranching might contribute to the greenhouse effect and global warming.

- -

4 In 2007, the IPCC declared that *the evidence for a warming trend during the 100 years ending in 2005 was 'unequivocal and amounted to a rise of 0.74 ± 0.18°C'.*
   a Explain what the phrase in italics probably means.
   b On what sort of evidence do you think it was based?
   c What implications does the IPCC declaration of 2 February 2007 have for global warming trends for the next 100 years?

OBJECTIVES

By the end of this spread you should be able to

- define leaching and eutrophication
- discuss the environmental issues arising from the use of fertilizers

Prior knowledge

- the nitrogen cycle (3.09)

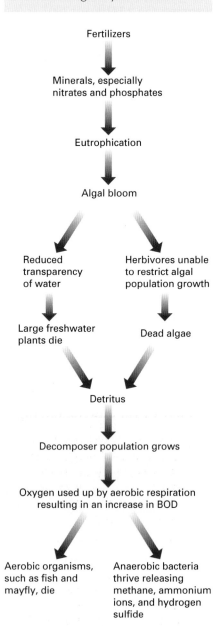

Fertilizers
↓
Minerals, especially nitrates and phosphates
↓
Eutrophication
↓
Algal bloom
↓
Reduced transparency of water / Herbivores unable to restrict algal population growth
↓
Large freshwater plants die / Dead algae
↓
Detritus
↓
Decomposer population grows
↓
Oxygen used up by aerobic respiration resulting in an increase in BOD
↓
Aerobic organisms, such as fish and mayfly, die / Anaerobic bacteria thrive releasing methane, ammonium ions, and hydrogen sulfide

Summary of eutrophication

Artificial fertilizers containing nitrates, phosphates, and potassium are often added to soils to maximize the growth of crops. Unfortunately, about half the minerals may **leach** from the soils into the groundwater, find their way into natural watercourses, and enter a freshwater habitat.

The addition of fertilizers can cause a body of water to become over-rich in nutrients, a process called **eutrophication**. One effect of eutrophication is to encourage the rapid growth of photosynthesizing organisms, especially algae. The dramatic, fast growth of algae is called an **algal bloom**. A bloom can smother large freshwater plants, reduce light intensity in the water, and produce toxins which kill fish. When the algae die, their decomposition by bacteria may lead to the complete deoxygenation of the water, causing the death of aerobic organisms.

Biological oxygen demand (BOD)

The addition of masses of dead algae to water stimulates the growth of microorganisms which feed on the dead organic material. As the density of microorganisms increases, their demand for oxygen also rises. This demand is called the **biochemical oxygen demand (BOD)**. It is measured as the mass (in mg) of oxygen used by 1 dm^3 of water stored in darkness at 20°C for 5 days. The BOD of unpolluted river water is about 15 mg O_2 dm^{-3}. The BOD of water that has had an algal bloom may be 100 times higher.

Oxygen depletion and eutrophication are not only caused by fertilizers. They may also be caused by any pollutant containing high concentrations of organic or inorganic nutrient, such as sewage, slurry (animal faeces and urine), or silage effluent which can leach off farmland and pollute water.

Indicator species

Indicator species can be used to monitor water quality and help detect incidents of water pollution by, for example, nitrate fertilizers.

An **indicator species** is a species that needs a particular environmental condition or set of conditions in order to survive. Whether the species is present, and in what numbers, provides information about the environment. Indicator species are used in a wide range of ecological investigations to find out about both the present and past conditions.

Biological monitoring of water quality

The leaching of large volumes of fertilizer into a freshwater habitat such as a river has effects similar to the addition of sewage, as illustrated here. There is an immediate impact on the environment at the point of entry and the pollutant reduces the quality of water for some way downstream. Changes in the abiotic environment are reflected by changes in the community of aquatic organisms.

The fall in oxygen content is the most dramatic effect of the addition of sewage or an algal bloom. Fauna requiring clean water and high oxygen levels are absent or in low numbers. Populations of sewage fungus (a slime consisting of filamentous bacteria, protoctists, and algae, as well as fungi), bacteria, and other organisms that thrive in organic waste are high. In extreme cases, the oxygen content may fall so low that conditions become anaerobic. There comes a point at which the organic material is completely broken down and the river fauna starts to recover. Populations of clean-water species increase and those adapted to the polluted conditions decline.

These changes show how, within limits, the activity of aquatic organisms, helped by dilution of the pollutant as it passes downstream, can purify the water. The distance over which the purification takes place varies and is affected by factors such as temperature, volume of water, and the existing populations of microorganisms.

Studies of many rivers show that there are characteristic communities of organisms associated with different levels of pollution in the river. Clean-water organisms include stonefly nymphs (larvae), mayfly larvae, and caddis fly larvae. These organisms are indicator species of unpolluted, well oxygenated water, and a community containing the organisms is an **indicator community** of well oxygenated water. As pollution increases and oxygen content falls, populations of species sensitive to a decrease in oxygen content fall and those of species tolerant to organic pollution, such as bloodworms, rat-tailed maggots, and tubifex worms, increase.

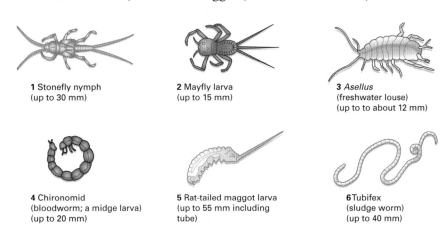

1 Stonefly nymph
(up to 30 mm)

2 Mayfly larva
(up to 15 mm)

3 *Asellus*
(freshwater louse)
(up to to about 12 mm)

4 Chironomid
(bloodworm; a midge larva)
(up to 20 mm)

5 Rat-tailed maggot larva
(up to 55 mm including
tube)

6 Tubifex
(sludge worm)
(up to 40 mm)

Biological indicators of organic pollution: these species are arranged in order of the level of pollution they tolerate, where 1 is the least polluted and 6 the most polluted water. Respiration among aquatic insect larvae varies with oxygen availability. Stonefly nymphs and mayfly larvae have delicate gills with which to acquire oxygen from well aerated waters; chironomid larvae contain haemoglobin to extract oxygen where levels are low – the lower the level, the greater the haemoglobin concentration. Rat-tailed maggots have long breathing tubes at the rear through which they can obtain air from above the water surface, allowing them to live in water totally depleted of oxygen.

In the UK, samples of freshwater organisms are routinely taken from rivers to monitor levels of pollution. A biologist can grade the quality of river water on a scale ranging from clean water with high species diversity to grossly polluted water with no organisms or only a few anaerobic ones.

In most situations, environmental quality is monitored both biologically and physicochemically. Biotic indices are used for routine, continuous monitoring of the environment and when they indicate a problem, physicochemical tests are carried out to identify the precise cause.

Check your understanding

1 Explain why you would not expect aerobic organisms such as mayfly larvae to occur in a body of water after an algal bloom.

2 Define eutrophication.

- - - - - - - - - - - - - - -

3 Why would the absence of stonefly nymph from a stream not necessarily indicate that its water was not of the highest quality?

OBJECTIVES

By the end of this spread you should be able to

- explain how ecosystems are dynamic and usually move from colonization to climax communities in the process of succession

- understand that at each stage of succession certain species may be recognized which change the environment so that it becomes more suitable for other species

- describe the changes in the abiotic environment that result in a less hostile environment and changing diversity

Prior knowledge

- species diversity (*AS Biology for AQA* spread 19.01)

Ecosystems are dynamic. They are constantly changing, not only in response to physical factors such as climate, but also in response to biological factors resulting from the activities of organisms within the community. Communities gradually change from one type to another. The sequence of change, from the initial colonization of a new area to establishing a relatively stable community, is called **ecological succession**. There are two main types: **primary succession** and **secondary succession**. Ecological succession happens in stages called **seres** or **seral stages**; each stage has its own community of organisms, and can usually be identified by the plants it contains.

Primary succession: colonizing a new area

Primary succession takes place in newly formed areas where no life previously existed, for example, volcanic islands such as Surtsey Island, glacial deposits of rock, new ponds or lakes, and sand dunes.

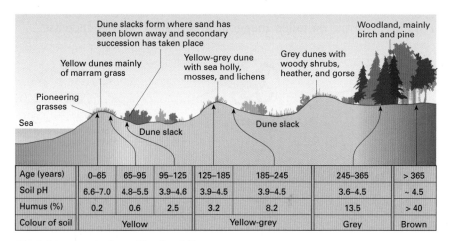

Age (years)	0–65	65–95	95–125	125–185	185–245	245–365	> 365
Soil pH	6.6–7.0	4.8–5.5	3.9–4.6	3.9–4.5	3.9–4.5	3.6–4.5	~ 4.5
Humus (%)	0.2	0.6	2.5	3.2	8.2	13.5	> 40
Colour of soil	Yellow			Yellow-grey		Grey	Brown

Sand dune succession at Studland Bay

On new rocky surfaces, the following types of plant communities usually occur in favourable climates:

Pioneer species → annuals → herbaceous perennials → shrubs → trees

This succesion may be a very slow process, taking hundreds of years.

In ponds and lakes, organic matter builds up from the dead remains of plants and animals and sediment is brought in by water running off the land. The water becomes shallower and richer in nutrients, allowing rooted plants to crowd the margins and extend further and further into the lake or pond. In doing so, they tend to trap more sediment and make the water even shallower. Ponds and small, shallow lakes might develop into marshes, or perhaps tree-dominated swamps. Given the right climate and ecological conditions, pond and lake succession would ultimately lead to woodland, like succession on bare rock. Many ecologists, however, believe that this rarely, if ever, happens.

The process of succession

Succession is a complex process driven by many factors acting simultaneously. In the early stages, abiotic factors are the most important in determining which organisms colonize a habitat. Only those organisms that can tolerate the hostile abiotic environment can establish themselves. These are called **pioneer species**.

Pioneer species arrive haphazardly. Those tolerant of the hostile conditions survive. In surviving, they may change the habitat in ways that favour other species more strongly than themselves. After the pioneer species have become established, biotic factors become increasingly important in the habitat. For example, early colonizers of rocky surfaces can often fix nitrogen, compensating for the poor nutrient content of the soils. When the colonizers die, they enrich the soil, making the habitat less hostile and allowing other species to become established. As succession continues, the **species diversity** of the community increases, and the food web of the ecosystem becomes more complex.

According to the **climax theory**, ecological succession leads to a more complex community with a high species diversity that is in equilibrium with its environment. This has been called the **climax community**. Some ecologists believe that there is a single type of climax community for each climatic region, called the **climatic climax**. Oak woodland, for example, is often stated to be the climatic climax of much of lowland Britain.

It is generally assumed that an increase in species diversity makes an ecosystem more stable because of the large number of alternative links between members of the community. For example, several species might be carrying out the same function, such as decomposition or primary production. If one species became extinct, the others would still be available to carry out that function.

Sere	Pioneer community	Subclimax community	Climax community
Species diversity	low	$\longrightarrow$	high
Number of links between community members	few	$\longrightarrow$	many
Resistance of ecosystem to change	fragile	$\longrightarrow$	resistant

Summary of the main changes that occur during ecological succession

Check your understanding

1 What is a climatic climax?

2 How do pioneer species change the humus content of sand?

3 Marram grass is not a pioneer species, but it is an early colonizer of sand dunes. Why is it absent on grey dunes, a later stage of sand dune succession?

- -

4 The box 'Primary succession in action' describes some of the changes that have taken place on Surtsey from 1963 to the mid-1990s. What do you think has happened since then? Suggest what the final outcome of ecological succession on Surtsey is likely to be.

Primary succession in action

In 1963, a new volcanic island called Surtsey emerged off Iceland's southern coast, rising to 150 m above the sea. As soon as the molten lava cooled, life began to colonize the bare rock. The pioneer species included lichens and mosses: hardy, drought-resistant organisms that established themselves and reproduced in hostile areas where other organisms could not live. These organisms acted on the rock to produce soil (lichens, for example, secrete acids which break down rock) and when they died they added humus to the soil. After time, the soil became deep and rich enough for plants to take root. The first plant was the sea rocket, a quick-growing annual. By the mid-1990s, there were nearly 50 other plant species flourishing on Surtsey. These plants produce large numbers of tiny seeds which are easily dispersed to colonize new areas. When fully established, the annual plants out-competed the pioneer species for nutrients and light, and replaced them. Some of the next plant arrivals were herbaceous perennials.

The formation of Surtsey Island

OBJECTIVES

By the end of this spread you should be able to

- distinguish between primary succession and secondary succession

- discuss how conservation of habitats frequently involves management of succession

Prior knowledge

- ecological succession I (4.03)

- reasons for conservation (*AS Biology for AQA* spread 19.03)

Secondary succession

Secondary succession takes place in areas where life is already present but has been altered in some way. The changes are similar to those in primary succession on a rock, but the process is usually faster as soil, often containing seeds, is present.

There are three main causes of secondary succession:

- natural catastrophes that remove existing vegetation, such as 'blow outs' on sand dunes, or fires

- human destruction of climax communities, for example, by the drainage of marshes and the burning of forests to increase farmland

- human management of habitats to maintain communities at an early successional stage

A typical sort of secondary succession occurs in lowland Britain on arable land that has been ploughed and then abandoned. The successional changes are similar to those in primary succession, but usually take place faster because soil already contains seeds.

To prevent secondary succession occurring on arable land, the field would need to be ploughed regularly.

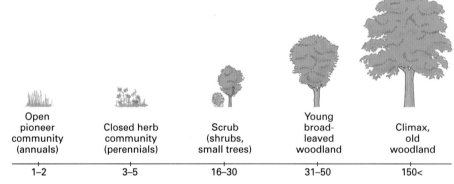

Open pioneer community (annuals)	Closed herb community (perennials)	Scrub (shrubs, small trees)	Young broad-leaved woodland	Climax, old woodland
1–2	3–5	16–30	31–50	150<

Time after ploughing (years)

Secondary succession: the succession from ploughed field to oak woodland is similar to primary succession on a rock, but the process is much more rapid because soil, often containing seeds, is already present.

Conservation biology and habitat management

Conservation biology is the scientific study of how biological diversity can be safeguarded. It involves studying human impacts on wildlife and developing ecologically sound strategies to deal with them. The strategies may range from the **ex-situ conservation** of individual endangered species within zoos to the **in-situ conservation** of whole ecosystems and landscapes.

One of the guiding principles of conservation biology is that individual species can survive only if they have suitable habitats in which to live. Therefore, a priority of many conservation projects is to protect and restore specific types of habitats. To do this requires a good understanding of ecological succession. For example, much of the grassland and heathland in Britain are **plagioclimaxes**: relatively stable climax communities arising from succession that has been deflected or arrested, either directly or indirectly, as a result of human activity. These habitats occur in Britain mainly because controlled burning, grazing, or some other factor prevents them from reaching their climatic climax. If the factor causing the plagioclimax is removed, these ecosystems would develop towards their climatic climax by ecological succession.

Conservation biologists use other ecological principles to guide their decision-making about habitat management. For example, a single large area of habitat with the potential to support a large population is generally more effective at conserving a species than several habitat fragments, each with the potential to support only a small population, especially if individuals are unable to move between the habitat fragments. Another guiding principle is that if it is necessary to have several small habitat areas, it is better if they are close together. Sometimes it is possible to create connecting corridors between habitat areas, making dispersal easier for the species.

It is also generally accepted that the presence of humans adversely affects many species, and areas inaccessible to humans are easier to conserve than those that are within easy reach. There is a general consensus among ecologists that it is more effective to conserve whole ecosystems in which many species live than to focus efforts on conserving individual species.

Dealing with biological conservation issues

When dealing with specific conservation issues, for example, whether to make a large clearing in a woodland to create a flower meadow plagioclimax, conservation biologists and policy-makers need to consider the many contributing elements. In theory, the process should involve the following steps.

- Scientific assessment: gathering information and identifying any problems – this may require investigations and experiments to be conducted and data to be gathered.

- Risk analysis: using the results of the scientific assessments, sometimes by constructing models, to predict the potential negative and positive consequences of carrying out a course of action (including ecological and economic costs).

- Public education and involvement: public awareness and support is often an essential part of dealing with conservation issues. When there are alternative actions possible, the public should be informed and, if possible, involved in the decision making.

- Action: selecting a course of action and carrying it out.

- Evaluation: monitoring the effects of the course of action to see whether it should be continued, stopped, or modified.

In practice, dealing with biological conservation issues is rarely as neat and tidy as the five steps described above imply.

Check your understanding

1 What type of succession occurs after a field is ploughed?

2 Distinguish between a climatic climax and a plagioclimax.

3 Suggest why dealing with real biological conservation issues does not always follow the five theoretically ideal steps of scientific assessment, risk analysis, public education and involvement, action, and evaluation.

5.01 Evolution in outline

OBJECTIVES

By the end of this spread you should be able to

- appreciate the importance of Charles Darwin's contribution to our understanding of how new species are formed

- understand that variation within a species and geographic isolation can lead to populations that differ from one another, and that this has the potential to form new species

Prior knowledge

- variation (*AS Biology for AQA* spreads 10.01–10.02)

- genes and alleles (*AS Biology for AQA* spread 11.03)

- the species concept (*AS Biology for AQA* spread 17.01)

One of the most fundamental questions in biology is: 'where do all living things come from?'. According to most biologists, the millions of species living on Earth today (including humans) are descended from other species that inhabited the world in the past. This change from one species to another has come about by a process called **evolution**. Evolution takes place when the genetic composition of different populations changes over successive generations. When the changes are sufficiently great and the populations are geographically isolated, a new species may be formed. (A **species** is a group of closely related organisms potentially capable of interbreeding to produce fertile offspring.)

Darwin's theory of evolution

Evolution is not a modern concept. In ancient times, philosophers such as Confucius in China and Aristotle in Greece suggested that complex species evolve from simpler pre-existing ones by a process of continuous and gradual change. However, it was not until the nineteenth century that scientists came up with plausible mechanisms for evolution. The mechanism that is most widely accepted among biologists today is based on the work of the nineteenth-century naturalist Charles Darwin.

Between 1831 and 1836, Darwin was the naturalist on board HMS *Beagle*, a research vessel engaged in mapping different parts of the world. After spending over three years surveying the coast of South America, the *Beagle* landed on the Galapagos Islands in the Pacific Ocean. Darwin compared the species he found in different locations. He observed that closely related species on the geographically isolated islands of the Galapagos resembled those on the South American mainland and they were probably formed from a common ancestor. This led him to develop his theory of evolution.

Darwin came to the conclusion that, over successive generations, a new species comes into being by slow and gradual changes from a pre-existing one. He believed that these changes are brought about by a process which he called **natural selection**.

Darwin's theory was based on three main observations:

1 Within a population are organisms with varying characteristics, and these variations are inherited (at least in part) by their offspring.

2 Organisms produce more offspring than are required to replace their parents.

3 On average, population numbers remain relatively constant and no population gets bigger indefinitely.

From these observations, Darwin came to the conclusion that within a population many individuals do not survive, or fail to reproduce. There is a 'struggle for existence'. For example, members of the same population compete to obtain limited resources, and there is a struggle to avoid predation and disease, or to tolerate changes in environmental conditions such as temperature. In this struggle for existence those individuals that are best adapted to their environment will have a **selective advantage**: they will be more likely to survive and produce offspring than less well adapted organisms. In Darwin's own words:

As many more individuals of each species are born than can possibly survive, and as, consequently, there is a frequently recurring struggle for existence, it follows that any being, if it vary however slightly in any manner profitable to itself under the complex and sometimes varying conditions of life, will have a better chance of surviving and thus be naturally selected. From the strong principle of inheritance, any selected variety will tend to propagate its new and modified form.

The Origin of Species

For more than 20 years, Darwin collected evidence to support his theory and refined his ideas. He delayed publishing his findings until 1858, when Alfred Russel Wallace sent him a letter describing a theory of evolution identical to Darwin's own.

Wallace was a British naturalist who had worked in the Malay Archipelago for eight years. He concluded from his research that some organisms live while others die because of differences in their characteristics, such as their ability to resist disease or escape predation. Darwin and Wallace published a paper jointly describing their theory of evolution by natural selection. However, Darwin's name has become more strongly linked with the theory because of a book he published on 24 November 1859. The book, entitled *On the Origin of Species by Means of Natural Selection or the Preservation of Favoured Races in the Struggle for Life*, has been called the most important biology book ever written. It not only gives a full description of the theory of evolution by natural selection, but also contains a huge mass of evidence to support the theory.

The reaction to Darwin

Many people found it difficult to accept Darwin's ideas, especially the idea that modern humans and apes are probably descended from a common ancestor. However, his theory is supported by so much evidence that the majority of biologists accept it. Evolution by natural selection has become a central theme which underpins much of modern biology. The modern theory of evolution is called **neo-Darwinism** (*neo* = new) because it incorporates modern scientific evidence, particularly from genetics and molecular biology, that was unavailable to Darwin. For example, we now know that the variations that are so important in natural selection come about by random and spontaneous changes in genes to form different **alleles**, particularly from mutations in reproductive cells. Despite modifications to Darwin's theory, in neo-Darwinism natural selection is still the driving force behind evolution.

(a) A tool-using, insect-eating finch that uses a small twig or cactus spine to probe for insects

(b) A cactus finch has a beak adapted to feed from the prickly pear flower or fruit.

(c) A small tree finch that uses its beak to grab insects

Of all the places he visited, the Galapagos Islands made the greatest impression on Darwin. His observations of the islands' animals and plants led him to his idea that new species could arise from an ancestral form by the gradual accumulation of adaptations to different environments. From studies made years after Darwin's voyage, biologists have concluded that this is what happened to the Galapagos finches. The Islands have a total of 14 species of closely related finches, some found on only a single island. The most obvious differences between them are their size and their beaks, which are adapted to the specific diets found on each island.

Check your understanding

1 Give the biological meaning of evolution.

2 How does neo-Darwinism differ from Darwin's original theory of evolution?

5.02 Inheritance

OBJECTIVES

By the end of this spread you should be able to

- understand the particulate nature of inheritance; that it depends on factors, which we call genes, that pass from one generation to the next unaltered

- recall the law of segregation in modern terms

Prior knowledge

- DNA and genes (*AS Biology for AQA* spread 11.03)

Early ideas of inheritance

Even the earliest humans must have noticed that although offspring differ from each other and from their parents in some respects, there are often remarkable similarities between parents and their offspring. They probably realized that some characteristics are passed on to the next generation while others appear to be lost. However, it was not until 1856, when Gregor Mendel carried out a large number of experiments, that **genetics** (the scientific study of inheritance) really began.

Before Mendel's experiments, most people believed that parental features were blended and transmitted to offspring as though they were fluids: for example, marriage between a short person and a tall person would produce children who would grow to an intermediate height. Even though Mendel had no knowledge of genes or chromosomes, his work showed that inheritance is particulate: it depends on the transfer of discrete (separate) factors from parents to offspring.

Mendel's experiments

Mendel was an Austrian monk who studied the inheritance of characteristics in garden peas (*Pisum sativum*), which he grew in the vegetable garden of his monastery. Mendel first had to establish pure-breeding plants, that is, plants which when self-fertilized produce identical offspring generation after generation. Mendel then selected for his experiments two pure-breeding plants with alternative expressions of a particular characteristic (for example, a tall plant and a dwarf plant). He crossed the plants by transferring pollen from one plant (the 'male' parent) to the stigma of a second plant (the 'female' parent). Mendel made sure that the 'female' plant could not be pollinated either by its own pollen or by any other by removing its anthers and covering its flowers with fine muslin bags.

Mendel collected the seeds produced by the 'female' parent and grew them the next year to give the first-generation offspring. (The offspring of pure-breeding parents are often called the **first filial generation** or the F_1.) He carefully recorded the characteristics of these plants and then crossed two plants from this generation. This type of cross involving plants of the same generation is called a **self-cross**. Again, the seeds produced were collected and grown the following year to give the second-generation offspring. (The offspring produced by a cross between F_1 parents is often called the **second filial generation** or the F_2.) The characteristics of each plant were again recorded.

- -

In his experiment involving a cross between pure-breeding tall and dwarf pea plants, Mendel found that all the F_1 plants were tall, but that the F_2 plants were a mixture of tall plants and dwarf plants in the approximate ratio of 3:1, as shown here. The most striking features of these experiments were that

- there were no plants of intermediate height

- there were no dwarf plants in the F_1 generation, though they reappeared in the F_2

P (parents)	Tall ♂	×	Dwarf ♀
F₁ (first generation)		Tall (self-crossed)	
F₂ (second generation)	Tall (787)		Dwarf (277)
Approximate ratio	3	:	1

Inheritance of a single characteristic: the results of one of Mendel's experiments. A pure-breeding tall male plant was crossed with a pure-breeding dwarf female plant.

From the first observation, Mendel concluded that characteristics are not blended together like different colours of paint, but that they are determined by definite, discrete particles which he called 'factors' and which we now call genes.

From the second observation, Mendel concluded that the factor for dwarfness must be carried in the F_1 plants, but that it is 'hidden' by the factor for tallness. The dwarf factor is expressed only in the absence of the tall factor. Therefore the plants must carry two factors for each characteristic, one factor coming from each parent. Because all the F_1 plants are tall, the factor for tallness must be **dominant** to the factor for dwarfness, which is the **recessive** factor.

Similar crosses involving other single characteristics produced similar results, all giving an approximate ratio of 3:1 in the F_2. From these results, Mendel proposed his first law of inheritance, the **law of segregation**. In modern terms, the law can be stated as follows.

> The characteristics of a diploid organism are determined by **alleles** which occur in pairs. Of a pair of such alleles, only one can be carried in a single **gamete**.

Mendel's success

Mendel was fortunate in choosing simple features controlled by single genes. However, his good fortune would have amounted to nothing without his hard work, good scientific method, and insight in interpreting the results of his breeding experiments. Mendel established the following important facts which formed the foundation of modern genetics:

- Inheritance is particulate; it depends on factors or particles.
- The factors are passed from generation to generation unaltered.
- The factors may be dominant or recessive, so that some factors are not expressed in every generation.
- Each organism has a pair of factors controlling a given characteristic.

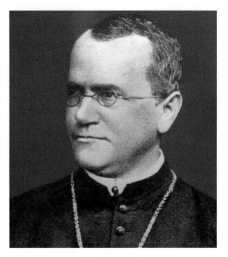

Gregor Mendel, 1822–84

Check your understanding

1 Explain how Mendel's experiments showed that inheritance depends on particles or factors and is not the result of blending.

2 State the law of segregation in modern terms.

- -

3 Mendel is now regarded as one of the most famous scientists and hailed as the 'father of genetics'. Mendel presented his results in a paper published by the Natural History Society of Brunn in 1866. It was a brilliant paper which reflected his genius for scientific experimentation and interpretation. However, the paper was ignored by the scientific community until 1900, 16 years after his death. Suggest why Mendel's ideas remained in obscurity for so long.

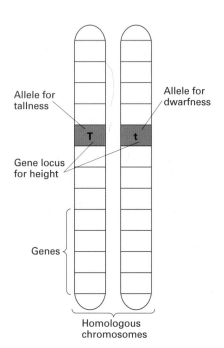

Outline of a pair of homologous chromosomes to show the distinction between genes and alleles

Genes and alleles

Mendel's breeding experiment with tall and dwarf plants (see spread 5.02) is an example of **monohybrid inheritance**: that is, inheritance involving a single characteristic determined by one gene.

In genetics, a **gene** can be regarded as a specific length of DNA which occupies a position on a chromosome called a **locus**. The gene determines a particular characteristic of an organism, in the case of Mendel's experiment, the height of the plant. Mendel recognized that the two ways in which height could be expressed (tall and dwarf) were determined by factors. We call these factors **alleles**. Alleles are different forms of the same gene which may alter the way in which a particular characteristic (such as the height of a plant) is expressed. More accurately, an allele is a particular sequence of **nucleotide bases** making up a gene.

The **genotype** of an organism is its genetic make-up. It refers to all of its alleles. An allele may or may not be expressed in the **phenotype** (the observable or measurable features of an organism). For example, a recessive allele will not be expressed if a dominant allele is also present. So in adult peas the gene for height is always found at the same locus on a particular pair of **homologous chromosomes**, but either the allele for tallness or the allele for dwarfness will be present at each locus.

Representing genetic crosses

By convention, a gene is represented by a letter or letters. If the letters are written by hand, they are chosen with care so that the upper and lower cases can be distinguished easily. The upper case (capital) letter represents the dominant allele and the lower case letter represents the recessive allele. In our example of height in pea plants,

- **T** represents the allele for tallness

- **t** represents the allele for dwarfness

Mendel recognized that diploid organisms have two factors (alleles) for each characteristic (gene), but only one of the factors is carried by each gamete. In the cross between pure-breeding tall and dwarf plants, all the offspring were tall, therefore the allele for tallness must be dominant to the allele for dwarfness. As the parent plants were pure breeding, they must have been homozygous, therefore the genotypes of the parents can be represented as **TT** (tall) and **tt** (dwarf).

The male tall parent plant produced gametes. Each carried one dominant allele. The female gametes each carried one recessive allele. The offspring were all heterozygous tall plants, **Tt** (they carried two different alleles, one for tallness, the other for dwarfness). We can summarize these points in a genetic diagram, as shown on the next page. A circle is drawn around the genotype of the gametes to reinforce the idea that each allele is discrete and can potentially join with alleles in any other gamete from the sexual partner.

In a self-cross between heterozygous tall plants there are three different possible genotypes in the offspring: **TT**, **Tt**, and **tt**. Assuming random fusion of gametes, a genetic diagram also shows us the probability of each genotype occurring in the offspring:

- there is a 1 in 4 (or 0.25) chance that a pea plant will have the genotype **TT**
- there is a 2 in 4 (or 0.5) chance that it will be **Tt**
- there is a 1 in 4 (or 0.25) chance that it will be **tt**

As all plants carrying at least one **T** allele are tall, 3 out of every 4 plants would be expected to be tall. That is, we would expect a ratio of 3 tall offspring to 1 dwarf offspring. In Mendel's experiment, out of 1064 offspring, 798 would be expected to be tall, and 266 dwarf. However, because these figures are based on chance and probabilities, it is unlikely that the actual number of offspring would match the expected value precisely. Mendel obtained 787 tall offspring and 277 dwarf offspring. A **chi-squared test** (a statistical test of the significance of the differences between observed numbers and expected numbers, see page 29) would show that these results do not differ significantly from the 3:1 ratio predicted from Mendel's law of segregation. A 3:1 ratio is typical of the results of a monohybrid cross between two heterozygous organisms, where one of the alleles of a gene is dominant and the other is recessive.

Test cross

If you have a tall pea plant, you can tell whether it is heterozygous (**Tt**) or homozygous (**TT**) by carrying out a **test cross**. An organism showing the dominant characteristic is crossed with one that is homozygous recessive. If homozygous, it will not produce any dwarf offspring; if heterozygous, it will produce dwarf and tall offspring in the expected ratio of 1:1.

(a) Cross between pure-breeding tall plants and dwarf plants

Parental phenotypes	Tall	×	Dwarf
Parental genotypes	TT		tt
Gametes	T		t

Female gametes: t

Offspring genotypes and phenotypes — Male gametes T — Tt — Tall

F₁ phenotypes — All tall

(b) Self-cross between F₁ offspring

Parental phenotypes	Tall	×	Tall
Parental genotypes	Tt		Tt
Gametes	T and t		T and t

Female gametes: T, t

Offspring genotypes and phenotypes — Male gametes T, t

	T	t
T	TT Tall	Tt Tall
t	Tt Tall	tt Dwarf

Expected phenotype ratio — 3 tall : 1 dwarf

*Diagrams showing a genetic explanation of Mendel's monohybrid cross. The circles show the alleles and the boxes (called **Punnett squares**) show the various outcomes resulting from random fusion of gametes.*

Parental phenotypes	Tall	×	Dwarf
Parental genotypes	Tt		tt
Gametes	T and t		t and t

Female gametes: T, t

Offspring genotypes and phenotypes — Male gametes t

	T	t
t	Tt Tall	tt Dwarf

Expected phenotype ratio — 1 tall : 1 dwarf

In this test cross the unknown plant turns out to be heterozygous.

Check your understanding

1 Distinguish between genotype and phenotype.

2 Is a dwarf pea plant homozygous or heterozygous? Give reasons for your choice.

3 What would be the result of a test cross of tall plants that turn out to be homozygous?

Codominance

Codominance is shown by two alleles that are neither dominant nor recessive to each other. In the heterozygous condition, they produce a phenotype that is different from the homozygous state of either allele. For example, in *Antirrhinum* (snapdragon), two alleles for flower colour are codominant. Homozygous individuals may be either red or white, whereas heterozygous individuals are pink. This is because in the heterozygous individual both alleles of the pair are expressed in the phenotype. The heterozygous *Antirrhinum* is part-way between the phenotypes of the two types of homozygous individuals. This is often the case with codominant alleles.

As neither allele is dominant, the different alleles are not represented by upper case and lower case letters. Instead, the gene is given an upper case letter and each allele is represented by a superscript upper case letter. For example, C^R could represent the allele for red flowers and C^W the allele for white flowers. The possible genotypes and phenotypes of snapdragon flowers are therefore as shown in the table.

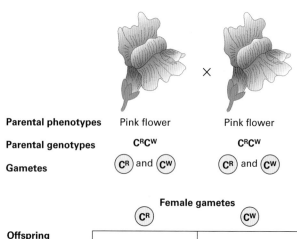

Parental phenotypes   Pink flower   Pink flower

Parental genotypes $C^R C^W$ $C^R C^W$

Gametes C^R and C^W C^R and C^W

Female gametes

	C^R	C^W
Offspring genotypes and phenotypes C^R	$C^R C^R$	$C^R C^W$
Male gametes C^W	$C^R C^W$	$C^W C^W$

Expected phenotype ratio   1 red flower : 2 pink flowers : 1 white flower

Codominance of flower colour in snapdragons

Genotype	Phenotype
$C^R C^R$	red flower
$C^R C^W$	pink flower
$C^W C^W$	white flower

The phenotypes and genotypes for flower colour in antirrhinum

If pollen from a white-flowered plant pollinates a red-flowered plant, the offspring will all have pink flowers. If these flowers are self-crossed, there is a 1 in 4 chance (0.25 probability) that the offspring will have red flowers, a 2 in 4 chance (0.5 probability) that they will have pink flowers, and a 1 in 4 chance (0.25 probability) that they will have white flowers. The figure illustrates this.

Multiple alleles

The human **ABO blood group** system involves codominance and another genetic condition called multiple alleles. **Multiple alleles** refer to the condition in which there are more than two alternative forms of a particular gene that occupy a gene locus. In some cases there are as many as 100 possible alleles for a particular gene, resulting in many slightly different phenotypes for the same trait. Of course, only two alleles can occur in any one diploid organism. By convention, in genetic diagrams the gene is given an upper case letter and each allele a superscript. For example, in the ABO blood group system:

- **I** (for immunoglobulin) represents the gene locus
- I^A, I^B, and I^O represent the three alleles

The I^A and I^B alleles are codominant, whereas both these alleles are dominant to I^O. Therefore, these three alleles give six possible genotypes and four possible phenotypes, the blood groups A, AB, B, and O, as shown in the table.

Phenotype	Possible genotypes
blood group A	$I^A I^A$ or $I^A I^O$
blood group AB	$I^A I^B$
blood group B	$I^B I^B$ or $I^B I^O$
blood group O	$I^O I^O$

The ABO blood group phenotypes and genotypes

Offspring of a mother who is heterozygous for blood group A and father who is heterozygous for blood group B might have any one of the four blood groups, as shown in this Punnett square.

	♀		♂
Parental phenotypes	Group A	×	Group B
Parental genotypes	$I^A I^O$		$I^B I^O$
Gametes	I^A and I^O		I^B and I^O

Female gametes

Male gametes	I^A	I^O
I^B	$I^A I^B$ Group AB	$I^B I^O$ Group B
I^O	$I^A I^O$ Group A	$I^O I^O$ Group O

Offspring genotypes and phenotypes

Inheritance of ABO blood groups in humans: possible offspring of a heterozygous blood-group A mother and a heterozygous blood-group B father

Check your understanding

1 Two heterozygotes for a particular condition produce offspring with three phenotypes in the ratio 1:2:1. What sort of alleles does this gene have?

2 Give the possible genotypes of a person who is blood group B.

3 There are some genes that have dozens of different alleles. How many alleles occur in a diploid cell?

- -

4 Codominance and multiple alleles are not the only conditions that result in non-Mendelian ratios. One classic example involves the gene for fur colour in mice. Wild mice have grey-coloured fur, a condition known as agouti. Occasionally, a mouse with yellow fur is born. A self-cross between mice with yellow fur produces offspring in the ratio of 2 yellow to 1 agouti. These results suggest that the allele for yellow (**Y**) is dominant to the allele for agouti (**y**). However, the ratio 2:1 is not a typical Mendelian ratio. Suggest an explanation. (Hint: examination of the uteri of female yellow mice made pregnant by a yellow male revealed dead yellow fetuses. No dead fetuses were found in yellow mice impregnated by agouti mice.)

OBJECTIVES

By the end of this spread you should be able to

- use fully labelled diagrams to predict the results of monohybrid crosses involving sex linkage

Prior knowledge

- monohybrid inheritance (5.03)

Colour blindness

Colour blindness refers to any condition in which colours cannot be distinguished. The most common form is red-green colour blindness in which a person cannot distinguish between red and green. This is caused by a recessive allele carried on the X chromosome. The pattern of inheritance of red-green colour blindness is therefore similar to that of haemophilia: men are more likely to show the defect (6–8% of Caucasian males are red-green colour blind, but less than 0.5% of women have this condition).

Sex chromosomes

Human sex is determined by a pair of **sex chromosomes** called **X** and **Y**. Because these chromosomes do not look alike, they are sometimes called **heterosomes**. All other chromosomes are called **autosomes**. Females have two X chromosomes (XX). Males have one X and one Y chromosome (XY). All mammals and some insects (including *Drosophila*, the fruit fly commonly used in genetic experiments) have similar sex-determining chromosomes. However, in birds females are the heterogametic sex and have the XY genotype. In some insects such as locusts, females have two X chromosomes (XX) while males have only one X and no Y chromosome. Their genotype is represented as XO.

X linkage

Genes that are located on one or other of the sex chromosomes are said to be **sex linked**. Sex linkage results in an inheritance pattern different from that shown by genes carried on autosomes. Theoretically, sex-linked genes could be on the X chromosome (X linkage) or the Y chromosome (Y linkage). However, the X chromosome is much larger than the Y chromosome, and all sex-linked genes so far confirmed in humans are located on the X chromosome.

In addition to carrying genes involved in determining female sex characteristics, the X chromosome also carries genes for non-sexual characteristics such as the ability to see particular colours and the ability to clot blood efficiently.

In **X linkage**, a recessive allele is expressed more often in males than in females because males carry only one allele for an X-linked gene whereas females have two alleles for the gene. So a female with a recessive allele on one of her X chromosomes and a dominant allele on the other will not show the recessive condition. However, a male with a recessive allele on his X chromosome has no dominant allele to counteract it, because his Y chromosome lacks the corresponding gene. The phenotype of the male will therefore show the recessive condition.

In animals such as *Drosophila* which are used in genetics experiments, X linkage is distinguished from autosomal linkage by making **reciprocal crosses**. In a reciprocal cross, parents are chosen such that the new male parent has the same phenotype as the female parent in the original cross, and the new female parent has the same phenotype as the original male parent. In X linkage, reciprocal crosses produce different results. So in the inheritance of X-linked eye colour in which the allele for red eye is dominant to the allele for white eye, a cross between a female with white eyes and a male with red eyes produces 50% red-eyed females and 50% white-eyed males. However, in a cross between a homozygous female with red eyes and a male with white eyes, all the offspring (male and female) are red eyed.

Haemophilia: an X-linked characteristic

Haemophilia is a condition in which the blood does not clot normally. Haemophilia A (the most common form of haemophilia) is a sex-linked character caused by a recessive allele carried on the X chromosome. Females have a pair of alleles for the gene that controls the production of a clotting factor, but males have only one alelle. Therefore, if a male inherits one allele for haemophilia A, he has the disease because he cannot possess another allele to mask its effects. On the other hand, a female can have the disease only if she inherits two recessive alleles, one from each parent. Heterozygous females are carriers of the disease.

The inheritance of haemophilia within a family can be shown in a **pedigree** (a chart of the ancestral history of a group of related individuals). One of the best established pedigrees for haemophilia is that of Queen Victoria's family, as shown below.

Parental phenotypes	Female carrier	×	Normal male
Parental genotypes	X^HX^h		X^HY
Gametes	X^H and X^h		X^H and Y

Female gametes

Offspring genotypes and phenotypes	Male gametes	X^H	X^h
	X^H	X^HX^H Normal female	X^HX^h Female carrier
	Y	X^HY Normal male	X^hY Haemophiliac male

A cross between a female carrier of haemophilia and a male without haemophilia. In sex-linkage studies, the sex chromosomes are always shown and dominant or recessive alleles are represented by superscript upper and lower case letters. In the transmission of haemophilia, X^HX^h would represent a female carrier and X^HX^H a normal female. The Y chromosome does not carry a gene for the clotting factor, so the male without haemophilia is shown as X^HY.

Key
- ■ Haemophiliac male
- □ Normal male
- ⬤ Carrier female
- ○ Female not known to be carrier
- ? Inadequate information

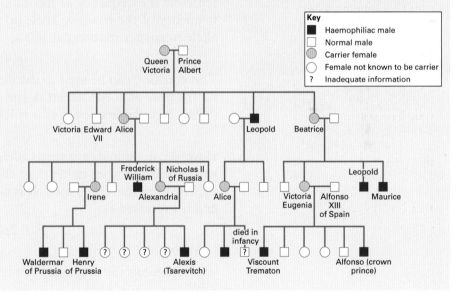

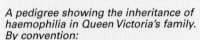

A pedigree showing the inheritance of haemophilia in Queen Victoria's family. By convention:

- *circles represent females; squares represent males*
- *shaded circles or squares represent affected individuals; unshaded circles or squares represent unaffected individuals*
- *two parents are linked by a horizontal line joining a circle and a square*
- *vertical lines run down from parents to children*
- *the children in one family are linked by a horizontal line above them*

Check your understanding

1 Why is sex linkage in humans due to genes carried on the X chromsome and not the Y chromosome?

2 What is the probability of a haemophiliac father and a carrier mother having a haemophiliac daughter?

3 Male-related baldness pattern is referred to as a sex-limited characteristic. How does this differ from sex linkage?

OBJECTIVES

By the end of this spread you should be able to

- understand that species exist as one or more populations

- define the concepts of gene pool and allele frequency

- state the Hardy–Weinberg principle and recall the conditions under which the principle applies

- calculate allele, genotype, and phenotype frequencies from appropriate data and from the Hardy–Weinberg equation

Prior knowledge

- the species concept
(*AS Biology for AQA* spread 17.01)

- the founder effect and genetic bottlenecks
(*AS Biology for AQA* spread 12.02)

Population genetics concerns the study of heredity and the genetic composition of **populations**. Members of a population all belong to the same species. There may be only a single population for a given species, but most have more. Two important concepts used in population genetics are gene pool and allele frequency.

- The **gene pool** is the sum of all the alleles of a population. The members of each population, therefore, share the same gene pool.

- **Allele frequency** refers to the proportion or percentage of a particular allele of a gene in a population, relative to other alleles of the same gene.

The Hardy–Weinberg equation

G. H. Hardy (an English mathematician) and W. Weinberg (a German biologist) devised an equation to calculate allele and genotype frequencies within a population. The equation is based on the fact that the total frequency of alleles of a gene must equal 100% (this is usually expressed as a decimal, where 1.0 represents 100%).

Hardy and Weinberg considered a population with a gene that has two alleles, **A** which is dominant and **a** which is recessive. They used p to represent the frequency of the dominant allele (**A**) and q to represent the frequency of the recessive allele (**a**) in the population. Because the total allele frequency must equal 100%,

$$p + q = 1.0$$

This equation can be used to calculate the allele frequency. For example, for the gene with two alleles **A** and **a**, if the frequency of **A** is 0.45 (45%), then the frequency of **a** will be

$$1 - 0.45 = 0.55 \ (55\%)$$

However, in diploid organisms alleles occur in pairs. In order to calculate genotype frequencies rather than allele frequencies, a second equation derived from the first is used. It is called the **Hardy–Weinberg equation** and is expressed as:

$$p^2 + 2pq + q^2 = 1.0$$

where

p^2 represents the frequency of the **AA** genotype

$2pq$ represents the frequency of the **Aa** genotype

q^2 represents the frequency of the **aa** genotype

For example, in a population where 14% of babies are born with a recessive condition

$$q^2 = 0.14$$
$$q = \sqrt{0.14} = 0.374$$

Since $p + q = 1.0$,

$$p = 1.0 - 0.374 = 0.626$$

Therefore $\qquad p^2 = 0.626^2 = 0.392$

and $\qquad 2pq = 2 \times 0.626 \times 0.374 = 0.468$

This means that 46.8% of the babies born are heterozygotes.

(Note that the frequency of the homozygous recessive phenotype, q^2, is needed in order to use the Hardy–Weinberg equation to work out genotype frequencies. Individuals showing the dominant phenotype may be either heterozygous or homozygous: the frequency of the dominant phenotype is $p^2 + 2pq$. This does not allow us to find either p or q separately.)

The Hardy–Weinberg principle

In 1908 Hardy and Weinberg, working independently, identified a mathematical principle which states that

The frequency of dominant and recessive alleles in a population will remain constant from generation to generation provided certain conditions are met.

These conditions are:

1 The population is large.

2 Mating is random.

3 No mutations occur.

4 There is no immigration into or emigration from the population.

5 All genotypes are equally fertile so that no natural selection is taking place.

If the first four conditions are met, a change in allele frequency will be due to natural selection. Measurements of allele frequency therefore provide a means of measuring the rate of evolutionary change.

Hardy and Weinberg recognized that allele frequencies are likely to remain constant only if populations are large. In small populations chance factors play a significant part in determining which alleles are passed on to the next generation. The smaller the population, the greater the probability that the allele frequency will differ between one generation and the next.

The name given to a change in allele frequency due to chance is **genetic drift**. It may occur when a few individuals from a large population colonize isolated areas such as an island (the so-called **founder effect**); or when a population decline results in a few individuals acting as the source of future generations. In either case, there will be less genetic variation within the small population than in the original population. In addition, some alleles from the large, original population may be absent in the small population and others may be disproportionately represented.

If a species has several populations that have undergone genetic drift and they become reproductively isolated from each other, they may evolve into new species by natural selection. This is the phenomenon that Charles Darwin observed on the Galapagos Islands.

Shells of the common snail Cepaea nemoralis *have several colour varieties, a condition known as* **polymorphism**. *Population genetics includes investigating why a population can have two or more forms coexisting in nature.*

Check your understanding

1 Define the term gene pool.

2 Under what conditions can changes in allele frequency be used to measure evolution by natural selection?

3 Use the Hardy–Weinberg equation to calculate the frequency of carriers if one in 2200 people in a population has cystic fibrosis, an inherited disorder caused by a single recessive allele carried on an autosomal chromosome.

- - - - - - - - - - - - - - -

4 In one founder population of wild flowers that exhibit codominance, the frequency of red flowering plants is 60% and that of pink flowering plants is 40%; there are no white flowering plants.

a Assuming there are two codominant alleles for white (C^W) and red coloration (C^R), calculate their allele frequency.

b Assuming that all the conditions for the Hardy–Weinberg principle apply, predict what will happen in subsequent generations.

5.07 Natural selection

OBJECTIVES

By the end of this spread you should be able to

- understand that differential reproductive success can affect the allele frequency within a gene pool
- distinguish between directional selection and stabilizing selection

Prior knowledge

- population genetics (5.06)

Selection pressures

Environmental factors that keep populations in check are called **selection pressures** or **environmental resistances**. These include

- disease
- competition for resources such as food and water, or a place in which to live
- predation
- unsuitable abiotic conditions, such as lack of light or oxygen, or unfavourable temperatures

Survival of the fittest

Darwin had the idea that **natural selection** is the mechanism that drives evolution after reading *An Essay on the Principle of Population* by Thomas Malthus, a clergyman and political economist. Malthus argued that, in time, the growth of human populations will outstrip the food supply, and that this will lead to 'famine, pestilence, and war'. Darwin applied this idea to populations of other animals and of plants. In his book on the origin of species, Darwin wrote

> There is no exception to the rule that every organic being naturally increases at so high a rate that, if not destroyed, the Earth would soon be covered by the progeny of a single pair.

Darwin observed that no single species had, despite their reproductive potential, completely over-run the planet; he concluded that populations are kept in check by a 'struggle for existence' as they compete for limited resources and are exposed to disease.

Those organisms best suited to the environmental conditions, with characteristics that give them an advantage in the 'struggle for existence', will have the best chance of surviving and producing offspring. Their high **natality** (birth rate) gives them a selective advantage. On the other hand, those with unfavourable characteristics are more likely to die. Their high **mortality** (death rate) gives them a selective disadvantage.

Darwin argued that this difference in natality and mortality results in natural selection. As environmental conditions change, certain characteristics within a randomly varying population are favoured, and natural selection occurs. This has become known as the '**survival of the fittest**'.

In evolutionary biology, **fitness** is defined as the ability of an organism to pass on its **alleles** to subsequent generations compared with other individuals of the same species. The 'fittest' individual in a population is the one that produces the largest number of offspring that survive to reproduce themselves. Natural selection by 'survival of the fittest' means that the genetic characteristics of a population gradually change from generation to generation in response to changes in the environment. Natural selection affects a **gene pool** by increasing the frequency of alleles that give an advantage, and reducing the frequency of alleles that give a disadvantage.

Three types of natural selection

Natural selection is not always a mechanism for change. There are three different types of selection: stabilizing selection, directional selection, and disruptive selection. These reflect the three different ways in which natural selection acts on the **phenotypes** in a population (the observable characteristics such as height or colour). Typically, the frequency in the population of each phenotype has a normal distribution, described by a bell-shaped curve.

- **Stabilizing selection** usually takes place in an unchanging environment. Extremes of the phenotype range are selected against, leading to a reduction in variation (more individuals tend to conform to the mean). Stabilizing selection occurs in the natural selection of birth mass in humans. The frequency of alleles carried by babies of intermediate birth mass will tend to increase in the gene pool.

- **Directional selection** favours one extreme of the phenotype range and results in a shift of the mean either to the right or to the left. This type of selection usually follows some kind of environmental change. The long neck of the giraffe is thought to have evolved in this way. Probably, when food was in short supply, only the tallest individuals could reach enough food to survive. They passed on their genes to the next generation. Directional selection results in alleles associated with one extreme of the phenotype range increasing in frequency in the gene pool.

- **Disruptive selection** selects against intermediate phenotypes and favours those at the extremes. This leads to a **bimodal distribution** (the distribution curve has two peaks or modes) and two overlapping groups of phenotypes. If the two groups become unable to interbreed, then each population may give rise to a new species. Disruptive selection may have contributed to the evolution of finches on the Galapagos Islands. Because there were fewer other competing birds, finches with short strong beaks had exclusive use of nuts as a food source, while those with long slender beaks had almost exclusive use of insects. Those finches with an average, unspecialized beak were more likely to have been in competition with other species of bird and would have reproduced less successfully. Disruptive selection results in alleles associated with both extremes of the phenotype range increasing in frequency in the gene pool.

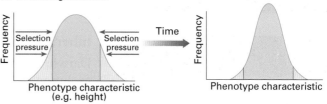

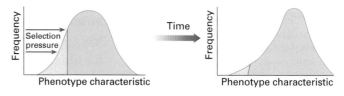

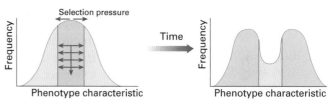

The three main types of natural selection. The areas shaded yellow indicate those individuals with a selective disadvantage that are being removed from the populations.

Check your understanding

1 What is meant by fitness in evolutionary terms?
2 What type of natural selection is thought to have brought about the long neck of giraffes?

- -

3 Suggest why the standard deviation around the mean birth mass is greater for babies born in economically developed countries than for those in less economically developed countries.

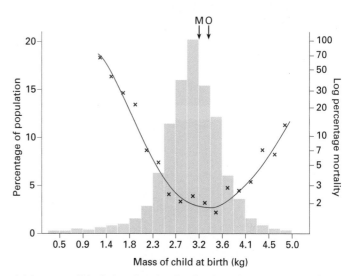

A histogram (blue) showing the distribution of mass at birth of children born in University College Hospital over a 12-year period, and a line graph (red) showing the percentage perinatal mortality (those failing to survive for four weeks) on a logarithmic scale. At the optimal birth mass (O), the percentage mortality is lowest. M is the mean birth mass.

OBJECTIVES

By the end of this spread you should be able to

- use examples and information to explain how selection produces changes within species
- interpret data relating to the effect of selection in producing change within populations

Prior knowledge

- antibiotics (*AS Biology for AQA* spread 18.01)
- mutations (*AS Biology for AQA* spread 18.02)
- antibiotic resistance (*AS Biology for AQA* spread 18.03)
- natural selection (5.07)

Pale and melanic forms of the peppered moth Biston betularia

Inherited variations

Darwin established that natural selection is the mechanism that drives evolution by causing differential mortality (unequal chances of dying), or 'survival of the fittest'. Within a population, individuals vary and those with favourable characteristics are more likely to survive than those with unfavourable characteristics. Darwin recognized that for evolution to occur, the characteristics of survivors have to be inherited by their offspring. However, he had no knowledge of genes.

We now know that inherited variations are determined by genes, and the ultimate source of these variations is **mutation**. A **gene mutation** is a random event resulting in a new allele. Most mutations that affect the phenotype are harmful. However, occasionally a mutation provides a new phenotype such that the mutant has a selective advantage over other individuals in a population. Over many generations, populations may gradually change so that individuals with the mutant allele become more frequent, a process called **microevolution**. Examples of such a change include the microevolution of antibiotic resistance in bacteria, pesticide tolerance in insects, heavy-metal tolerance in grass, and industrial melanism.

Antibiotic resistance

Antibiotic resistance is a severe problem throughout the world. For example, some strains of the common bacterium *Staphylococcus aureus* are resistant to antibiotics such as penicillin and methycillin (strains resistant to methycillin are described as MRSA: methycillin-resistant *Staphylococcus aureus*). Penicillin resistance has probably evolved in the following way:

- By chance, a mutation produces an individual bacterium that can produce an enzyme, penicillinase, which deactivates penicillin.
- This bacterium is immediately resistant to penicillin. (Bacteria have only one strand of DNA and one copy of each gene, so the mutant allele is expressed immediately, unmasked by a dominant allele.)
- If the population is exposed to penicillin, the mutant will survive and reproduce whereas those without the mutant allele will be killed.

The penicillin applies a selection pressure that results in **directional selection** and a large population of penicillin-resistant bacteria evolves. Because bacteria can reproduce very rapidly under ideal conditions (such as inside a warm human body), the single mutant can produce many millions of antibiotic-resistant offspring within 24 hours.

Pesticide resistance and heavy metal tolerance in plants have evolved in a similar manner.

Industrial melanism

The peppered moth *Biston betularia* has two main forms with different wing colours. One form has pale wings with dark markings; the other form is called **melanic** because the wings contain large amounts of **melanin** (a black pigment) which makes them almost black.

Until 1849, collections of moths from Manchester contained only the pale form, but by 1895, 98% of moths caught in Manchester were melanic. This change in phenotype frequency was called **industrial** melanism because the increase in melanics was associated with the industrialization that occurred in Manchester. By 1895, Manchester had become so highly industrialized that soot from factory chimneys had blackened trees and fences in and around the city.

A simple Darwinian explanation of industrial melanism proposes that melanic forms, resting against a sooty industrialized background, are difficult to see. They escape predation by birds and survive to reproduce. In rural areas, the opposite would occur.

Observations and field experiments conducted in the 1950s by H. B. D. Kettlewell seem to confirm this explanation. Kettlewell bred pale and dark moths, marked them, released both forms in two separate areas, one in Dorset (unpolluted) and the other in Birmingham (polluted), and recorded the percentage recaptured. His results are shown in the table. Kettlewell had observed that moths resting on tree trunks during the day were hunted by birds that rely on sight to catch their prey. He therefore concluded that these birds produced the main selection pressure by feeding on moths differentially according to their background.

Whether a peppered moth is pale or melanic is determined by a single gene with two alleles, usually represented as **C** and **c**. Moths with genotypes **CC** and **Cc** are melanic, and moths with genotype **cc** are pale.

In 1958, the frequency of melanic moths was high in industrialized areas or upwind of them. Since the 1970s, air pollution has decreased sufficiently for trees and fences to have become cleaner so that lichens can grow on them. Consequently, the frequency of the pale forms (and therefore the **c** allele) has increased.

It is important to realize that a mutation of the **c** allele to the **C** allele occurred spontaneously; it was not produced by pollution. A change in an environmental factor, such as air pollution, changes the frequency of an allele in a population, but it does not affect the rate at which such an allele arises by mutation. Before 1849, melanic moths probably occurred but were too rare to have been collected; they would almost certainly have been eaten by birds before the moths could reproduce.

Although industrial melanism of the peppered moth demonstrates natural selection in action, the different forms (morphs) have remained members of the same species because they can interbreed. If the moths in polluted and unpolluted areas were prevented from interbreeding for long enough, they would probably become sufficiently different to evolve into two distinct species.

Polluted region		
	Pale form	Melanic form
Marked and released	137	447
Recaptured	18	123
% recaptured	13.1	27.5

Unpolluted region		
	Pale form	Melanic form
Marked and released	496	473
Recaptured	62	30
% recaptured	12.5	6.3

Data from Kettlewell's mark-release-recapture experiment

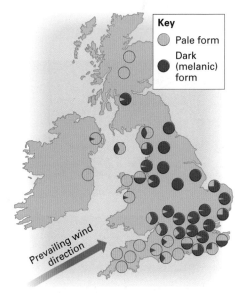

Key
- Pale form
- Dark (melanic) form

Prevailing wind direction

The distribution of pale and melanic forms of the peppered moth in the British Isles in 1958

Check your understanding

1 Why are antibiotic-resistant bacteria rare where antibiotics are not used?

2 What effect did industrial pollution have on
   a the frequency of the **C** (melanic) allele within a population of peppered moths
   b the rate of mutation of the **c** allele to the **C** allele?

- -

3 Which statistical test could you carry out to test the significance of Kettlewell's results? What would be the null hypothesis?

OBJECTIVES

By the end of this spread you should be able to

- explain how geographic isolation of populations of a species can result in the accumulation of differences in the gene pools

- appreciate the importance of geographic isolation in the formation of a new species

Prior knowledge

- the species concept (*AS Biology for AQA* spread 17.01)

- natural selection (5.07)

The coyote (Canis latrans) and the wolf (Canis lupus) do not normally interbreed, although biologically they are capable of doing so. They are reproductively isolated.

Evolution occurs whenever the inherited characteristics of a population or of a species change over a period of time. When these changes lead to the formation of one or more new species, **speciation** has taken place.

A definition of species

A species can be defined as 'a group of organisms with similar features which can interbreed to produce fertile offspring, and which are reproductively isolated from other species'.

Organisms that do not interbreed to produce fertile offspring under normal circumstances are regarded as **reproductively isolated**, and they belong to separate species.

Reproductive isolation: becoming unable to interbreed

Rarely do all members of a species exist in a single large population. They usually live in a number of local interbreeding populations called **demes**. Each deme has its own distinct gene pool, but two or more demes will belong to the same species as long as some individuals from the different demes continue to interbreed and produce fertile, hybrid offspring. On the other hand, if **reproductive isolation** occurs between the demes (the members of different demes are unable to interbreed at all), the demes may over a long period of time evolve by natural selection into new species.

Mechanisms that prevent breeding between populations and which can eventually lead to speciation are called **isolating mechanisms**. The most important isolating mechanism is thought to be **geographical isolation**, in which two populations originally of the same species are separated from each other by a physical barrier such as a mountain, river, or ocean.

When geographical isolation leads to new species being formed, **allopatric speciation** is said to have occurred. (Allopatric means literally 'different countries'.) Any physical barrier that prevents members of different populations from meeting must inevitably prevent them from interbreeding.

Hypothetically, allopatric speciation could take place when rising sea levels divide one island into two, as shown below. Although geographical isolation would be the original cause of allopatric speciation, the two isolated populations would have to diverge so much from each other that if reunited they would be unable to interbreed. Other isolating mechanisms would keep the two species from breeding together, as shown in the table.

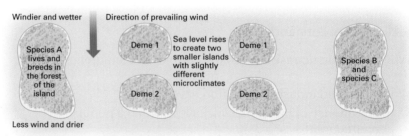

1 Species A inhabits the forests of an island, forming a single interbreeding population

2 The sea separates the two islands and isolates the population of species A into two demes which adapt independently to their environments

3 Over a long period of time, demes 1 and 2 have evolved different physiological and anatomical adaptations

4 The two islands rejoin to form one island with a single forest. Physiological and anatomical differences between deme 1 and deme 2 prevent them from interbreeding. They are now two new species, B and C, each with its own gene pool.

A hypothetical example of allopatric speciation

Type of isolation	Reason for isolation
Geographical isolation	Organisms isolated by a physical barrier, such as a mountain, river, or ocean.
Temporal isolation	Organisms breed at different times of year.
Ecological isolation	Organisms live in different habitats within the same area.
Behavioural isolation	Organisms have different behaviour patterns, for example, they use different behaviour to attract a mate. In the fruit fly *Drosophila*, for example, normal mating involves males performing a ritualized 'dance' that has a definite sequence of wing and body movements. Closely related species will not normally mate because the courtship dances of the males are different. But experiments have shown that, in some cases, mating will occur if the antennae of the female are removed. Presumably, the female is unable to detect the wrong courtship dance and permits mating.
Mechanical isolation	Organisms cannot mate because of anatomical differences which make it impossible for gametes to come together.
Genetic isolation	Genetic or physiological incompatibility between different organisms prevents hybrids forming, for example, pollen may fail to grow on a particular stigma with incompatible genes.
Hybrid isolation	Different organisms interbreed but offspring do not survive or are infertile.

Isolating mechanisms

Check your understanding

1 Name three different types of geographical isolation.

2 Three populations of butterflies, A, B, and C, live separately on three oceanic islands, as shown here. The butterflies all feed on nectar, but have slightly different wing colouring. The islands are swept through the year by strong prevailing winds from the north-west.

   a What type of speciation is likely to have occurred?

   b Describe the isolating mechanism involved.

   c Two populations can interbreed and produce fertile offspring. Which are they likely to be?

   d Two populations can mate, but the offspring are infertile. Which are they likely to be?

   e When brought together, two populations do not attempt to mate, even in captivity. Which are they likely to be?

- - - - - - - - - - - - - - - - - -

3 When the coyote and wolf are kept in captivity they can interbreed and produce fertile offspring. However, they are regarded as two separate species. Why?

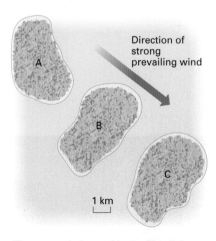

Three populations of butterflies living on three islands

OBJECTIVES

By the end of this spread you should have a better understanding of the significance of the following biological principles and be able to apply them to a range of contexts:

- random sampling results in the collection of data which is unbiased and suitable for statistical analysis
- ATP is the immediate source of energy for biological processes
- limiting factors
- energy is transferred through ecosystems
- genetic variation occurs within a species and geographic isolation leads to the accumulation of genetic difference in populations

Prior knowledge

- random sampling (1.03)
- ATP (2.01)
- limiting factors (2.05)
- energy transfer (3.01–3.02)
- genetic variation and geographic isolation (5.01 and 5.07–5.09)

Unit 4 Populations and environment covers a number of key biological principles. In chapters 1–5, these have been dealt with in specific contexts. However, you are expected to appreciate their wider significance and be able to apply them to a range of biological contexts.

Random sampling

Random sampling is a basic technique used to select a group of units (a **sample**) from a population. In this context, a **population** is any entire collection of people, organisms, places, or things from which we may collect data. In simple **random sampling**, each individual is chosen entirely by chance and each member of the population has an equal chance of being included in the sample.

Before making a random sample, it is important that the population is carefully and completely defined. This should include a description of the members to be included in the sample. For example, the population for a 2008 study of infant health might be all children born in the UK between 2000 and 2007. The sample might be all babies born on 18 April in any of the years.

Random sampling is designed to select a representative, unbiased sample of the population so that meaningful generalizations can be made about the population from which the sample has been taken. It is an essential requirement for many statistical analyses, enabling a sample statistic to provide reliable information about a corresponding population parameter. For example, the mean for a set of data from a random sample gives information about the overall population mean. Statistical techniques, such as calculating standard deviations and standard errors from sample means, assume random sampling has been used.

Biased, precise	Biased, imprecise	Unbiased, imprecise	Unbiased, precise
	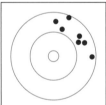	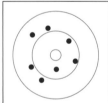	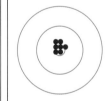

A target analogy showing the effects of bias on precise and imprecise measurements. Random sampling minimizes bias. In an unbiased sample errors from chance will tend to cancel each other out with many measurements; those from a biased sample will not. Measurements taken of an unbiased sample, even if not very precise, can provide a sample mean that estimates the true population mean within known confidence limits.

ATP

ATP, the mononucleotide **adenosine triphosphate**, appears to be as fundamental to life as DNA. All known organisms use ATP as their only direct source of chemical energy. They all require a continual supply of energy to carry out the processes that keep them alive. Some processes, such as cellular repair, occur continually. Others, such as the production of light in fireflies, occur only at certain times. Hydrolysis of ATP releases energy used to drive all of these vital biological processes.

Fireflies emitting light use energy from ATP.

Plants generate ATP by **photophosphorylation** using radiant energy from the Sun. All organisms generate ATP by **cellular respiration**. During respiration, chemical energy stored in nutrients such as carbohydrates, lipids, and proteins, is released so that ATP can be synthesized from ADP and inorganic phosphate.

Limiting factors

The principle of limiting factors is based on Liebig's law of the minimum. This law was originally applied to crop growth. It states that

> Growth is controlled not by the total of the resources available to a crop, but by the resource in scarcest supply.

The **principle of limiting factors** builds on this concept. It states that

> The functioning of an organism is limited by the essential factor or combination of factors present in the least favourable amount.

This wording of the principle recognizes that some biological processes such as photosynthesis can be controlled by too much of a factor (such as heat) rather than too little (see spread 2.05). It also recognizes that other processes, such as the growth of a population, are controlled by a complex interaction of a number of factors rather than by just one limiting factor. It is important to appreciate that whereas some factors might be essential throughout the life of an organism, others may be essential only at critical periods in the life cycle or at specific times of the year.

Energy transfer

The principle of energy transfer through an ecosystem (spread 3.01) is similar to that of energy transfer in *any* biological system from the level of the cell to the biosphere. All such energy transfers normally obey the laws of thermodynamics.

The first law, the law of conservation of energy, states that energy can neither be created nor destroyed. Consequently, the total energy entering a system always equals the total energy leaving it.

The second law of thermodynamics relates to entropy, a measure of the disorder in a system. It states that in any self-contained thermodynamic system entropy increases. An important consequence of this is that whenever energy is transformed from one type to another, some is dissipated as heat. This means that transformations from, for example, the chemical energy of food to the kinetic energy of locomotion, are not 100% efficient. Also, biological processes that involve energy transformations always generate some heat.

Genetic variation and geographic isolation

Genetic variation within a species and **natural selection** acting on geographically isolated populations can lead to a change in the genetic composition (**allele frequency**) of the species over successive generations. The changes may be so great that new species evolve that are adapted to different environments and ways of life. This type of evolution is one of the unifying themes of modern biology. When biologists study the characteristics of an organism, they invariably ask: how has it evolved? what is its function? to what is it adapted?

Active organisms are continually transforming one type of energy to another. In the process, heat is dissipated. It is this heat that infrared cameras detect.

Check your understanding

1. Distinguish between the term 'population' in a statistical context and in an ecological context.

2. What type of energy is always generated when energy transformations occur?

- - - - - - - - - - - - - - - - -

3. The extinction of animal and plant species is of concern today because it is accelerated by human activities. However, extinction has occurred since the dawn of life. The 20–30 million species on the Earth today represent a small proportion of all the species that have ever existed.

   a. Suggest why more than 99.9% of all species that ever evolved have become extinct by natural processes.

   b. Explain why the highest rates of extinction in recent times have occurred among species that are geographically isolated on small oceanic islands.

OBJECTIVES

By the end of this spread you should be able to

- interpret an investigation into the effect of a specific limiting factor such as light intensity, carbon dioxide, or temperature on the rate of photosynthesis

- use knowledge and understanding about the effects of specific limiting factors on the rate of photosynthesis

- analyse and interpret data related to an investigation into the effects of specific limiting factors on the rate of photosynthesis

- evaluate the methodology, evidence, and data related to the effect of specific limiting factors such as light intensity, carbon dioxide, or temperature on the rate of photosynthesis

Prior knowledge

- photosynthesis (2.02)
- limiting factors (2.05)

A: Temperature and photosynthesis

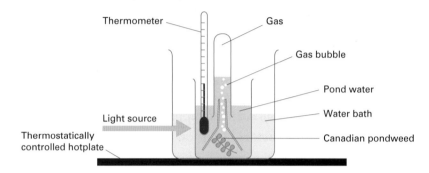

Measuring the evolution of oxygen from Elodea

A convenient way of investigating the effect of a limiting factor on the rate of photosynthesis is to measure the evolution of oxygen from an aquatic plant. In one investigation into the effect of temperature on photosynthesis, *Elodea* (Canadian pondweed) was used. A freshly cut stem was placed, bubbling end upwards, under a glass filter funnel in a beaker of 0.5% sodium hydrogencarbonate pondwater solution. The funnel was raised off the bottom of the beaker on pieces of Plasticine. The beaker was put into a thermostatically controlled water bath and the plant was allowed to come into equilibrium with each temperature as it was investigated. After equilibration, the number of gas bubbles emerging from the cut stem of *Elodea* were counted for a timed period of 5 minutes. During the investigation, the light intensity was kept constant by using the same light source at a fixed distance from the plant. The investigation was repeated twice and the mean rate of bubble production calculated for each temperature. The table shows the results.

Temperature (°C)	Mean rate of bubble production (number of bubbles per minute)
5	12
10	19
15	26
20	31
25	28
30	23
35	15
40	3

Questions

1 In this investigation, it was assumed that the gas bubbles evolved from Canadian pondweed contained oxygen. How could this assumption be tested?

2 Why is counting bubbles not a very reliable estimate of the volume of gas production?

3 Suggest two methods of measuring more accurately the gas production from Canadian pondweed.

4 How was light intensity kept constant?

5 Why was a 0.5% solution of sodium hydrogencarbonate used in the investigation?

6 Suggest why the filter funnel was raised off the bottom of the beaker.

7 This investigation only estimates the *apparent* rates of photosynthesis. Would the true rate be higher or lower than the apparent rate? Give reasons for your answer.

8 Describe and explain the observed effects of temperature on the rate of photosynthesis in *Elodea*.

9 Suggest a factor other than carbon dioxide, light, or temperature that might have an effect on the rate of photosynthesis.

B: Light intensity and photosynthesis

In an investigation to study the effect of light intensity on the rate of photosynthesis, discs from the leaves of young cabbage plants were used. Batches of five discs of equal area were punched using a cork borer and placed in a 5 cm³ transparent plastic syringe containing 5 cm³ of 0.2 M sodium hydrogencarbonate solution. By means of a simple procedure, the gas filling the intercellular spaces in the discs was removed and replaced with the sodium hydrogencarbonate so that all the discs sank to the bottom of the syringe. Syringes containing batches of five discs were then placed in different light intensities. All of the discs rose to the surface. The time taken for each disc to rise to the surface was recorded and the mean time at each light intensity was calculated. The table shows the results.

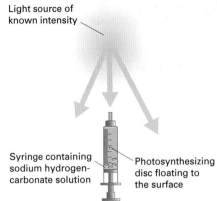

Investigating the effect of light intensity on the rate of photosynthesis

Questions

1 State one important consideration when selecting areas of leaf for cutting the discs.
2 Why do the leaves float?
3 Using the data, comment on the changes in the rate of net photosynthesis with changing light intensity.
4 When a black cover was placed over the syringe after the discs had risen to the surface, the discs sank. Explain why.

Light intensity (arbitrary units)	Net rate of photosynthesis [(1/mean time in minutes) $\times 10^3$]
20	15
40	27
60	75
80	220
100	215

C: Carbon dioxide exchange and light intensity

Study the graph showing carbon dioxide exchange for two plants at varying light intensities.

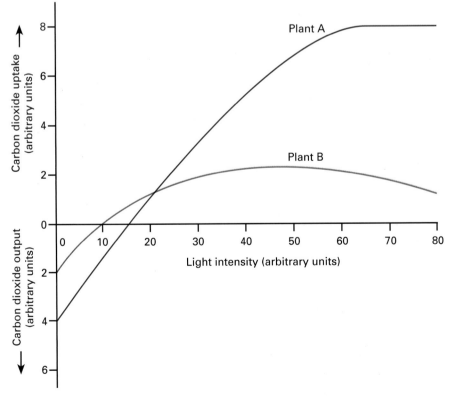

Carbon dioxide exchange and light intensity

Questions

1 Suggest an explanation for the differences between the two plants.
2 Define the carbon dioxide compensation point.
3 For each plant, give its compensation point.

OBJECTIVES

By the end of this spread you should be able to

- interpret an investigation into the effect of a specific variable such as substrate or temperature on the rate of respiration of a suitable organism

Prior knowledge

- cellular respiration (2.07–2.10)

Questions

1 Draw a graph to summarize the results.

2 Answer the following questions:

   a What causes the methylene blue to lose its blue colour during this experiment?

   b Calculate the initial rate of respiration for the yeast cells in glucose solution. Give your answer as % transmission min^{-1}. Show how you made the calculation.

   c Comment on the rate of respiration during the 20 minutes of the experiment.

   d State one way in which the rate of respiration of yeast in glucose solution differs from that in sucrose solution.

   e Give one reason for the shape of the lactose curve.

   f Explain why the tubes were kept covered during the investigation.

A: The effect of different substrates on the rate of respiration

An investigation was carried out to study the effect of different substrates on the rate of respiration of yeast cells. Three carbohydrate substrates were used: glucose, sucrose, and lactose.

The investigation uses the fact that at certain stages during cellular respiration, oxidation occurs by the removal of hydrogen atoms from the substrate. These hydrogen atoms are normally picked up by NAD. However, in this investigation another hydrogen acceptor, a dye called methylene blue, is used to indicate the rate of respiration. In its fully oxidized form, the dye is a deep blue colour. As it becomes reduced by picking up hydrogen atoms, it loses its colour. When fully reduced, it is colourless.

A colorimeter was used to measure the intensity of light passing through carbohydrate solution containing a culture of yeast cells. In this investigation, the colorimeter was calibrated so that in the results 100% transmission of blue light means that the methylene blue is fully oxidized; 0% transmission means that it is fully reduced.

In the first part of the investigation, a test tube was set up containing 10 cm^3 of a suspension of yeast cells in 0.5% glucose solution made up at a temperature of 30°C. 1 cm^3 of 0.1% methylene blue solution was added to the tube. The tube was covered and placed in a water bath at 30°C for a period of 20 minutes during which the percentage of light transmitted through the tube was measured by a colorimeter.

The investigation was repeated, first using a new batch of the same volume of yeast in 0.5% lactose solution and then in 0.5% sucrose solution. The results of the investigation are shown in the table.

Time (min)	Percentage of blue light transmitted		
	Substrate: glucose	Substrate: lactose	Substrate: sucrose
0	100	100	100
2	75	100	94
4	53	100	88
6	37	100	82
8	27	100	76
10	21	100	70
12	18	100	64
14	16	100	58
16	15	100	52
18	14	100	46
20	14	100	40

The effect of different substrates on the rate of respiration

B: The rate of respiration in a small invertebrate

Respirometers are devices that measure the rate of respiration, typically by measuring oxygen consumption and carbon dioxide output. Respirometers vary in complexity. Some incorporate sophisticated computer technology that automatically measures the volumes of gases exchanged and draws off small samples to analyse the proportions of oxygen and carbon dioxide in the gases. Others, designed to measure respiration rates in small invertebrates, may be simple.

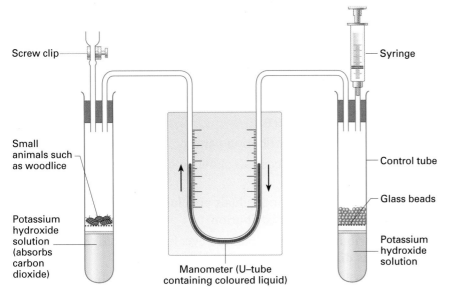

A simple respirometer

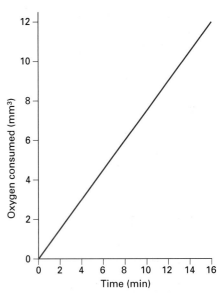

The oxygen consumed by the woodlice in a respirometer over time

In one investigation, woodlice had been placed in a simple respirometer consisting of a boiling tube connected to a U-tube. The U-tube acted as a manometer (a device for measuring pressure changes). The other end of the U-tube was connected to a control tube which was treated in exactly the same way as the first tube, except that it had no woodlice. Instead, glass beads which take up the same volume as the woodlice were placed in the tube. The two boiling tubes (but not the manometer) were kept in a water bath at a constant temperature. The U-tube contained a coloured liquid which moved up and down according to the pressure exerted on it by the gases in the two boiling tubes. Both tubes contained potassium hydroxide solution which absorbed any carbon dioxide produced.

When the woodlice respire aerobically, they consume oxygen, causing the liquid to move in the U-tube in the direction of the arrows.

The rate of oxygen consumption can be estimated by timing how long it takes for the liquid to rise through a certain height. Or, if the internal radius of the manometer tube is known, the volumes of oxygen consumed can be calculated using the equation:

$$\text{Volume} = \pi r^2 h$$

where r is the internal radius of the tube and h is the distance moved by the liquid.

Questions

1 What is a respirometer?

2 What is the function of the potassium hydroxide in the respirometer?

3 One end of the U-tube was connected to a tube containing woodlice. Why did the other tube to which it was connected contain glass beads?

4 The graph above shows the volume of oxygen consumed by the woodlice over a 16 minute period at 15°C. The total mass of woodlice was 10 g. Calculate the rate of oxygen consumption per gram of woodlouse. Show your workings.

5 What would you expect to happen to the rate of respiration if the temperature was increased to 25°C? Give your reasons.

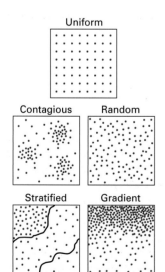

Types of distribution patterns

As part of your A-level studies, you are expected to undertake fieldwork involving the use of frame quadrats and line transects and the measurement of a specific abiotic factor in at least one habitat. Then, after collecting the data, you are expected to apply elementary statistical analyses to the results. The fieldwork may be undertaken in one or several habitats and on one or more occasions. It may be completed as a class exercise or as an individual project. The biology fieldwork you undertake will probably be concerned with investigating the effect of one or more abiotic factors, such as pH or light intensity, on the distribution and abundance of living organisms. No matter how it is undertaken, the fieldwork must be carefully planned if it is to be effective. Such planning must include a detailed **risk assessment** (see spread 1.05).

Types of distribution pattern

There are several types of distribution pattern. These include

- **random distribution** – individuals show no clear pattern of distribution

- **contagious distribution** – individuals are clustered together (also called **clumped** or **patchy distribution**)

- **uniform distribution** – individuals are evenly distributed (also called **homogeneous distribution**)

- **stratified distribution** – individuals are uniformly distributed in different sub-areas of the habitat

- **gradient distribution** – the distribution of individuals varies smoothly over the sampling area

It is possible to determine whether a population has a random, contagious, or uniform distribution by analysing the quantitative data obtained when a large number of frame quadrats are placed at random in the study area and the number of individuals in each quadrat are counted. If a population is uniformly distributed, there should be about the same number of individuals in every quadrat counted. If a population has a contagious distribution, there should be some quadrats containing a large number of individuals and many quadrats that are empty. If a population is randomly distributed, there will be no clear pattern in the counts from the quadrats.

In statistical analyses, the **variance to mean ratio** is used to measure population dispersion. Variance (s^2) is the square of the standard deviation.

- $s^2/\text{mean} > 1.0$ indicates a clumped distribution
- $s^2/\text{mean} = 1.0$ indicates a random distribution
- $s^2/\text{mean} < 1.0$ indicates a uniform distribution

The significance of the differences from 1.0 will depend on the number of quadrats sampled. The larger the sample size, the more significant any difference will be.

Sampling protocol

Before starting any fieldwork investigation, you should decide on a sampling procedure. In simple **random sampling** a two-dimensional coordinate grid is superimposed on the investigation area. The required sampling coordinates are then obtained using random numbers. This procedure is best where there is a uniform distribution of individuals.

In **systematic sampling**, the first sampling point is chosen at random but all others are taken at a fixed distance from this. This procedure is often used when mapping the distribution of organisms but it suffers from the disadvantage of having the potential to be biased. Also, there is no reliable method of estimating the standard error of the sample mean.

Stratified random sampling is carried out in habitats which have clearly distinguishable sub-areas, such as the zones on a tidal rocky shore. The area to be studied is divided into sub-areas within which random sampling is carried out. The number of sampling points in each sub-area may be constant or may be weighted according to the relative size of each sub-area (for example, if a total of 20 samples are taken in an area that has two sub-areas, one occupying 60% and the other 40% of the total area, 12 samples will be taken in the first area and 8 in the second).

An element of randomization can be used in **transect sampling** by identifying alternative start and end points and choosing them at random.

Relationships between two variables

In many fieldwork investigations, measurements of one or more abiotic factors are taken at the same time as organisms are sampled. When possible, the data collected should be quantitative, precise, and accurate.

Quantitative data collected for any two variables can be presented as a **scattergram** to reveal the type of relationship between them. Sometimes a scattergram shows no relationship between the two variables plotted. Some pairs of variables may show a strong negative or positive relationship which may be linear (straight lined) or non-linear (for example, curvilinear). The statistical significance of linear relationships can be tested using **Spearman's rank correlation coefficient** (see spread 1.06). Even if a strong correlation is confirmed, it must be remembered that the test does not prove a causal relationship between the variables.

In some circumstances, a **chi-squared test** (see spread 1.06) can be carried out to see if an environmental variable is affecting the distribution and abundance of an organism.

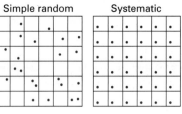
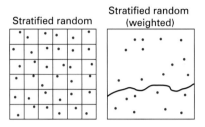

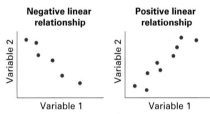

Four basic methods of sampling

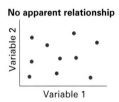

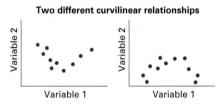

Scattergrams showing five relationships between two variables

---

Questions

1 When analysing the results of quadrat sampling, a student found that some quadrats contained a large number of individuals, but many quadrats were empty. What type of distribution pattern did this indicate?

2 Which type of sampling technique is best to use when the distribution pattern of the individuals being studied is uniform?

3 Which statistical technique can be used to test the significance of the correlation between two variables which appear on a scattergram to have a positive linear relationship?

4 With reference to the sampling protocol and analysis of the data collected, how would you use frame quadrats to determine whether a particular species has a clustered, random, or uniform distribution pattern?

The Heathland Restoration Project, near Honiton, Devon, has been running since 1995. (a) The woodland clearance. (b) The control strip five years later. The aim of the project is to restore a lowland heath habitat on the site, while comparing different management techniques for the restoration process. Details of the project can be found on the Woodland Education Centre website.

Question 1: Sampling flowering plants

The photos show a habitat management experiment. A similar project was carried out to compare different land management techniques to restore a lowland heath habitat in a woodland ecosystem. At the start of the project an area was cleared of mixed woodland, with an understory of rhododendron. The area was treated with a biodegradable herbicide to clear the site of all existing vegetation. The site was then divided into a number of different strips and was allowed to regenerate naturally. Different management methods were used on the strips with the aim of determining which was the best for regenerating heathland.

One of the strips was left untouched to act as a control. This strip measured 10 m by 5 m. The table shows the mean density values of plant species from ten 1 m² quadrat samples taken within the control strip.

Species	Mean number of plants per m²
birch seedlings (*Betula* spp)	12
common bent (*Agrostis capillaris*)	5
European gorse (*Ulex europaeus*)	4
hazel seedlings (*Corylus avellana*)	1
heather (*Calluna vulgaris*)	7
rhododendron seedlings (*Rhododendron ponticum*)	2
sycamore seedlings (*Acer pseudoplanatus*)	3
violets (*Viola riviniana*)	10
Yorkshire fog (*Holcus lanatus*)	6

Mean density of different plants as measured in 10 quadrat samples in the control strip

a Distinguish between an ecosystem and a habitat. [2]

b **i** Describe how the quadrat sites could be selected at random. [2]

　ii Why was it necessary to collect data from more than one quadrat? [2]

　iii Use Simpson's species diversity index formula to calculate the species diversity of the plants. Show your working. [2]

　iv Distinguish between species diversity and species richness. [1]

　v The control strip was left for 5 years. What would you expect to happen to the species diversity? Give reasons for your answer. [4]

c What effect would this change in plant diversity have on the diversity of animals living in the same ecosystem? Explain your answer. [2]

[Total marks = 15]

Question 2: Cod and grey seals

Grey seals are marine mammals that live in the seas of the North Atlantic. They are voracious carnivores feeding on fish, including cod. In the summer, grey seals come ashore to breed at haul-out sites. According to a Canadian Fisheries and Oceans report, Atlantic cod in the seas off the eastern Scotian Shelf and Sydney Bight were at their lowest abundance on record in 2005. They had a restricted distribution

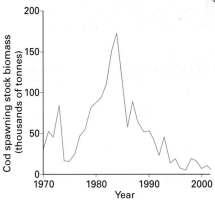

Changes in cod spawning stock biomass

and exhibited poor growth and condition, early maturation, and poor recruitment. Mortality was very high despite a cessation of directed fishing in these areas since 1993 and the return of oceanographic conditions closer to long-term averages after a severe cold spell in the late 1980s to early 1990s.

Fisheries scientists suspected that the decline in cod stocks might be linked to an increase in the grey seal population: in addition to eating juvenile cod, seals are also the primary host to a parasite called the sealworm. The cod is the intermediate host of this parasite; high sealworm infection rates could be contributing to mortality. Most grey seal pupping occurs on Sable Island where an increase in annual production has been observed over the last 40 years, resulting in an exponential increase in the population, as the graph shows. Cod represents only 1% of the grey seal diet, but most predation is on juvenile cod. Whales and fish are also important predators of cod. Nevertheless, seals were found to be the single most important predators of juvenile cod in the late 1990s. Although seals eat cod, they also eat other fish that prey on cod. Seals and cod exist in a complex ecosystem, which makes it difficult to provide simple solutions to problems such as the lack of recovery of cod stocks.

a With reference to the top graph, describe the changes in the cod spawning stock biomass between 1970 and 2000. [3]

b Spawners of age 5 and older, that is, experienced as distinct from first-time spawners, comprised more than 50% of the spawning biomass prior to 1995. However, the proportion of these older spawners declined substantially in the late 1990s and early 2000s. Suggest why these older spawners probably make a larger contribution to the production of larval cod. [2]

c What type of increase has occurred in the grey seal population in the last 40 years? [1]

d Calculate the percentage increase in the total grey seal population between 1975 and 2000. Show your working. [4]

e Give reasons why predation by grey seals could be an important factor in preventing cod recovery even though cod represents only 1% of the seal diet. [2]

f Suggest why reducing the grey seal population may not increase cod stocks. [3]

[Total marks = 15]

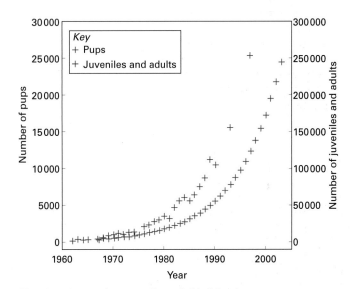

The abundance of grey seals on Sable Island

OBJECTIVES

By the end of this spread you should have answered questions on

• photosynthesis and respiration

Prior knowledge

• photosynthesis and respiration (2.01–2.10)

Stretch and challenge

These long structured questions on photosynthesis and respiration are posed in the style of the AQA examination questions. As in the AQA examinations, these questions are designed with an incline of difficulty such that the later subquestions offer a genuine challenge to the most able students.

Question 1: Photosynthesis

Study the figure which outlines the light-independent reactions of photosynthesis. (Note: this is based on the outline of the Calvin cycle in spread 2.04, but the numbers of all molecules have been doubled.)

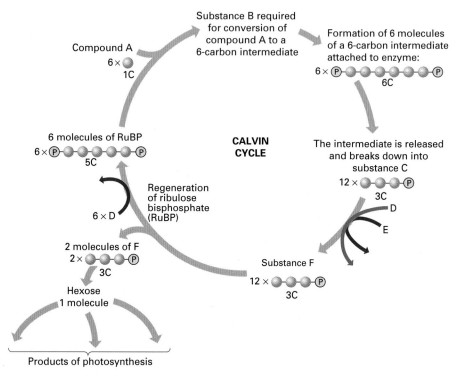

Outline of the Calvin cycle

a Name three groups of biological molecules formed from hexose. [3]

b Explain how the light-independent reactions of the Calvin cycle bring about the continual production of hexose sugars. [6]

Study the figure showing the Z scheme.

c The light-dependent reactions involve electron transport chains. Describe the role of an electron transport chain. [6]

[Total marks = 15]

An outline of the Z scheme, light-dependent reactions that involve electron transport chains

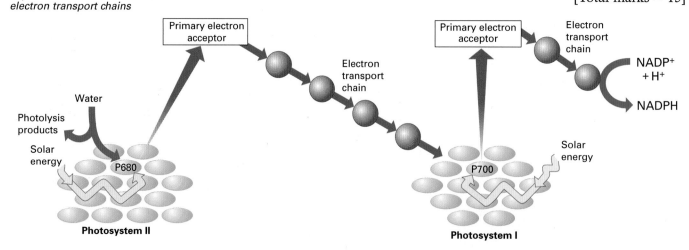

Question 2: Respiration

a Study the diagram opposite showing the main stages of anaerobic respiration in yeast.

  i Process X results in the conversion of glucose to pyruvate. What name is the process given? [1]

  ii Give one piece of evidence from the diagram that suggests that the conversion of pyruvate to ethanol involves decarboxylation. [1]

  iii Although the conversion of pyruvate to ethanol does not provide any more ATP directly, it is important indirectly in allowing further production of ATP in anaerobic respiration. Explain why. [2]

b Compare and contrast the anaerobic respiration of glucose in yeast with the anaerobic respiration of glucose in a muscle cell. [4]

c The effect of temperature on the rate of anaerobic respiration in yeast was studied using the fermenter shown in the diagram below. The yeast suspension was mixed with glucose solution and the increase in gas pressure after 5 minutes was recorded and analysed on the computer to give the total volume of gas evolved from the yeast.

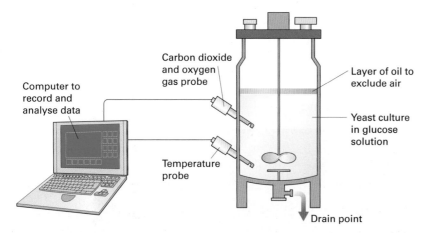

A fermenter used to study respiration in yeast

  i The experiment was repeated three times and the results were pooled. Explain the advantages of collecting more than one result. [2]

  ii At 20°C, the following results were obtained.

Volume of gas (cm³)	result 1	result 2	result 3
	38.5	29.7	30.9

Calculate the mean rate of gas production in $cm^3 \ s^{-1}$. Show your working. [2]

  iii If the investigation was conducted in an aerobic environment at the same temperature, how would the results differ from those obtained in the anaerobic environment? Explain your answer. [3]

[Total marks = 15]

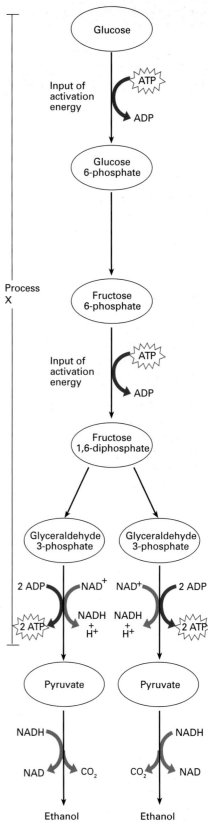

The main stages of anaerobic respiration in yeast

113

OBJECTIVES

By the end of this spread you should have answered questions on

• energy and food production

Prior knowledge

• energy and food production (3.01–3.09)

Stretch and challenge

These long structured questions on energy and food production are posed in the style of the AQA examination questions. As in the AQA examinations, these questions are designed with an incline of difficulty such that the later subquestions offer a genuine challenge to the most able students.

Question 1: Energy transfer

a The following table shows the mean values for primary productivity in four ecosystems.

Ecosystem	Region	Primary production $(kJ\ m^{-2}\ y^{-1})$
grassland on an intensive farm ecosystem	temperate	30 000
deciduous woodland	temperate	25 000
natural grassland	temperate	14 000
rainforest	tropical	40 000

i Present the results in a suitable graph. [4]

ii Explain why the primary production is higher in the tropical rainforest than in the temperate deciduous woodland. [2]

iii Why is the productivity measured in units of energy rather than units of biomass? [2]

iv The photographs show a natural grassland and grassland on an intensive farm. Suggest two reasons why the grassland on the farm has a much higher primary productivity than that of the natural grassland. [2]

Natural grassland and grassland on an intensive farm

b Study the diagram of energy flow through a freshwater ecosystem.

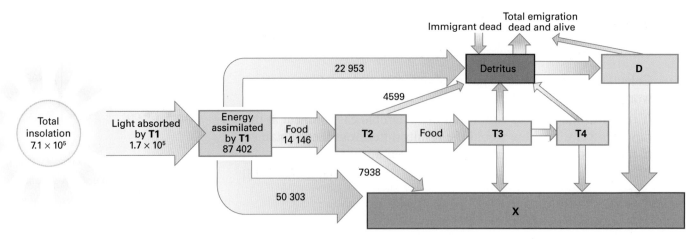

Energy flow through a freshwater ecosystem. All figures are in $kJ\ m^{-2}\ y^{-1}$.

i What does X represent? [1]

ii What type of organisms are in box D? [1]

iii What type of organisms are represented by T2? [1]

iv Calculate the amount of energy transferred as food from T2 to T3. [2]

[Total marks = 15]

Question 2: Food production – fertilizers and pesticides

a i Look at the photograph of the roots of a leguminous plant. What are the structures labelled Z, and how do they enable the leguminous plant to survive in soils with a low nitrate ion concentration? [3]

ii Suggest why waterlogged soil may have a low concentration of nitrate ions. [2]

iii Natural organic and artificial inorganic fertilizers add nitrates to the soil, increasing crop yield. Other than financial cost, give one advantage and one disadvantage of using an organic fertilizer. [2]

iv The graph shows the effects of applying an NPK fertilizer at different rates to a crop of wheat. Explain how the graph shows the law of diminishing returns. [2]

b i There are three major groups of pest management methods. Chemical and biological are two. Name the third. [1]

ii Resistance to pesticides has increased rapidly since 1950. Using information in the graph below, calculate the percentage increase in resistance between 1970 and 1990. Show your working. [2]

Roots of a leguminous plant

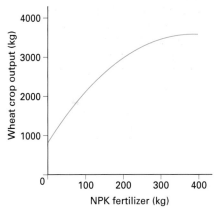

The effects of applying an NPK fertilizer at different rates to a wheat crop

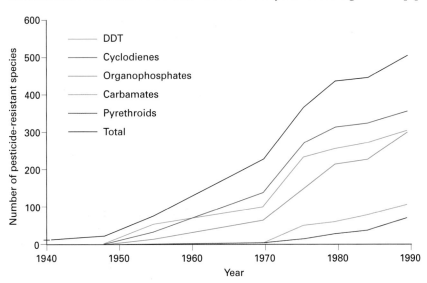

Graph showing the resistance of major pests to five major pesticides

iii Explain how an insect population can become resistant to a pesticide. [3]

[Total marks = 15]

OBJECTIVES

By the end of this spread you should have answered questions on

• environmental issues

Prior knowledge

• environmental issues (4.01–4.04)

Stretch and challenge

These long structured questions on environmental issues are posed in the style of the AQA examination questions. As in the AQA examinations, these questions are designed with an incline of difficulty such that the later subquestions offer a genuine challenge to the most able students.

Question 1: Pollution

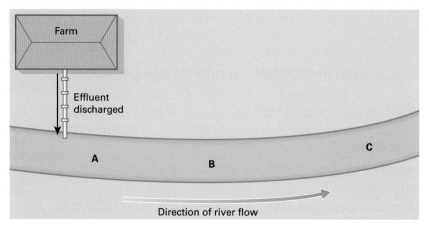

The discharge of effluent into a river

	A	B	C
BOD (biochemical oxygen demand) (mg dm^{-3})	38.0	17.0	3.0
Ammonium ions (mg dm^{-3})	0.4	0.8	0.1
Nitrate ions (mg dm^{-3})	5.0	50.0	8.0

Biochemical analyses of water at sites A, B, and C

a Study the diagram that shows part of a river into which farm effluent has been discharged. The table shows the results of analyses of water samples taken at sites A, B, and C along the river.

  i Define the biochemical oxygen demand (BOD) and describe how it is measured. [3]

  ii Explain why the BOD decreases from site A to site C. [3]

  iii Explain the increase in ammonium and nitrate ions from site A to site B. [4]

b In addition to chemical analyses, indicator species are used to monitor water quality.

  i Explain what indicator organisms are. [2]

  ii One indicator organism commonly monitored is *Escherichia coli*. Suggest why water downstream of the effluent discharge is likely to have higher population levels than water upstream of the discharge. [3]

[Total marks = 15]

Question 2: Ecological succession

Study the diagram showing stages of ecological succession in a pond.

1 Water fills the depression – microscopic algae colonize

2 Layer of sediment forms on the bottom – submerged plants appear

3 Plants grow on the surface and edges of the pond – submerged plants die out

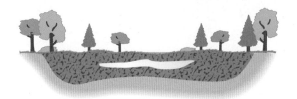

4 Mud and organic matter build up to form a marsh

5 The marsh dries out and a community of land plants becomes established

Pond succession

a Explain how the diagrams illustrate the features of an ecological succession. [6]

b Suggest how management of the pond may result in a plagioclimax rather than a climatic climax. [4]

c The final stage of succession is known as the climax community. During this stage species diversity and the size of each population remains fairly constant. Explain what limits the size of popul tions in a climax community. [5]

[Total marks = 15]

OBJECTIVES

By the end of this spread you should have answered questions on

• variation and evolution

Prior knowledge

• variation and evolution (5.01–5.09)

Stretch and challenge

These long structured questions on variation and evolution are posed in the style of the AQA examination questions. As in the AQA examinations, these questions are designed with an incline of difficulty such that the later subquestions offer a genuine challenge to the most able students.

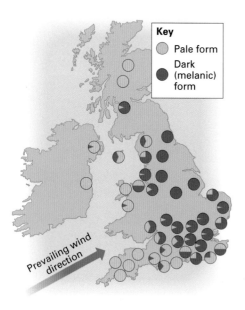

The distribution of pale (typica) and melanic (carbonaria) forms of the peppered moth in the British Isles in 1958

Question 1: The peppered moth

The peppered moth, *Biston betularia*, has three main colour morphs determined by a gene that has three alleles designated C^c, C^i, and C^t.

• C^c is the allele for the carbonaria morph; it is dominant to the other two alleles

• C^i is the allele for the insularia morph; it is dominant to the allele C^t

• C^t is the allele for the typica morph

*Biston betularia: **(a)** the carbonaria morph; **(b)** the insularia morph; **(c)** the typica morph*

a i Using appropriately annotated genetic notation show all the possible crosses that can occur between two carbonaria moths. [5]

ii The three types of peppered moth are classified as varieties of the same species. How is it possible to show that the three varieties belong to the same species? [2]

b In the 1950s, H. B. D. Kettlewell studied the general distribution of two different colour morphs of *Biston betularia* in the British Isles. He made a special study of their distribution in an area of rural Dorset free from industrial pollution and an area of the city of Birmingham where industrial pollution was high. Typically, in the Dorset study area the barks of trees such as birch were free of soot and covered with lichen whereas those in the Birmingham study area were heavily covered in soot with few if any lichens growing on them. The map summarizes his results.

i What does the map indicate about the relative frequencies of the two morphs in Dorset and Birmingham? [1]

ii Suggest how the process of natural selection might have led to the different frequencies of the varieties of the peppered moth seen in Dorset and Birmingham. [4]

iii Over a long period of time, varieties of one species can develop into separate species. Suggest how this could happen in moths such as the peppered moth. [3]

[Total marks = 15]

Question 2: Guinea pig species

a The guinea pig is a small mammal. In an examination, a student wrote its scientific name as:

cavia Porcellus

Rewrite its scientific name correctly. [2]

b In South America, there are several species of guinea pig. Suggest how they might have arisen by allopatric speciation. [1]

c Using the information below, construct a genetic diagram to explain why repeated crossings of two guinea pigs resulted in offspring of four different genotypes, all of which had short, black hair.

- The male and female starter guinea pigs both had short, black hair.
- The male starter guinea pig was homozygous for hair length.
- The female starter guinea pig was homozygous for hair colour.
- Two genes on separate chromosomes control hair length and hair colour.
- The hair may be either long or short and its colour either black or brown. [4]

Guinea pigs showing two different coat phenotypes

d In a breeding experiment, the same female starter pig was twice mated with another male which had long, brown hair. Of the 15 offspring, 11 had short, black hair and 4 had long, black hair. The experimenters assumed that the genes were not linked and expected the phenotype ratio of the offspring to be 1:1.

i State a suitable null hypothesis for a chi-squared test of the results. [1]

ii Copy and complete the table below constructed to carry out the chi-squared test and give the chi-squared value. [4]

Phenotype	Observed (O)	Expected (E)	(O − E)	(O − E)²	$\frac{(O-E)^2}{E}$
long, black hair					
short, black hair					
Chi squared $= \Sigma \frac{(O-E)^2}{E} =$					

Chi-squared test table

iii Use the probability table below to determine whether the null hypothesis is supported by the data. Explain fully how you arrived at your answer. [3]

[Total marks = 15]

Degrees of freedom	Probability value					
	0.99	0.95	0.01	0.05	0.01	0.001
1	0.0002	0.0039	2.71	3.84	6.63	10.83
2	0.020	0.103	4.61	5.99	9.21	13.82
3	0.115	0.352	6.25	7.81	11.34	16.27
4	0.297	0.711	7.78	9.49	13.28	18.47

Probability values for chi-squared test

OBJECTIVES

By the end of this spread you should

- know which biological principles you are expected to understand for the synoptic assessment
- understand how the synoptic element of the AQA Specification is to be assessed

What is synoptic assessment?

According to the AQA Specification for AS and A2 Biology, synoptic assessment of science is defined as follows:

> Synoptic assessment requires candidates to make and use connections within and between different areas of the subject at AS and A2: for example, by
>
> - applying knowledge and understanding of more than one area to a particular situation or context
> - using knowledge and understanding of principles and concepts in planning experimental and investigative work and in the analysis and evaluation of data
> - bringing together scientific knowledge and understanding from different areas of the subject and applying them

Biological principles which you are expected to understand

When you have completed each of the units of the AQA Specification for AS and A2 Biology, you are expected to have an understanding of a number of key biological principles, shown in the table opposite. As part of the synoptic assessment, the examiners may draw on an understanding of these principles in the assessments of subsequent units. Therefore,

- the assessment of Unit 2 may include the biological principles of Unit 1
- the assessment of Unit 4 may include the biological principles of Units 1 and 2
- the assessment of Unit 5 may include the biological principles of Units 1, 2, and 4

When will synoptic assessment occur?

Although there is an element of synoptic assessment in all units, synoptic assessment features more strongly in the A2 units. According to the AQA Specification

> … it is assessed by two types of question: structured questions carrying 15 marks and an essay question carrying 25 marks. The structured question on Unit 4 develops a How Science Works context and requires candidates to use subject matter and skills from Unit 4 and the AS Specification. The structured question in Unit 5 tests data handling skills and requires candidates to apply knowledge, understanding, and skills from the AS and A2 Specification. In the essay in Unit 5, candidates are required to use knowledge and understanding from across all units.

Practising synoptic assessments

Spreads 6.11–6.12 have questions in the style of the structured question on Unit 4 which develops a How Science Works context and requires candidates to use subject matter and skills from Unit 4 and the AS Specification.

Spreads 12.07–12.08 have questions in the style of the structured question on Unit 5 which tests data-handling skills and requires candidates to apply knowledge, understanding, and skills from the AS and A2 Specification.

Spread 12.09 considers the essay questions in Unit 5 for which candidates are required to use knowledge and understanding from across all units.

Biological principles in the AS Specification	Unit
Proteins and polysaccharides are made up of monomers that are linked by condensation.	1
Many of the functions of proteins may be explained in terms of molecular structure and shape.	1
Enzymes are proteins and their rates of reaction are influenced by a range of factors: temperature, the presence of inhibitors, pH, and substrate concentration.	1
Substances are exchanged by passive or active transport across exchange surfaces. The structure of plasma membranes enables control of the passage of substances across exchange surfaces.	1
A species may be defined in terms of observable similarities and the ability to produce fertile offspring.	2
Living organisms vary and this variation is influenced by genetic and environmental factors.	2
The biochemical basis and cellular organisation of life is similar for all organisms.	2
Genes are sections of DNA that contain coded information as a specific sequence of bases.	2
During mitosis, the parent cell divides to produce genetically identical daughter cells.	2
The relationship between size and surface area to volume ratio is of fundamental importance in exchange.	2
Biological principles in the A2 Specification	
Random sampling results in the collection of data which is unbiased and suitable for statistical analysis.	4
ATP is the immediate source of energy for biological processes.	4
Limiting factors	4
Energy is transferred through ecosystems.	4
Chemical elements are recycled in ecosystems.	4
Genetic variation occurs within a species and geographic isolation leads to the accumulation of genetic difference in populations.	4
Organisms regulate their internal environment and so maintain optimum conditions for their metabolism.	5
Animals respond to their internal and external environment as a result of stimulus perception, chemical and electrical coordination, and a response by effectors. Plants respond to their external environment as a result of specific growth factors that regulate cell growth.	5
The genetic code is held in the base sequence of nucleic acids and determines the amino acid sequence of polypeptides produced by a cell.	5
Regulating gene expression enables a cell to control its own activities and development.	5
Scientists are able to manipulate gene expression for many agricultural, industrial, and medical purposes.	5

The biological principles studied in AS Biology and A2 Biology

A heathland ecosystem dominated by heathers and gorse

Heathland restoration

In 2003, a restoration project began on a 2 hectare site on Exmoor. Historically, the site had been a rich heathland ecosystem that had been grazed by sheep and cattle, but it had been left unmanaged for several years before the restoration project started. As a result the community typical of a heathland ecosystem had been replaced by that of scrubland. Heaths are generally dominated by dwarf shrubs, with heathers and gorse usually forming a major part of the plant community, whereas 80 per cent of the scrubland was dominated by bramble and blackthorn.

In 2003, the scrub was completely cleared except for one area measuring 10 metres by 10 metres. The cleared land was divided into eight blocks of equal area. The blocks were subjected to different management techniques to see which was the best method to restore heathland. Each year since 2003, an ecologist has been monitoring the changes in the different blocks, as well as in the area of scrub left uncleared. The ecologist used two different methods: random quadrat sampling and a belt transect. The number of samples taken by using random quadrat samples in each block was twice the number of samples taken by the belt transect sampling. In all the blocks managed by controlled grazing, two species of heather occurred: common heather (*Calluna vulgaris*) and bell heather (*Erica cinerea*).

Erica cinerea is a xerophyte. Its leaves are dark green in colour, shiny, and arranged in whorls, in the axils of which are small leafy shoots. The stomata are situated on the lower surface which is covered with fine hairs. In addition the edges of the leaves are rolled inwards until they nearly meet, leaving a groove between the two opposing surfaces. The stomata are protected by the inrolled surfaces of the leaves.

Question 1

a With reference to the heathland restoration project, distinguish between primary succession and secondary succession. [2]

b Heathland is a plagioclimax. How does this differ from a climatic climax? [2]

c Suggest why the ecologist monitored the site using two methods rather than just one method. [1]

d Describe how the random quadrat sampling could have been carried out. [3]

e With reference to handling the data collected by the ecologist, explain the significance of

   i the number of random quadrat samples being twice the number of belt transect samples

   ii the monitoring being carried out by the same ecologist rather than by different ecologists. [2]

f In what sort of conditions would you expect *Erica cinerea* to grow? Suggest how it is adapted to these conditions. [5]

[Total marks = 15]

Moss models for desiccation

Plants are exposed to widely fluctuating environments, and they have evolved adaptations to these fluctuations that limit cell damage. The adaptations of flowering plants include anatomical structures and rapid physiological and biochemical responses. Mosses lack many of the specialized structures that characterize flowering plants, but in some cases they show a remarkable resistance in stress situations, having evolved efficient molecular mechanisms. This makes mosses a good model for studying the role and the effectiveness of these molecular mechanisms, and for understanding how plants use them to resist stress factors.

Mosses differ from flowering plants in that they obtain most of their water from the atmosphere (for example, from precipitation of rain onto their leaves). As atmospheric sources can vary greatly, most mosses have become adapted to a continually moist habitat or have evolved the ability to tolerate desiccation. This makes mosses suitable model organisms for studying desiccation tolerance. Desiccation in mosses is a rapid process because they are unable to slow the water loss. Desiccation tolerance is expressed in the ability to rehydrate and recover normal metabolic processes, especially photosynthetic activity.

In one study, mosses including *Mnium affine* were collected from their natural habitat and maintained in room conditions (20°C and natural light period, avoiding direct sunlight) in plastic tubs in the soil they were growing in. They were watered daily. After 2 weeks 2 g of fresh moss leaves were used to prepare an enzyme extract before desiccation of the mosses. The remaining plant material was desiccated simply by stopping watering. After 12 months the mosses were rehydrated by adding water to the plastic tub. Immediately before rehydration, leaf samples were collected to measure water loss from them, and to prepare an enzyme extract from dry mosses. The degree of water loss in desiccated mosses was determined by obtaining 0.35 g of leaves from the desiccated moss and measuring the same moss leaves after 3 hours of heating at 115°C. Plant material was collected 20 minutes after rehydration and 16 hours after rehydration.

Catalase was one of the enzymes analysed. It is an enzyme that breaks down hydrogen peroxide to water and oxygen. Changes in the activities of catalase in two moss species – the drought-tolerant *Tortula ruralis* and drought-sensitive *Cratoneuron filicinum* – have been studied. A positive correlation between increased enzyme activities of catalase and dehydration is found in *Tortula ruralis*. The opposite occurred in the sensitive moss.

Question 2

a How does water uptake in mosses differ from water uptake in flowering plants? [2]

b Explain how the researchers determined when the moss was completely dehydrated. [2]

c Present the results from the table showing the rate of photosynthesis of mosses at different states of hydration in a suitable graph. [4]

Photosynthesis measured in mosses	Rate of photosynthesis (nmol CO_2 g^{-1} dry mass s^{-1})
before desiccation	6.5
when desiccated	0
20 minutes after rehydration	6.3
16 hours after rehydration	8.7

The rate of photosynthesis in Mnium affine *measured at the same light intensity. The units are nanomoles of carbon dioxide per gram of dry mass per second. (Adapted from Christov, K. and Bakardjieva N. (1997) Peroxidase, catalase, and superoxide dismutase in long-term desiccated moss* Mnium affine *and their role in the process of rehydration. In* Plant Peroxidase Newsletter, *No. 11.)*

d Excluding experimental error, suggest an explanation for the difference in the rate of photosynthesis of the mosses before desiccation and the mosses 16 hours after rehydration. [2]

e What is catalase? [1]

f Explain how catalase activity may be decreased by dehydration. [2]

g Describe the relationship between catalase and drought resistance in mosses. [2]

[Total marks = 15]

The mountains of Madeira, where the land snails belonging to the genus Discula *live*

Discula, *the Madeiran land snail*

Scientists working together

Discula is a genus of terrestrial pulmonate gastropod molluscs. The genus is of particular interest because several species living on Madeira Island are among Europe's most endangered animals. For example, *Discula testudinalis* is listed in the Red Data Book as 'critically endangered'. In order to establish their conservation status, taxonomists, ecologists, and ecological geneticists have been working together to understand the taxonomy and distribution of these molluscs.

Madeira Island is part of an archipelago about 650 km (400 miles) off the coast of Africa. Originally, all its *Discula* species were thought to have originated from the same founder species which was carried to the island on floating logs or by birds. However, recent evidence suggests that colonization may have occurred more than once and that there might have been two founder populations. The evidence comes from comparative DNA analysis by DNA–DNA hybridization. Using this and other evidence, taxonomists have split the genera so that four species (which we will call A, B, C, and D) are now in two subgenera: species A and B are in one subgenera, and species C and D in another.

Ecologists studying the ecological niches of these species found that all four species had slightly different food requirements. They also revealed that species A has two subspecies, which occupy separate mountains.

Ecological geneticists have been monitoring variations in the allele frequency of the two subspecies and seeing if they are changing with time.

Question 1

a Explain what is meant by the following terms:

   i genus   **ii** species   [4]

b Explain why different species found in the same habitat would be expected to have different diets.   [2]

c Define DNA–DNA hybridization.   [2]

d **i** Explain what allele frequency is.

   ii Suggest what is likely to happen to the allele frequency of the two subspecies of A and explain how these changes might occur.   [5]

e Suggest why competition might be greater between species A and species B rather than between species A and species D.   [2]

[Total marks = 15]

The life of a marine biologist

Dr Alan Southward (1927–2007) was one of the most influential marine biologists of the twentieth century. For more than 50 years he worked for the Marine Biological Association (MBA) in Plymouth. Listed below are just some of the projects in which he was involved.

• **Barnacle biology and taxonomy**

With D. J. Crisp, Southward revised early work from Charles Darwin's monograph on barnacles. The highly variable species of this group had prompted some of Darwin's early thoughts on evolution.

Dr Alan Southward carrying out marine biological research

- **Ecological interactions of rocky shore organisms**

In 1951, at Port Erin on the Isle of Man, Southward pioneered intertidal field experiments in which species were removed or transplanted to look at interactions in the region between high and low tide marks.

- **Effects of oil pollutants**

Southward's work on shores became invaluable after the Torrey Canyon oil spill, when the break-up of a supertanker contaminated most west Cornish shores in 1967. Vast amounts of dispersants were used to remove the oil from rocky shores, but these were far more toxic than the oil itself, killing major grazers such as limpets. Recovery took 10–15 years on many shores overtreated with dispersant, compared with only two or three years on shores where nature was left to take its course.

- **Biology and nutrition of animals from hydrothermal vents**

Southward, working with his wife Eve, studied the gutless worms called pogonophorans that inhabited hydrothermal vents discovered in the 1970s. They found that the worm relied on chemosynthetic symbionts. They also showed that similar modes of nutrition occurred in shallow-water organisms from methane and sulfide-rich environments.

- **Effects of climate change in marine ecosystems**

From 1953, Southward took responsibility for the long-term study of zooplankton and fish such as sardines and herring in the English Channel. His research indicated that climatic fluctuations were the most likely explanation for the inconsistency of fish stocks in the Channel's ecosystem. By using historical records of fish catches, he showed that fluctuations between cold-water herring and warm-water pilchard (sardines) had been occurring since the middle ages. His longitudinal studies were interrupted in the 1980s through lack of funding, but were restarted in the 1990s amid concern about global warming. His results are proving to be vitally important in distinguishing the effects of long-term human-driven climate change from other factors.

Question 2

a Suggest why the high variability of barnacle species made them particularly interesting to Charles Darwin. [2]

b Suggest why many of the rocky shores of Cornwall became carpeted in green weeds in the aftermath of the Torrey Canyon. [2]

c What are chemosynthetic symbionts of pogonophoran worms and how do they obtain their energy? [4]

d Suggest factors other than long-term human-driven climate change that might cause fluctuations in fish stocks in the English Channel. [4]

e At Port Erin, Southward studied the interaction between limpets and seaweeds on the rocky shore. His results showed a negative correlation: when limpets were abundant, there were few seaweeds; when seaweeds were abundant, there were few limpets.

  i What type of competition occurs between individuals of two different species? [1]

  ii Suggest reasons for the negative correlation between the two species. [2]

[Total marks = 15]

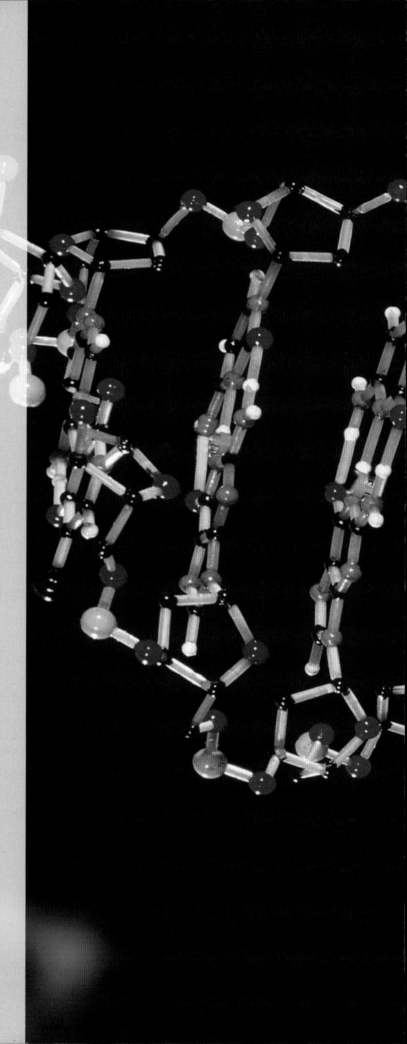

Unit 5: Multicellular organisms are able to control the activities of different tissues and organs within their bodies. They do this by detecting stimuli and stimulating appropriate effectors: plants use specific growth factors; animals use hormones, nerve impulses, or a combination of both. By responding to internal and external stimuli, animals increase their chances of survival by avoiding harmful environments and by maintaining optimal conditions for their metabolism.

Cells are also able to control their metabolic activities by regulating the transcription and translation of their genome. Although the cells within an organism carry the same genetic code, they translate only part of it. In multicellular organisms, this control of translation enables cells to have specialized functions, forming tissues and organs.

The sequencing and manipulation of DNA has many medical and technological applications.

Consideration of control mechanisms underpins the content of this unit. Students who have studied it should develop an understanding of the ways in which organisms and cells control their activities. This should lead to an appreciation of common ailments resulting from a breakdown of these control mechanisms and the use of DNA technology in the diagnosis and treatment of human diseases.

AQA Approved Specification (July 2007)

Control in cells and organisms

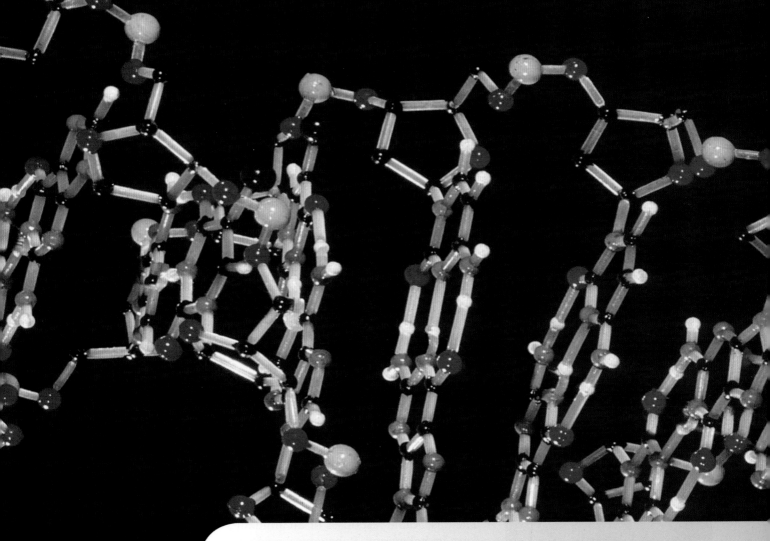

A three-dimensional model of DNA. 'The sequencing and manipulation of DNA has many medical and technological applications.'

Sensitivity: responding to stimuli

OBJECTIVES

By the end of this spread you should be able to

- appreciate that all organisms increase their chances of survival by responding to changes in their environment
- understand the nature and function of tropisms
- distinguish between taxes and kineses

Responding to change

All living organisms must be able to detect changes in their environment and respond appropriately. Changes in the environment are called **stimuli** (singular stimulus). A stimulus may be either external (outside the organism) or internal (inside the organism).

Sensitivity, the ability to respond appropriately to stimuli, is one of the characteristic features of life. Each organism has its own specific type of sensitivity that improves its chances of survival. A single-celled amoeba, for example, can move away from a harmful stimulus such as very bright light, and move towards a favourable stimulus such as food molecules. However, it can only distinguish between a limited number of different stimuli.

In an amoeba, the detection of a stimulus and the response to it must both take place in a single cell. However, in large multicellular organisms stimuli are usually detected in **sense organs** which contain specialized cells called **receptors** that are particularly sensitive to specific stimuli.

Structures that bring about a particular response to a stimulus are called **effectors**. The sense organs and effectors may be in quite different parts of the body. Responses usually involve the coordinated actions of many different parts of the body. To achieve this coordination, one part of the body must be able to pass information to another part.

Plant responses

Descriptions of large terrestrial plants moving from one place to another are the stuff of science fiction. Nevertheless, plant parts do appear to move in response to environmental stimuli. For example, the fruits of some flowering plants explode when they become dry, the leaves of *Mimosa* (the sensitive plant) curl when touched, and the petals of daisies open when exposed to sunlight.

Tropisms

Tropisms are responses to directional stimuli that help maintain the roots and shoots of flowering plants in a favourable environment. The plant body part responds to a stimulus by growing either towards a directional stimulus (a **positive tropism**) or away from it (a **negative tropism**). If the directional stimulus is light, the response is called a **phototropism**; if gravity, it is called a **geotropism**, and if water, it is called a **hydrotropism**.

Animal responses

The observable responses of an animal to the environment around it form its **behaviour**. Behavioural responses may involve movements of the whole body or one or more body parts. They range from the very simple to the extremely complex. Two relatively simple forms of behaviour that involve the locomotion of organisms or cells in response to specific external stimuli are called **kineses** and **taxes**.

Kineses and taxes

Kineses and taxes are both forms of simple behavioural responses which help to keep an individual animal, such as a woodlouse, in a favourable

Mimosa *before and after being touched*

environment (see diagram below). A **kinesis** is a random movement in which the rate is related to the intensity of the stimulus, but not to its direction. There are two main types:

- **orthokinesis** involves changes in speed of movement
- **klinokinesis** involves changes in the rate of turning (the lower the rate of turning, the more likely it is that the animal will leave an unfavourable area)

In a choice chamber which is divided into dry and damp areas, woodlice tend to move faster and turn less in the dry areas. This tends to orientate them towards the damp areas.

A **taxis** is a movement in response to the direction of a stimulus. Movements towards a stimulus are positive; those away from a stimulus are negative. Woodlice generally show **negative phototaxis**: they move away from light and tend to be found in shaded or dark places, such as under stones and leaves.

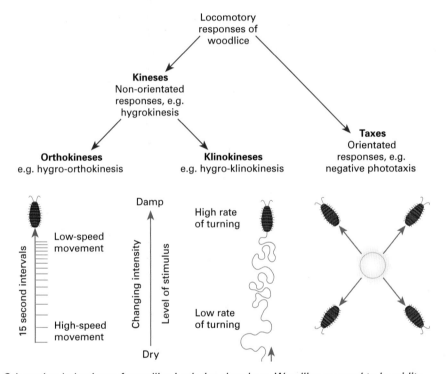

Orientation behaviour of woodlice in choice chambers. Woodlice respond to humidity levels by changing their rate of turning (hygro-klinokinesis) and their speed of movement (hygro-orthokinesis). They also move away from light (negative phototaxis).

Check your understanding

1 What is a stimulus?
2 Explain what is meant by a shoot being 'positively phototropic'.
3 Distinguish between a taxis and a kinesis.

- -

4 When a single light source provides the stimulus, a woodlouse may move its whole body axis in a straight line away from the light source. The woodlouse has bilateral light receptors that can make a simultaneous comparison of the light on either side of its body. *Euglena* is a single-celled organism that has a single light-sensitive area. Suggest how it might orientate itself towards or away from a single light source.

By the end of this spread you should be able to

- describe a simple reflex arc involving three neurones

- discuss the importance of simple reflexes in avoiding damage to the body

- distinguish between a reflex action and stereotyped behaviour that incorporates fixed action patterns

Prior knowledge

- sensitivity: responding to stimuli (7.01)

The simplest type of animal behaviour is a **reflex action** in which a stimulus brings about an automatic, involuntary, and **stereotyped response** (that is, the response to the same stimulus is always similar). This is an **unlearned behaviour**. Reflex actions are **innate** and depend on nerve pathways that are inherited. Many reflex actions, such as withdrawal of the hand from a hot object, are protective.

The nerve pathway that transmits information rapidly from a receptor to an effector is called a **reflex arc**. Reflex arcs vary in complexity. The one that removes a hand from a hot object, shown in the diagram below, has three neurones:

- a **sensory neurone** that carries information from sensory receptors to the **central nervous system**

- a **motor neurone** that carries information to an **effector organ**

- an **interneurone** in the central nervous system between the sensory neurone and motor neurone. The interneurone is also called a **relay neurone** or an **association neurone** because it makes synaptic connections with nerves that pass upwards into the association areas of the brain. These connections enable the brain to modify a reflex action, but it cannot stop it completely because information is transmitted along the reflex arc too quickly.

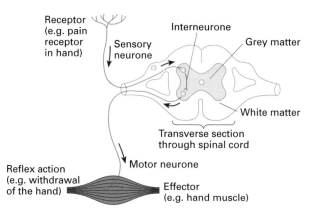

A simple reflex arc consisting of three neurones: a sensory neurone, an interneurone, and a motor neurone

Testing reflexes

One of the best known human reflexes is the knee-jerk response. It is used by doctors to test neural responses. It has a **monosynaptic reflex arc**: the sensory neurone has a single connection (called a **synapse**) with the motor neurone; there is no interneurone. This type of pathway is adapted for responses that need to happen as quickly as possible. Doctors test the **knee-jerk reflex** (also called the patellar reflex) by tapping the tendon just below the kneecap.

In a healthy person, the tap should cause a reflex contraction of the quadriceps muscle in the upper thigh so that the lower leg kicks. In this reflex pathway, a stretch receptor in the tendon is stimulated by the tap and a sensory neurone carries a nerve impulse to the spinal cord. Here the impulse passes across a synapse to a motor neurone, which runs from the spinal cord to the quadriceps, which extends the leg. Contraction of this

A doctor testing the knee-jerk reflex

muscle brings about the knee-jerk response. Failure of the leg to respond to the tap may indicate disease or damage to part of the reflex arc.

A reflex contraction of a muscle in response to its being stretched is called a **stretch reflex**. Stretch reflexes are important in maintaining body posture.

Even the knee-jerk reflex is not as simple as it might seem. When the stretch receptors within the quadriceps are stimulated, in addition to bringing about the stretch reflex they also send information to antagonistic muscles ensuring that they are relaxed. This process is called **reciprocal inhibition**.

Stereotype behaviour incorporating fixed action patterns

Fixed action patterns (FAPs) are complex forms of stereotype behaviour characterized by relatively fixed patterns of coordinated movements. FAPs occur during nest-building in birds, the courtship dance of stickleback fish, the food-begging behaviour of gull chicks, and the suckling response of newborn babies. They are brought about by specific stimuli called **sign stimuli**.

(a)

(b)

Spot colour	Percentage effectiveness
Red	100%
Black	86%
Blue	85%
White	71%
No spot	30%

*Feeding behaviour of gull chicks. (**a**) Normal behaviour of a chick on the nest. It pecks at the red spot on the parent's beak to stimulate the parent to regurgitate food. (**b**) Laboratory experiments using hand-held cardboard models of a gull's head compared the number of pecks at each model with the number of pecks at the red spot model. The feeding response involves fixed action patterns.*

FAPs are largely innate responses but they differ from reflex actions in that they may vary according to the precise conditions in which the sign stimuli are presented, and they can also be modified by experience. For example,

- The calls of many birds involve FAPs that are greatly influenced during early development by the songs of nearby members of the same species.

- The begging response of a herring gull chick is triggered by the red spot on its parent's beak, but experiments show that the response varies with the size of the spot and other features of the beak, as shown above.

- Egg retrieval by ground-nesting birds may vary according to the size of the stimulus – gulls and geese retrieve eggs when they roll out of the nest. Egg retrieval involves a series of FAPs triggered by the sight of the egg rolling away. Experiments show that the larger the egg, the stronger the response. An abnormally large model egg provides a **supernormal stimulus** which triggers a greater than normal response.

Check your understanding

1 List the main components of a reflex arc involving three different neurones.

2 Name one function of a reflex action.

3 Distinguish between a reflex action and a fixed action pattern.

- - - - - - - - - - - - - - - -

4 Using information in the text about FAPs, suggest why a willow warbler may feed a cuckoo chick in preference to its own offspring.

OBJECTIVES

By the end of this spread you should be able to

- discuss the role of receptors, the autonomic nervous system, and effectors in controlling the heart rate

Prior knowledge

- the heart (*AS Biology for AQA* spread 8.01)
- the cardiac cycle (*AS Biology for AQA* spread 8.02)
- cardiac output (*AS Biology for AQA* spread 8.03)
- the reflex arc (7.02)

The human heart is an amazing organ. Every second of every day of a person's life, it pumps life-giving blood around the body. Its cells have the peculiar property of having their own inherent rhythm of contraction. This enables them to contract without any external stimulation, a condition referred to as **myogenic**.

The **heart rate** (the number of ventricular contractions per minute) is normally determined by **pacemaker** cells in the **sinoatrial node**. The pacemaker initiates a heartbeat by passing waves of electrical stimulation first over the **atria** and then, via the **atrioventricular node**, over the **ventricles**. A heart with no nervous or hormonal stimulation contracts at a rate of about 50 beats per minute. However, nervous stimulation coordinated by the **cardiovascular centre** in the **medulla oblongata** (the brain stem) normally overrides this inherent rhythm.

Nervous control of the heartbeat

Nervous control of the heartbeat is mainly by **reflex actions** which involve **reflex arcs** originating in **sensory receptors**.

- **Baroreceptors** (pressure receptors) are located in the **carotid sinus** (a swelling of the carotid artery) and the walls of the heart.
- **Chemoreceptors** sensitive to changes in pH and the concentration of blood carbon dioxide and blood oxygen are located in the walls of the heart, the carotid and aortic arteries, and the medulla oblongata.

These receptors pass their sensory information via sensory neurones to the cardiovascular centre from which two nerves, the vagus nerve and the sympathetic nerve, lead to the pacemaker in the sinoatrial node. These nerves are part of the **autonomic nervous system**, that part of the peripheral nervous system that regulates body activities that are usually involuntary and performed unconsciously. A person's actual heart rate depends in part on the relative activity of these two nerves which act antagonistically:

- The **vagus nerve** (part of the parasympathetic nervous system) runs directly from the cardiovascular centre to the pacemaker region in the sinoatrial node. It has an inhibitory effect – stimulation of the vagus nerve causes the heart to slow down.
- The **sympathetic nerve** (part of the sympathetic nervous system) goes to the sinoatrial node via the spinal cord. It is an excitatory nerve – stimulation of the sympathetic nerve accelerates the heart rate.

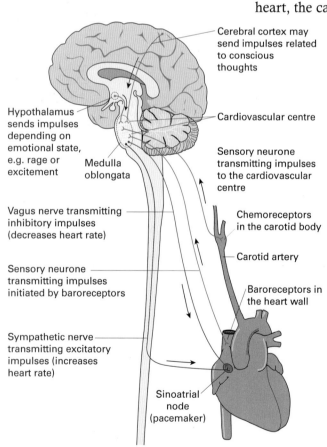

Cerebral cortex may send impulses related to conscious thoughts

Hypothalamus sends impulses depending on emotional state, e.g. rage or excitement

Medulla oblongata

Cardiovascular centre

Sensory neurone transmitting impulses to the cardiovascular centre

Vagus nerve transmitting inhibitory impulses (decreases heart rate)

Chemoreceptors in the carotid body

Carotid artery

Sensory neurone transmitting impulses initiated by baroreceptors

Baroreceptors in the heart wall

Sympathetic nerve transmitting excitatory impulses (increases heart rate)

Sinoatrial node (pacemaker)

A diagram showing some of the nerve connections between sense organs, higher centres of the brain, the cardiovascular centre, and the sinoatrial node in the heart

Heart rates vary

Resting heart rates vary with the age and sex of a person. Although there is considerable individual variation, adult males have an average heart rate of about 70 beats per minute, adult females 75 beats per minute, and infants about 100 beats per minute. Well trained endurance athletes tend to have very low resting heart rates, often below 40 beats per minute.

Modification by nervous stimulation and hormones enables the heart rate to meet the demands of particular circumstances. During exercise, the heart rate increases in proportion to the workload and oxygen uptake, and reaches up to 200 beats per minute at maximal activity. The increase in heart rate is part of a **negative feedback mechanism** that controls the blood carbon dioxide concentration, blood pH, and, indirectly, the blood oxygen concentration.

Exercise leads to an increase in the blood carbon dioxide concentration and a lowering of blood pH. This is detected by chemoreceptors located in the **carotid body**, an area of glandular tissue close to the carotid artery. Sensory information is passed to the cardiovascular centre, causing the sympathetic nerve to pass excitatory impulses to the sinoatrial node. The heart rate increases, thereby increasing the rate at which carbon dioxide is transported to the lungs for exhalation. When the blood carbon dioxide concentration falls, excitation of the pacemaker decreases, and the heart rate returns to normal.

Several other factors affect heart rate:

- At times of excitement or danger, nerve impulses from higher centres in the brain stimulate the sympathetic nerve in the cardiovascular centre, causing the heart rate to increase. The sympathetic nervous system also stimulates the release of the hormone **adrenaline** (epinephrine) from the adrenal glands. Adrenaline increases the heart rate.

- A temperature increase of 1°C raises the heart rate about 10 beats per minute. This explains why the pulse increases during a fever.

- Drugs such as caffeine and amphetamines increase the heart rate while others such as **beta blockers** tend to slow it down.

- If the blood pressure is high, pressure receptors in the carotid artery and aorta are stimulated. These send nerve impulses to the cardiovascular centre, inhibiting the activity of the sympathetic nerve, causing the heart rate to decrease.

- At the start of exercise, when a muscle contracts, stretch receptors in the muscle transmit nerve impulses to the cardiovascular centre, causing the heart rate to increase in anticipation of the increased metabolic demands.

Artificial pacemakers

In some people, the conduction of electrical impulses from the natural pacemaker of the heart (the sinoatrial node) is impaired. This condition, known as heart block, can be remedied by inserting an artificial pacemaker under the muscle in the upper thorax. The battery-powered pacemaker stimulates the heart through a wire connected to the ventricle or the heart lining. New pacemakers fitted with complex electronic circuits can sense changes in breathing, body temperature, and movement so that the heart rate is adjusted accordingly.

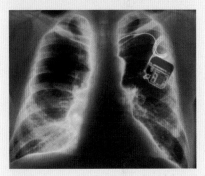

X-ray showing an artificial pacemaker fitted into the chest cavity

Check your understanding

1 What would happen to the heart rate if the vagus nerve was cut?
2 Make a simple flow diagram showing what happens when blood carbon dioxide concentrations increase.
3 What is adrenaline and what is its effect on the heart rate?

- -

4 Distinguish between the structure and function of the carotid sinus and the carotid body.

An animal's internal and external environments are continually changing. A detectable change is called a **stimulus**. To some extent all cells are sensitive to stimuli, but in animals certain cells have become especially sensitive to particular stimuli. These are called **receptor cells**. They act as transducers, converting the energy of the stimulus into the electrical energy of a nerve impulse which is transmitted along the sensory neurone to the central nervous system. The frequency of nerve impulses conveys information about the stimulus, enabling the animal to respond in an appropriate way. Receptor cells may act individually, or as a group in a tissue or organ (a **sense organ**).

The Pacinian corpuscle

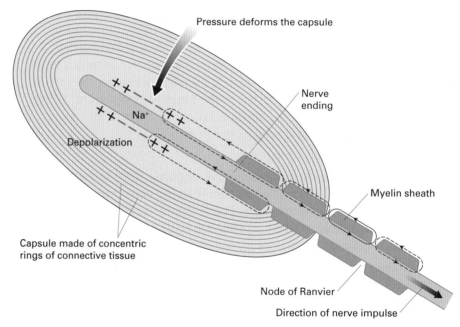

Pressure deforms the capsule

Na⁺

Depolarization

Nerve ending

Myelin sheath

Capsule made of concentric rings of connective tissue

Node of Ranvier

Direction of nerve impulse

A Pacinian corpuscle

The **Pacinian corpuscle** is a receptor sensitive to changes in pressure and touch. As such, it is called a **mechanoreceptor**. Pacinian corpuscles occur in the skin, tendons, muscles, and joints of mammals. Each corpuscle has a single sensory neurone, one end of which consists of a capsule made of concentric rings of connective tissue that acts as a pressure-sensitive pad. Application of pressure against the connective tissue deforms stretch-mediated sodium ion **channel proteins** in the plasma membrane. This causes an influx of sodium ions into the end of the sensory neurone. The sudden influx of sodium ions reverses the potential difference across the plasma membrane. The inside becomes temporarily positively charged, a process called **depolarization**. The electrical potential created by the deformation is proportional to the stimulus intensity. This graded potential is known as the **generator potential**. If the deformation stimulus is above a critical level, the graded potential is high enough to trigger the transmission of a nerve impulse along the sensory neurone. If the stimulus intensity is below the threshold, no impulse is transmitted (see left-hand graph, next page).

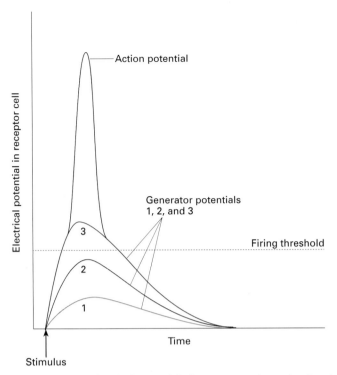

Changes in the electrical potential of a receptor when stimulated by three separate stimuli. Only the third stimulus produces a generator potential high enough to trigger a nerve impulse.

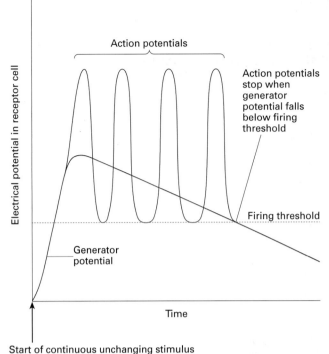

Sensory adaptation: the generator potential gradually declines in response to a constant stimulus. When it falls below the threshold, the nerve impulses stop.

The Pacinian corpuscle, like other receptors, responds only to a specific stimulus. It responds only to mechanical deformation. It does not respond to other stimuli such as temperature changes. The body has separate receptors to detect these.

Sensory adaptation

The Pacinian corpuscle, like other receptors, is adapted to detect potential harmful or beneficial changes in the environment. When given an unchanging stimulus, it stops responding so that the sensory system does not become overloaded with unnecessary or irrelevant information. Loss of responsiveness is brought about by a process called **sensory adaptation**, shown above right. An unchanging stimulus results in a decline in the generator potential of a sensory receptor such as a Pacinian corpuscle. This means that the nerve impulses transmitted in sensory neurones become less frequent and may eventually stop. The mechanism of sensory adaptation is not fully understood but in Pacinian corpuscles it probably involves changes in channel proteins in the plasma membrane of the nerve endings.

Check your understanding

1 Which part of a Pacinian corpuscle is deformed by the application of pressure?

2 After deformation of a Pacinian corpuscle, how do sodium ions create a generator potential?

3 Suggest why a person after a period of time becomes insensitive to the touch of coarse clothing on the skin.

OBJECTIVES

By the end of this spread you should be able to

- explain why differences in sensitivity and visual acuity are due to differences in the distribution of rods and cones and the connections they make in the optic nerve

Prior knowledge

- the reflex arc (7.02)
- receptors (7.04)

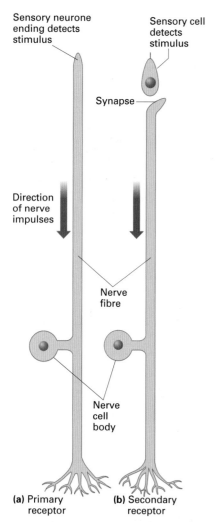

Sensory neurone ending detects stimulus

Sensory cell detects stimulus

Synapse

Direction of nerve impulses

Nerve fibre

Nerve cell body

(a) Primary receptor

(b) Secondary receptor

Two types of receptor cell: (a) a primary receptor (such as a Pacinian corpuscle) (b) a secondary receptor (for example, rods and cones)

The mammalian eye is a complex organ that transduces light (visible frequencies of electromagnetic radiation) into patterns of nerve impulses. The **transduction** takes place in the retina, a layer of photosensitive cells at the back of the eye. It contains two types of receptors called rods and cones and their nerves.

The structure of the retina: rods and cones

Rods and cones are classified as **secondary receptors** – they are more complex than **primary receptors** like Pacinian corpuscles. A primary receptor consists of a single neurone, one end of which detects changes in a particular stimulus. In secondary receptors, a modified epithelial cell detects changes in the stimulus and passes this information on to a separate neurone which transmits the information as nerve impulses. The diagram on the left shows this difference.

Pacinian corpuscles function individually, while rods and cones are part of a sense organ, the eye. **Sense organs** are complex stimulus-gathering structures consisting of grouped sensory receptors.

Up to 150 rods may connect via synapses to one neurone, a characteristic known as **convergence**. This enables a group of rods to function as a photosensitive unit, gathering light from a relatively wide area and combining its stimulatory effects. Convergence enables rods to provide photosensory information at low light intensities, but this increased sensitivity is at the expense of **visual acuity** (the ability to discriminate fine details). Consequently, in dim light when cones are not functioning, the rods provide only enough information for an ill-defined image that lacks colour. Also, because rods are the only type of photoreceptor at the outermost edge of the retina, something seen 'out of the corner of the eye' lacks detail. However, the rods are very sensitive to changes in light intensity caused, for example, by something moving. This ability is potentially vital because that 'something' could be an enemy or predator.

Transduction of light energy by rods

A rod cell has in its outer segment up to 1000 vesicles, each containing a photosensitive pigment called **rhodopsin**. Rhodopsin is made up of the protein opsin and retinal, a derivative of vitamin A. Light causes retinal to change shape. As a result, retinal and opsin break apart; a process called bleaching. This triggers a series of events which alters the permeability of the rod's plasma membrane and contributes to the formation of a generator potential in the sensory neurone. The generator potential results from the combined effects of all the rods serving a particular sensory neurone. When it exceeds a critical threshold level, a nerve impulse is transmitted. The nerve impulse passes first along the sensory neurone and then, via a fibre of the optic nerve, to the brain. The pattern of nerve impulses transmitted along different neurones is interpreted in the brain as patterns of light and dark.

Cones

Cones enable us to see colour and fine details. They are thought to work in a similar way to rods. However, there are three types of cone, each containing a different form of the pigment iodopsin which breaks down only in bright light. The three pigments together are sensitive over the visible spectrum, but one is most sensitive to blue light, one to green light, and the other to red light. According to the trichromatic theory, different colours are perceived by mixing the information from the different types of cone.

Whereas rods function in groups connected to one sensory neurone, a single cone may have its own sensory neurone. If light of sufficiently high intensity from two separate sources falls on two such cone cells, two separate impulses are transmitted along two sensory neurones and the brain receives two separate images. Thus cones provide much more detailed sensory information than rods, but they can function only in high light intensities as there is only a little convergence.

The fovea and blind spot

Over most of the retina, rods and cones are buried under a layer of blood vessels and nerve fibres which lead into the optic nerve. However, the **fovea**, a small depression in the retina opposite the lens, consists only of cones. The layer of nerve fibres here is thin and there are no capillaries. This means that light falling on the fovea produces a clear, well defined visual image in colour. When a person wants to examine the fine details of an object, the eyes move automatically so that light from the object falls on the fovea.

Adaptation to varying light intensities

The eyes adapt to different levels of brightness by varying their sensitivity. A prolonged period (more than about half an hour) in a darkened room results in photosensitive pigments being formed much faster than they are being broken down, thus increasing the sensitivity to light. This process is called **dark adaptation**.

A prolonged period in bright light results in photosensitive pigments in rods and cones being broken down, reducing sensitivity to light. This process is called **light adaptation**.

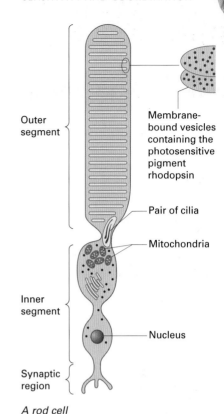

Outer segment

Membrane-bound vesicles containing the photosensitive pigment rhodopsin

Pair of cilia

Mitochondria

Inner segment

Nucleus

Synaptic region

A rod cell

Inside of eye

Pigmented epithelium

Cone

Rod

Synapse

Bipolar neurone cell body

Optic nerve fibre cell body (ganglion cell)

Optic nerve fibres

Light rays

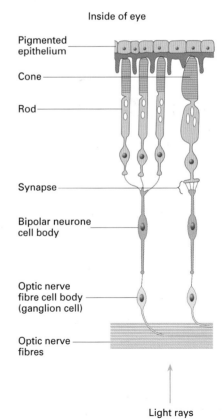

Simplified diagram of part of the retina. Note that light has to pass through the optic nerve fibres before it reaches the photoreceptor cells which are at the back of the retina.

Check your understanding

1 Which are more sensitive to light, rods or cones? Give reasons.

2 Which type of receptor is most responsible for seeing the fine detail in a photograph?

- -

3 Suggest why objects that appear brightly coloured in a well lit room appear in dim light to be either white or black, or various shades of grey.

OBJECTIVES

By the end of this spread you should be able to

- recall that coordination may be electrical or chemical in nature
- distinguish between the action of nerve impulses and mammalian hormones on target cells
- recall that nerve cells pass electrical impulses along their length and stimulate their target cells by secreting chemical neurotransmitters directly on to them, and that this results in rapid, short-lived, and localized responses

Prior knowledge

- sensitivity: responding to stimuli (7.01)

A multicellular organism functions as an integrated unit because it has specialized systems that coordinate the activities of its different parts. In animals, there are two main systems: one, the nervous system, is electrical; the other, the endocrine system, is chemical.

The nervous system

Nervous systems range from the simple nerve nets of jellyfish and sea anemones, which have no brain and relatively few interconnections, to the nervous systems of humans with brains of staggering complexity. The human brain contains many millions of cells each of which may communicate with thousands of other nerve cells. Their interconnections form circuits which enable us to control our muscles, think, remember, and even study our own brains.

All the various animal nervous systems are fast-acting communication systems containing nerve cells, **neurones**. These convey information in the form of **nerve impulses** which are electrical in nature.

Neurones take various forms but each has a **cell body**, containing a nucleus, and **nerve fibres**, long extensions that transmit nerve impulses rapidly from one part of the body to another. Fibres carrying impulses away from the cell body are called **axons**; those carrying impulses towards the cell body are called **dendrons**. Apart from the main nerve fibre, there may be small dendrons (**dendrites**) extending from the cell body.

In mammals, **sensory neurones** carry messages from peripheral sense organs to a **central nervous system** (CNS) consisting of the brain and spinal cord. The CNS acts as an integration centre and processes information from many sources. **Motor neurones** convey instructions from the CNS to effector organs (mainly muscles and glands).

The effect of a neurone on an effector depends on chemicals called **neurotransmitters** that are released by nerve endings directly onto target cells. Some neurones are excitatory – they secrete neurotransmitters that make an effector more active. Other neurones are inhibitory, secreting neurotransmitters that make an effector less active. Whether excitatory or inhibitory, the effect of the neurotransmitters is usually rapid, short lived, and localized.

Organization of the mammalian nervous system

Structurally, the mammalian nervous system consists of a **central nervous system** (the brain and spinal cord) and the **peripheral nervous system** (all the nerves and nervous tissue outside the brain and spinal cord). A nerve is a thread-like structure containing a bundle of nerve fibres (axons and dendrons). A single nerve may contain both sensory and motor neurones.

Functionally, the mammalian nervous system is divided into the **somatic nervous system** and the **autonomic nervous system**.

The somatic nervous system includes sensory neurones which transmit impulses from peripheral receptors to the central nervous system, and motor neurones which send impulses to skeletal muscles. It is sometimes called the **voluntary nervous system** because many of its actions are under conscious control.

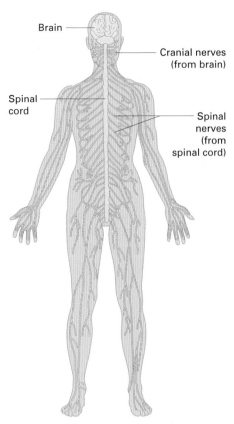

General plan of the human nervous system. The brain and spinal cord form the central nervous system; the cranial nerves and spinal nerves form the peripheral nervous system.

The autonomic nervous system is sometimes called the **involuntary nervous system** because it usually enables internal organs to function properly without any conscious control. It has two parts: the **parasympathetic nervous system** and the **sympathetic nervous system**. These generally act antagonistically (that is, they have opposing actions). While the parasympathetic nervous system maintains the body in non-threatening situations, the sympathetic nervous system prepares the body for action during periods of excitement or danger. Hence the vagus nerve slows down the heart rate and the sympathetic nerve speeds it up.

Check your understanding

1 What is an axon?
2 In what form is information conveyed in
   a   the nervous system
   b   the endocrine system?
3 What are neurotransmitters?

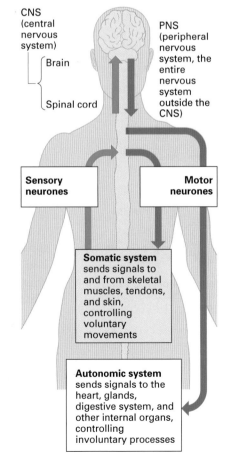

How the mammalian nervous system is organized

OBJECTIVES

By the end of this spread you should be able to

- distinguish between the action of nerve impulses and mammalian hormones on target cells

- recall that mammalian hormones are substances that stimulate their target cells via the blood system and that this results in slow, long-lasting, and widespread responses

- recall that histamine and prostaglandins are local chemical mediators released by some mammalian cells which affect only cells in their immediate vicinity

Prior knowledge

- immune system (*AS Biology for AQA* spread 9.01)

- sensitivity: responding to stimuli (7.01)

- chemical and electrical coordination I (7.06)

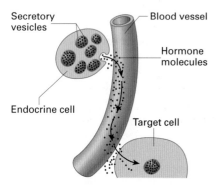

A hormone from an endocrine cell is secreted into the blood, which carries it to its target cell.

Typically, the nervous system is adapted to convey messages rapidly between specific locations so that quick responses can be made. Nerve impulses conveyed along a neurone are electrical, but communication between one neurone and another or between a neurone and an effector depends on the release of special chemicals called neurotransmitters. These are released at the nerve endings and act locally.

The endocrine system

In contrast to the nervous system, the endocrine system is adapted to carry information from one source to many destinations to bring about long-lasting responses.

The endocrine system consists of a number of glands that secrete **hormones** (organic chemicals, usually proteins or steroids). The glands of the endocrine system are called **endocrine glands** or **ductless glands** because they secrete their hormones directly into the bloodstream. Once inside a blood vessel, a hormone is carried in the bloodstream so that it can reach almost any cell in the body. However, each hormone has its own **target cells** on which it acts. Therefore, although all the hormones are transported together in the bloodstream, each has its own specific effect on the body. In some cases, a target cell has specific receptor molecules on its cell surface membrane which bind with the hormone molecule. Once bound onto the membrane, the hormone brings about its response.

Endocrine glands occur at strategic points around the body. Their hormones regulate a wide range of activities, including blood glucose concentration, gastric secretion, heart rate, metabolism, growth rate, reproduction, and water balance.

Local chemical mediators

The nervous and endocrine systems are not the only means of coordination in a multicellular organism. There are many local chemical mediators that affect only cells in their immediate vicinity. Two such chemical mediators are histamine and prostaglandins.

Histamine is part of the mammalian **non-specific immune system**. It is produced by white blood cells and mast cells (large mobile cells found in the matrix of connective tissue) when the skin is irritated by a foreign substance such as poison from a stinging nettle. Release of histamine into the vicinity of the irritation causes an acute inflammation involving pain, heat, redness, swelling, and sometimes loss of function of the affected part. Antihistamines act by inhibiting the action of histamine.

Prostaglandins are modified fatty acids, often derived from plasma membranes. They were first discovered in components of the semen produced by the human prostate gland – hence their name. In semen, prostaglandins stimulate contraction of the uterine wall, helping to convey sperm to the egg. They have another reproductive function. When secreted by cells in walls of the placenta, they make nearby muscles of the uterus more excitable, helping to induce labour during childbirth.

Their more general function is as local regulators in the mammalian defence system. During a viral or bacterial infection or after a physical trauma, various prostaglandins help induce fever and inflammation and also intensify feelings of pain. These unpleasant effects act as an alarm, informing the organism that something is wrong. Ibuprofen and aspirin inhibit the secretion of prostaglandins.

Coordinating systems compared

The table below summarizes and compares the action of the mammalian nervous system, endocrine system, and local chemical mediators.

Nervous system	Endocrine system	Local mediators
communication electrical in nature	communication chemical in nature	communication chemical in nature
response usually localized	response may be widespread and often involves different parts of the body	response localized
acts quickly	may take several minutes or even longer to act	acts more quickly than hormones
effects are usually short lived	hormones may have effects that are long lasting	response lasts as long as chemical is secreted and its action is not inhibited
nerve impulses are targeted on specific cells (e.g. muscles or glands) reached by nerve fibres or other neurones	hormones are targeted on specific cells some distance from the endocrine gland	chemicals are targeted on cells within the vicinity of secretion of the chemical
nerve impulses in motor neurones may excite or inhibit the action of a gland or muscle	hormones usually cause changes in metabolic activity of target cells	local mediators have different effects – histamines and prostaglandins are involved in the inflammatory and pain responses

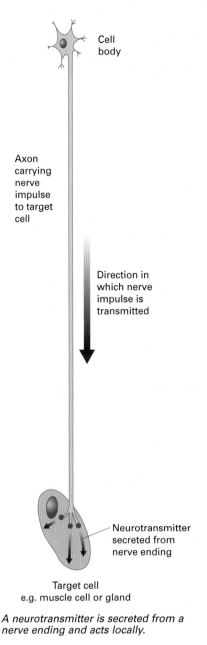

A neurotransmitter is secreted from a nerve ending and acts locally.

Check your understanding

1 Which coordinating system is slow to act and has long-lasting effects and widespread responses?

2 How are hormones conveyed to their target cells?

3 Why are histamines and prostaglandins not classified as hormones?

7.08 Plant growth factors

OBJECTIVES

By the end of this spread you should be able to

- recall that in flowering plants specific growth factors diffuse from growth regions to other tissues and, among other things, regulate growth in response to directional stimuli

- discuss the role of indoleacetic acid (IAA) in controlling tropisms in flowering plants

Prior knowledge

- water transport and transpiration in plants (*AS Biology for AQA* spreads 16.04–16.06)

- plant tissues (*AS Biology for AQA* spread 14.08)

- plant responses (7.01)

- chemical and electrical coordination (7.06–7.07)

Plant growth factors are chemicals that occur naturally at extremely low concentrations. They regulate various aspects of plant growth and development from seed formation and germination through to the ageing and death of a plant. They also coordinate many plant responses to environmental stimuli (for example, phototropism and geotropism).

Often, plant growth factors, like animal hormones, carry information from one part of the organism to another. Plant growth factors are transported in the transpiration stream of the **xylem** and by **mass flow** in the **phloem**. From the xylem and phloem, they usually reach their target cells by **diffusion**.

Unlike hormones, plant growth factors are not synthesized in special organs, and they sometimes act in the immediate vicinity of their production. **Ethene**, for example, usually acts on the tissues from which it is released.

Ethene is a gas released from ripening fruits, nodes of stems, ageing leaves, and flowers. It is involved in seed dormancy, fruit ripening, and **leaf abscission** (the process by which leaves are shed). Other plant growth substances include **gibberellins** and **cytokinins** which are growth promoters while **abscisic acid** (ABA) is a powerful growth inhibitor that often acts antagonistically to the growth promoters. ABA appears to promote dormancy in some seeds, and it stimulates the closing of stomata.

The main plant growth factor involved in phototropism is **indoleacetic acid** (IAA) which is synthesized mainly in the shoot tips. Its major effect is to promote growth by increasing the rate of cell elongation.

Phototropisms

Most of our knowledge about phototropisms has been gained from experiments on **coleoptiles**, the specialized, protective sheaths around the germinating shoots of grasses. Coleoptiles have been used because their response to light is easy to observe, they are small, and they are easy to grow in large numbers.

If a grass seedling is exposed to light from one side, it grows towards that light. Microscopic examination reveals that the cells on the shaded side of the coleoptile are significantly longer than those on the lit side. It seems that coleoptiles bend towards light by differential growth: cells on the shaded side elongate faster and become longer than those on the lit side. Early experiments on phototropism involved whole coleoptiles, decapitated coleoptiles, and coleoptiles shaded in various ways, as shown in the first diagram. Later experiments, shown in the second diagram, used agar blocks to collect a chemical diffusing downwards from shoot tips.

In the 1930s, the chemical messenger responsible for phototropic responses was extracted and given the name **auxin** (from the Greek *auxein*, to increase). Auxin was later found to consist mainly of indoleacetic acid (IAA).

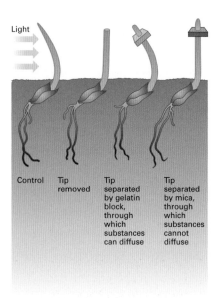

Some early experiments carried out on coleoptiles

Light

Control | Tip removed | Tip separated by gelatin block, through which substances can diffuse | Tip separated by mica, through which substances cannot diffuse

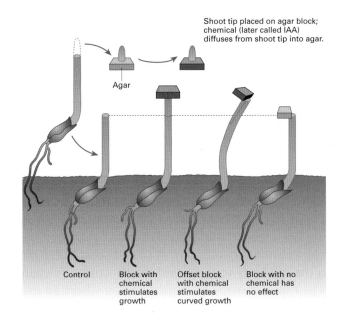

Shoot tip placed on agar block; chemical (later called IAA) diffuses from shoot tip into agar.

Agar

Control

Block with chemical stimulates growth

Offset block with chemical stimulates curved growth

Block with no chemical has no effect

Some experiments that show that phototropisms involve a chemical messenger. The agar blocks collect IAA which diffuses from the cut ends of shoot tips. By placing the blocks in various positions on coleoptiles with their tips cut off, it has been shown that the IAA stimulates cell growth. If more IAA is made to pass down one side of a coleoptile, that side will grow quicker than the other, causing the coleoptile to bend.

Coleoptile tips appear to produce equal amounts of IAA in the dark or the light, but its distribution can vary. It seems that when light strikes one side of a coleoptile, a receptor triggers the redistribution of IAA so that more travels down the shaded side of the coleoptile. The increased IAA concentration causes cells on the shaded side to grow longer than those on the lit side.

IAA is thought to stimulate cell elongation by causing target cells to secrete protons (hydrogen ions) into their cell walls. The resulting increase in acidity is thought to weaken the bonds between cellulose microfibrils, allowing the cell wall to expand when the cell takes in water.

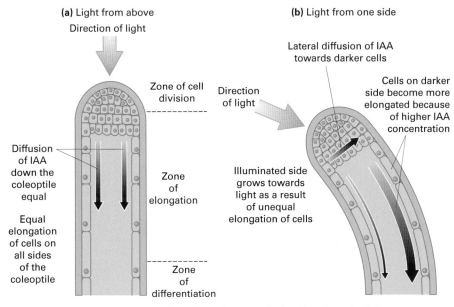

(a) Light from above
Direction of light

Zone of cell division

Diffusion of IAA down the coleoptile equal

Zone of elongation

Equal elongation of cells on all sides of the coleoptile

Zone of differentiation

(b) Light from one side

Lateral diffusion of IAA towards darker cells

Direction of light

Cells on darker side become more elongated because of higher IAA concentration

Illuminated side grows towards light as a result of unequal elongation of cells

Positive phototropism: growth of a coleoptile towards the direction of a light source.

Check your understanding

1 Name the specific plant growth factor that is involved in phototropisms of coleoptiles.

2 What effect does this factor have on cellular growth?

- - - - - - - - - - - - - - - - -

3 Most research on phototropism has used coleoptiles. Suggest the limitations of this research.

By the end of this spread you should be able to

- describe the structure of a myelinated motor neurone

Prior knowledge

- cell ultrastructure (*AS Biology for AQA* spreads 4.05–4.06)
- the reflex arc (7.02)
- chemical and electrical coordination (7.06)

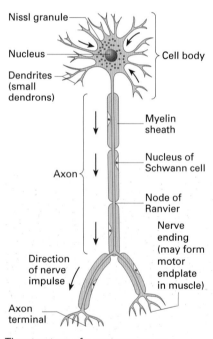

The structure of a motor neurone

Labels: Nissl granule — Nucleus — Dendrites (small dendrons) — Cell body — Myelin sheath — Nucleus of Schwann cell — Node of Ranvier — Nerve ending (may form motor endplate in muscle) — Axon — Direction of nerve impulse — Axon terminal

A mammalian motor neurone can generate and conduct impulses. In doing so, it conveys information rapidly over considerable distances. A single nerve impulse may be transmitted from the spinal cord to the feet in a few milliseconds.

The structure of a motor neurone

A typical motor neurone has a **cell body** containing a nucleus and other organelles. In the cytoplasm of the cell body are **Nissl granules** – these contain parallel rows of the **endoplasmic reticulum** and associated **ribosomes**. **Dendrites**, highly branched cytoplasmic extensions about 1 mm long, receive incoming information from other cells and conduct nerve impulses towards the cell body. A fine process called the **axon** extends from the cell body to the target cell. It varies in length from about 20 µm to 1 m or more. The conical region where an axon joins the cell body is called an **axon hillock**. The axon hillock integrates the incoming information from the dendrites and initiates a nerve impulse. The axon transmits outgoing nerve impulses from the cell body to the target cell. Axons and dendrites are often referred to as **nerve fibres**.

At the target cell, the axon divides into a number of nerve endings. The tip of each nerve ending has a swelling called the **axon terminal**. A narrow gap called the **synaptic cleft** separates the membrane of an axon terminal from the membrane of a target cell. The junction between one neurone and another cell is called a **synapse**.

The myelin sheath

The axons of many mammalian motor neurones are enclosed along most of their length by a thick insulating material called the **myelin sheath**. The myelin sheath is produced by special supporting cells called **Schwann cells**. The sheath is essentially a series of cell membranes, each produced by a single Schwann cell and wrapped many times around the axon. Gaps between the membranes of one Schwann cell and the next are called **nodes of Ranvier**. They play a key role in the fast transmission of nerve impulses.

Fast transmission enables mammals to respond almost instantaneously to stimuli. Nerve impulses can be directed along the nerve fibres to specific points in the body so that responses can be very localized.

Invertebrate neurones are not myelinated. The speed of conduction of their nerve impulses depends, among other things, on the diameter of the neurone.

The table summarizes the structure and functions of the organelles in a motor neurone.

Organelle	Structure and functions
dendrites	• receive incoming information • branched and numerous allowing many synaptic connections with other neurones
cell body	• has large volume to accommodate a nucleus and organelles • densely packed with mitochondria to provide the ATP needed to generate a nerve impulse • Nissl granules manufacture proteins
axon	• carries outgoing nerve impulse to target cell • long and thin to conduct impulses over a long distance rapidly
Schwann cells	• produce a myelin sheath that acts as an electrical insulator for the axon
nodes of Ranvier	• gaps in the myelin sheath that allow a nerve impulse to jump from one node to the next, speeding up the conduction of a nerve impulse
axon terminals	• highly branched to increase the contact area between a motor neurone and its target cell • bulbous end contains neurotransmitter which is released when a nerve impulse reaches it

The structure and functions of a motor neurone

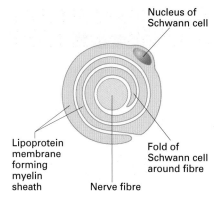

Transverse section through an axon showing the myelin sheath

Check your understanding

1 What is the name of the fine process that carries nerve impulses from a neurone cell body to a target cell?

2 What is the function of a Schwann cell?

3 Squids such as *Loligo vulgaris* shown in the photo can escape from danger because they have giant nerve fibres. These fibres can conduct nerve impulses very rapidly, since speed of conduction is directly related to the diameter of the fibre. Squids have nerve fibres of normal diameter to control their slow cruising movements, but giant nerve fibres control their rapid escape response. When danger threatens, giant nerve fibres carry information from the brain down the body, causing circular muscles to contract and force a jet of water out of the body, enabling the squid to make a quick backward escape. Suggest why squids have giant nerve fibres only for rapid escape responses.

The squid Loligo vulgaris *inhabits coastal waters and shallow seas in large numbers. Its giant nerve fibres enable it to respond quickly to danger.*

OBJECTIVES

By the end of this spread you should be able to

- define the resting potential

- explain how a resting potential is established in terms of differential membrane permeability, electrochemical gradients, and the movement of potassium and sodium ions

Prior knowledge

- diffusion (*AS Biology for AQA* spread 5.03)

- active transport (*AS Biology for AQA* spread 5.06)

- chemical and electrical coordination (7.06)

Defining the resting potential

A resting neurone is so called because it does not convey a nerve impulse, not because it is inactive. On the contrary, a resting neurone expends much energy in maintaining a potential difference across its membrane. This potential difference is called the **resting potential** and it is defined as the potential difference that is maintained across the membrane of an axon when a neurone is not conducting an impulse. The inside of the membrane is negative relative to the outside and the potential difference measures about -70 mV (millivolts).

Differential membrane permeability

During the resting potential, the inside of a neurone is negative relative to the outside because of an unequal distribution of charged ions. This is due mainly to the difference in permeability of the membrane to sodium and potassium ions. Sodium ions (Na^+) are present in higher concentrations outside the cell than inside. By contrast, the inside of the cell has a higher concentration of potassium ions (K^+).

This unequal distribution of ions results from a combination of **active transport** and **diffusion** of sodium and potassium ions across the cell membrane. A sodium–potassium pump actively transports sodium ions out of the neurone and potassium ions in. For every three sodium ions pumped out, only two potassium ions are pumped inwards. On its own, this would result in only a slight potential difference across the membrane. However, this difference is amplified by the membrane being about 50 times more permeable to potassium ions than to sodium ions. Potassium ions are able to diffuse freely back out of the cell down their concentration gradient, but the sodium ions diffuse back into the **axoplasm** (cytoplasm) of the neurone only very slowly. This results in the total number of positively charged ions on the outside of the membrane being greater than the total number inside, and creates a negative electrical charge inside compared with outside. Without active transport, an equilibrium would eventually be reached and there would be no potential difference across the membrane.

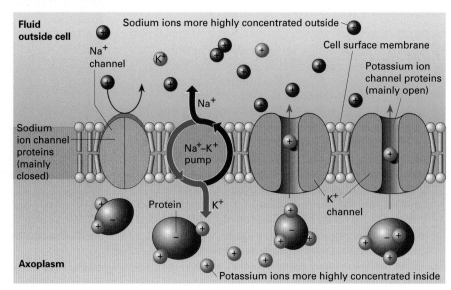

The combined effect of the differential permeability of the membrane to sodium and potassium ions and the action of the sodium–potassium pumps maintain a resting potential across the membrane of a neurone.

Electrochemical gradients

The diffusion of ions across the membrane of a neurone is due to a combination of electrical and chemical gradients. Sodium and potassium ions are positively charged and therefore tend to move down an electrical gradient towards a negatively charged region. The ions will also diffuse down a chemical gradient from a region where they are at a high concentration to a region where they are at a low concentration.

Movement of sodium and potassium ions

The rate of diffusion of sodium and potassium ions down a chemical gradient depends on **channel proteins** that are specific to each ion. Some channel proteins allow sodium ions to diffuse through the membrane into the axoplasm; others allow potassium to diffuse out. These proteins are **voltage gated** to control their permeability, that is, the opening and closing of gated channel proteins is controlled by changes in the potential difference across the cell membrane. When the gates are open, the ions can pass throught the membrane; when the gates are closed, they cannot. Membrane permeability to an ion depends on the proportion of gates open or closed. In a 'resting' axon, relatively more potassium gates are open than are sodium gates. This explains why the membrane is so much more permeable to potassium ions than to sodium ions, and why more potassium ions move out of the axoplasm than sodium ions move in.

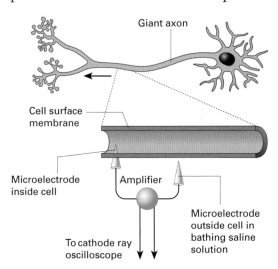

Measuring the potential difference across the membrane of a giant axon

Investigating nerve impulses

Nerves convey information rapidly from one part of the body to another, enabling animals to respond quickly to changes in their external and internal environments. The information is carried in the form of electrical signals called **nerve impulses**. Most of our understanding of the nature of nerve impulse comes from work done on giant axons of squids. These are the nerve fibres responsible for the rapid escape movements of squids. Their large diameter (up to 1 mm) makes it possible to measure the electrical activity in a giant axon when it is at rest and when it is carrying a nerve impulse.

A fine glass microelectrode is inserted inside an axon, and the voltage (potential difference) between it and a reference electrode on the surface of the axon can be displayed on a cathode ray oscilloscope, as shown opposite. By convention, the potential difference of the inside of the cell is always measured relative to that on the outside, so that the outside potential is taken as zero.

• • • • • • • • • • • • •

Check your understanding

1 What are the main mechanisms that maintain the resting potential of a neurone?

2 State whether most voltage gated sodium ion channel proteins are open or closed when a neurone is not transmitting a nerve impulse.

- -

3 Although the resting potential results mainly from the unequal distribution of potassium and sodium ions, other ions also play a part. On the outside of the membrane, in addition to sodium ions, there are more chloride ions (Cl^-) and calcium ions (Ca^{2+}). On the inside of the membrane, in addition to having more potassium ions, there are more protein anions. What charge will these anions have? Why do they not diffuse down a concentration gradient from inside to outside?

OBJECTIVES

By the end of this spread you should be able to

- explain how changes in membrane permeability lead to depolarization and the generation of an action potential

- define the all-or-none principle

- describe how the passage of an action potential along an unmyelinated or myelinated axon results in a nerve impulse

Prior knowledge

- generator potential (7.04)

- nerve impulses: resting potential (8.02)

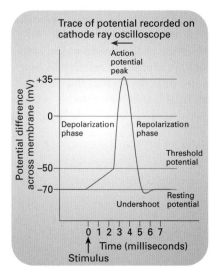

The changes in potential difference across the neurone membrane that make up an action potential

The effects of a stimulus

A nerve impulse occurs only when a neurone has a sufficiently high stimulus. A **stimulus** is any disturbance in the external or internal environment. It may be chemical, mechanical, thermal, or electrical, or it may be a change in light intensity.

When a stimulus is applied, the axon becomes **depolarized**; that is, the inside becomes temporarily less negative. A sub-threshold stimulus results in a **graded potential**, a change in electrical potential which is proportional to the stimulus intensity. However, if the stimulus is strong enough (if it exceeds the threshold level), an **action potential** occurs. There is a complete reversal of the charge across the nerve cell: the interior becomes positively charged relative to the outside. Typically, the action potential reaches a peak of about +35 mV. The potential difference then drops back down, undershoots the resting potential and finally returns to it. The return of the potential difference towards the resting potential is called **repolarization**. The entire action potential takes about 7 milliseconds.

Channel proteins and action potentials

The action potential results from changes in the permeability of cell membranes to ions. At rest, the membrane is more permeable to potassium ions than to sodium ions because most sodium ion channel proteins are closed. When a stimulus is applied, the sodium ion channels open, sodium ions move in, and the inside becomes more positively charged. If the stimulus reaches the threshold level, an action potential occurs.

When the action potential reaches its peak, the sodium ion channels close. Sodium ions stop moving into the axoplasm but more potassium ion channels open and potassium ions diffuse rapidly out. These changes cause the potential difference to drop back down, undershoot the resting potential, and finally return to it.

Transmission in an unmyelinated neurone

An individual action potential at any one point in a neurone is a short-lived, localized event. It is transmitted along the neurone as a nerve impulse because it causes a small current to flow in the axoplasm and the extracellular tissue fluid. In an **unmyelinated neurone**, this local current acts as the stimulus for the next part of the nerve membrane, causing further depolarization, and so on along the neurone. A nerve impulse is therefore transmitted as a self-propagating wave of depolarization with one portion of the neurone repolarizing as the next portion depolarizes.

Transmission in a myelinated neurone: saltatory conduction

In myelinated nerve fibres, the myelin sheath acts as an effective electrical insulator. Consequently the local flow of current can only be set up between adjacent nodes of Ranvier. There is no myelin sheath at these nodes, therefore the neurone membrane is exposed to the extracellular fluid. Also, there are many more sodium ion channels at the nodes of Ranvier than in the myelinated parts of a neurone. The nodes are about 1 mm apart. The local current set up by the depolarization of one node depolarizes the next node, and so on. The nerve impulse 'leaps' from

(a)

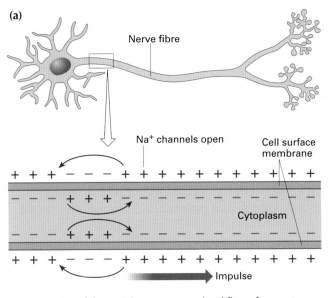

Depolarization of the membrane causes a local flow of current

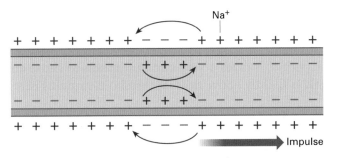

Local current stimulates the next region of the membrane, causing further depolarization

(b)

The impulse 'leaps' from one node to the next, which greatly speeds up the rate of propagation.

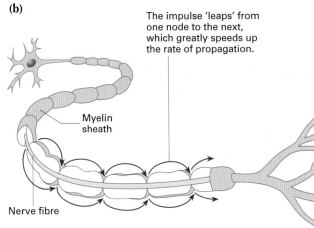

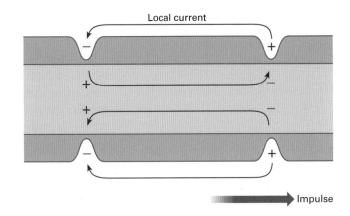

node to node. This type of nerve impulse transmission is called **saltatory conduction** (*saltus* is the Latin verb 'to leap' or 'to jump'). Saltatory conduction allows nerve impulses to be transmitted very quickly. It is also highly efficient because relatively few ions cross the membranes at the nodes, minimizing the need for active transport.

The all-or-nothing principle and frequency coding

In both unmyelinated and myelinated nerve fibres, action potentials obey the **all-or-nothing principle**. This means that no matter how strong the stimulus, the size of an action potential is always the same. Therefore, information about the strength of a stimulus is carried along a neurone only as changes in the frequency of impulses, rather than their size. Frequencies as high as 1000 impulses per second have been recorded in some mammalian neurones.

Frequency coding would provide limited information about changes in the environment if sense organs were connected to only one neurone. However, sense organs usually have many neurones that transmit information to the brain or another part of the body. Different neurones have different threshold levels. Therefore, further information about the nature of a stimulus is provided by the number, type, and location of the neurones affected by a stimulus.

Transmission of a nerve impulse: (a) in an unmyelinated neurone; (b) saltatory conduction along a myelinated neurone. Action potentials occur only at the nodes of Ranvier because local currents in one node reach the next node, starting a new action potential there – the impulse jumps from node to node.

Check your understanding

1 Puffer fish produce a highly potent neurotoxin called tetrodotoxin. This selectively blocks the entry of sodium ions across the membrane of a neurone. What effect will it have on the transmission of nerve impulses?

2 What happens to **a** the frequency **b** the amplitude (size) of an action potential when a stimulus is increased above the threshold level?

3 Define saltatory conduction.

The refractory period

Neurones can generate nerve impulses over a wide range of frequencies, from one or a few per second to more than 100 per second. The frequency is limited by the absolute refractory period, during which the neurone is completely inexcitable, and the relative refractory period, during which it is less excitable than normal.

During the **absolute refractory period** the sodium ion channels are open. During this period, which lasts about a millisecond or less, a second stimulus will not trigger a second action potential no matter how strong the stimulus is.

The **relative refractory period** follows the absolute refractory period. It corresponds to the time when the extra potassium channels open to repolarize the membrane and potassium ions flood out of the axoplasm, causing the membrane to become briefly more negative than the normal resting potential (a condition known as **hyperpolarization**). During this period, which lasts several milliseconds, a greater than normal stimulus is needed to initiate an action potential. However, the membrane becomes progressively easier to stimulate as the relative refractory period proceeds. At the beginning of the period, it takes a very strong stimulus to cause an action potential, but only a slightly above normal threshold stimulus near the end.

By limiting the maximum frequency of nerve impulses, the refractory period enables the nervous system to distinguish separate stimuli and make coordination possible. It ensures that each nerve impulse is separated from the next, with no overlapping signals, an essential feature for a system conveying frequency coded information.

How fast are nerve impulses?

A nerve impulse is the wave of depolarization that passes along an axon. Speeds, which range from 0.1 to 100 m s^{-1}, depend on three main factors: whether the axons are myelinated or unmyelinated, the axon diameter, and temperature.

Myelination

Only vertebrates have a **myelin sheath** surrounding neurones. **Saltatory conduction** increases the speed of propagation dramatically. Unmyelinated neurones transmit impulses at a maximum speed of about 1 m s^{-1}, while myelinated neurones have speeds of up to 100 m s^{-1}.

Axon diameter

Nerve fibres vary in diameter from about 0.5 to 1000 µm (1 mm). Unmyelinated axons with wide diameters can transmit nerve impulses faster than those with small diameters. The normal sized unmyelinated axons of a squid have conduction speeds of about 4 m s^{-1}, while its much wider unmyelinated giant axons conduct at speeds of 35 m s^{-1} or more. These giant axons are specifically adapted for quick escape responses.

The conduction speeds of myelinated fibres also increase with axon diameter, but the advantage of myelination means that there is no need for giant axons. A myelinated axon of a cat has to be only 4 μm in diameter to conduct nerve impulses at speeds of 35 m s^{-1}, whereas the giant axon of the squid needs a diameter 250 times greater to achieve this speed. The greatest advantage of myelinated axons comes from their small size, which allows a highly complex nervous system with high conduction speeds that does not take up much space.

Temperature

The conduction speed of a nerve impulse is strongly affected by temperature. Within limits, the higher the temperature, the faster the speed. This is mainly because the propagation of an impulse involves **diffusion** of ions, and the rate of diffusion increases with temperature as a result of the increased kinetic activity of the ions. Temperature also affects the integrity of membranes and the action of enzymes involved in the active transport required to maintain the resting potential. At very high temperatures, membranes may be damaged and enzymes denatured, resulting in disruption of the nerve conduction.

An important consequence of the relationship between temperature and conduction speed is that **homoiotherms** (so-called warm-blooded animals) generally have fast responses irrespective of the environmental temperature, while **poikilotherms** (cold-blooded animals) such as snakes can only respond quickly when their bodies have been warmed by, for example, the sun. There is more about this on spread 9.02.

Check your understanding

1 Distinguish between the absolute refractory period and the relative refractory period.
2 Explain why, within limits, the conduction speeds of nerve impulses become faster with temperature increases.

- -

3 The graph shows the relationship between conduction velocities in myelinated and unmyelinated nerve fibres. It shows that while conduction speed is linearly related to fibre diameter in myelinated fibres, in unmyelinated fibres it is proportional to the square root of the diameter.

  a What does this mean for fibres with diameters of less than 1 μm?

  b Unmyelinated fibres occur in the brains of mammals. Suggest reasons for this.

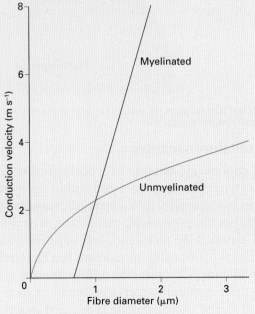

Conduction velocities in myelinated and unmyelinated nerve fibres

Nerve conduction velocity tests

A doctor may carry out a nerve conduction velocity test to see how well a nerve conducts an impulse. In this test, two small patch-like electrodes are placed a fixed distance apart on the surface of the skin overlying a nerve. At one electrode, a low intensity electric current stimulates the nerve and generates nerve impulses in its neurones. The other electrode records the compound action potential resulting from stimulating the nerve. The recording apparatus measures the time it takes for an impulse to be conducted along the fixed distance. This distance divided by the time gives the conduction velocity.

Nerve conduction velocity tests are used in the diagnosis of neurological diseases such as multiple sclerosis. A healthy nerve conducts impulses faster than a damaged nerve.

Multiple sclerosis (MS) is a major neurological disease that results from the gradual deterioration (demyelination) of the myelin sheath around neurones. Hardened scars replace the sheath. These scars interfere with the transmission of nerve impulses, slowing them down and causing a gradual loss of motor activity.

• • • • • • • • • • • • • • •

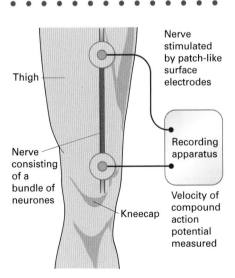

The nerve conduction velocity test

OBJECTIVES

By the end of this spread you should be able to

- describe the detailed structure of a synapse

- explain unidirectionality, temporal and spatial summation, and inhibition

- describe the sequence of events involved in the transmission of an impulse across a cholinergic synapse

Prior knowledge

- nerve impulses: resting potential and action potential (8.02 and 8.03)

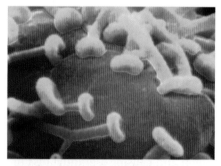

A scanning electron micrograph of presynaptic neurones terminating in synaptic bulbs (blue) which release neurotransmitter to postsynaptic membranes (×1000)

The nervous system transmits impulses throughout the body. Neurones are linked with other neurones and with effectors through specialized junctions called **synapses**. Neurones do not actually make contact with their target cell. They are separated by a narrow gap called the **synaptic cleft**.

Except for a few specialized electrical synapses, information passes across the synapse in the form of chemicals called **neurotransmitters**. Different neurones release different types of neurotransmitter which may stimulate or inhibit the activity of the target cell. **Excitatory presynaptic cells** release neurotransmitters that decrease the membrane potential of the target cell, making it more excitable and, if it is another neurone, more likely to generate nerve impulses. **Inhibitory presynaptic cells** release neurotransmitters that increase the membrane potential, making the target cell less excitable and less likely to generate nerve impulses.

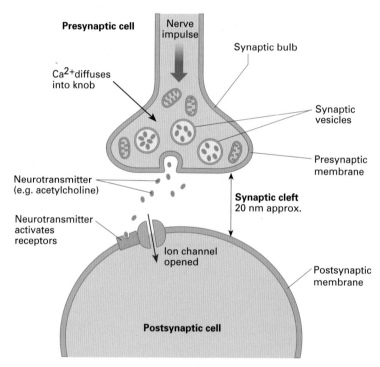

A chemical synapse, in which the neurotransmitter changes the excitability of the postsynaptic membrane

Excitatory synapses

A **cholinergic synapse** uses **acetylcholine** as the neurotransmitter. Acetylcholine is synthesized within the **synaptic bulb** and stored in special organelles called **synaptic vesicles**. When an action potential reaches the presynaptic membrane, it depolarizes the membrane, that is, it makes the membrane less negative than at rest. This depolarization triggers the opening of **calcium ion channels** in the presynaptic membrane. Calcium ions diffuse into the synaptic bulb, causing the vesicles containing acetylcholine to migrate and fuse with the presynaptic membrane. Acetylcholine is released into the synaptic cleft and diffuses across the synapse. Then it binds to specific protein receptor molecules on the postsynaptic membrane, a process known as **receptor activation**.

Inhibitory synapses

Receptor activation by inhibitory neurotransmitters causes other effects on postsynaptic membranes. Usually it opens chloride ion channels which makes the postsynaptic membrane more negative than normal (it becomes hyperpolarized) and less likely to depolarize sufficiently to generate an action potential.

Recycling the neurotransmitter

After a neurotransmitter has affected a postsynaptic membrane, two important processes take place:

- enzymes break down the neurotransmitter molecules in the synaptic cleft
- enzymes bring about the synthesis of neurotransmitter to refill the vesicles within the neurone from which they were originally released

The breakdown of neurotransmitter prevents further, unwanted effects. Acetylcholine, for example, dissociates from its receptor and is broken down by the enzyme acetylcholinesterase. The breakdown products diffuse back into the synaptic bulb where they are resynthesized into acetylcholine, using energy from ATP. The high density of mitochondria in the synaptic bulbs ensures that there is plenty of ATP available for the synthesis of the neurotransmitter.

Combining the effects of more than one impulse

A typical postsynaptic cell receives information from hundreds or even thousands of presynaptic neurones. The numerous synaptic connections allow the cell to combine different sources of information before responding. Its response will depend on the sum of all the excitatory and inhibitory postsynaptic potentials produced by spatial and temporal summation.

Spatial summation occurs when a single synapse does not release enough neurotransmitter to start an action potential on its own. But an action potential is fired when sufficient neurotransmitter builds up from several different synapses acting together. In this way the graded potentials produced by these several synapses can combine to trigger an action potential.

In **temporal summation**, a postsynaptic membrane fails to generate an action potential after a single impulse reaches the presynaptic membrane, but does so when two or more impulses arrive in quick succession from the same synapse. In this case, the graded potentials produced by successive impulses add together, generating an action potential in the postsynaptic neurone.

Transmission of nerve impulses through chemical synapses has several advantages.

- It enables information from different parts of the nervous system to be integrated.
- It provides a mechanism for filtering out trivial or non-essential information.
- It ensures that nerve impulses are unidirectional, passing only from presynaptic membranes to postsynaptic membranes.
- It allows the synapses to act as switches, so that nerve impulses can pass along one of several separate pathways in the nervous system.

Check your understanding

1 Explain why synaptic bulbs have a high density of mitochondria.

2 Which mineral ion causes synaptic vesicles to fuse with the presynaptic membrane?

3 What is a cholinergic synapse?

4 Distinguish between spatial and temporal summation.

5 How does a synapse ensure unidirectionality of a nerve impulse?

- - - - - - - - - - - - - - - -

6 In an electrical synapse, the presynaptic and postsynaptic membranes are close together with pores made of protein linking the cytoplasm of the two cells. This enables synaptic transmission to be very fast. Suggest what type of responses electrical synapses might be involved in.

The neuromuscular junction

Typically, as an axon approaches a muscle it loses its myelin sheath and branches extensively to form several areas of contact with different muscle fibres. Each point of contact between the motor neurone and one muscle fibre is a special plate-like synapse called a **neuromuscular junction** or **motor end plate**. It lies in a shallow much infolded depression on the surface of the muscle fibre. Synaptic vesicles are clustered in groups opposite the infolded regions of the **sarcolemma** (the muscle fibre membrane). As with a neurone-to-neurone synapse, a neuromuscular junction has a small gap between the membrane of the neurone and the membrane of the muscle fibre called a **synaptic cleft**.

Transmission across the neuromuscular junction

A neuromuscular junction functions in a similar way to a cholinergic synapse. Acetylcholine is always the neurotransmitter and, in vertebrate skeletal muscles, is always excitatory.

The following is a summary of the main events that take place when a nerve impulse passes along a motor nerve to a neuromuscular junction:

- When an action potential reaches a neuromuscular junction, calcium ion channel proteins open and calcium ions diffuse into the synaptic cleft.

- The diffusion of calcium ions causes synaptic vesicles to move to the junction membrane and fuse with it.

- Acetylcholine is released from the vesicles into the synaptic cleft.

- Acetylcholine diffuses across the cleft and attaches onto receptor molecules on the sarcolemma.

(a) Muscle fibres and their nerve supply

(b) Detail of neuromuscular junction

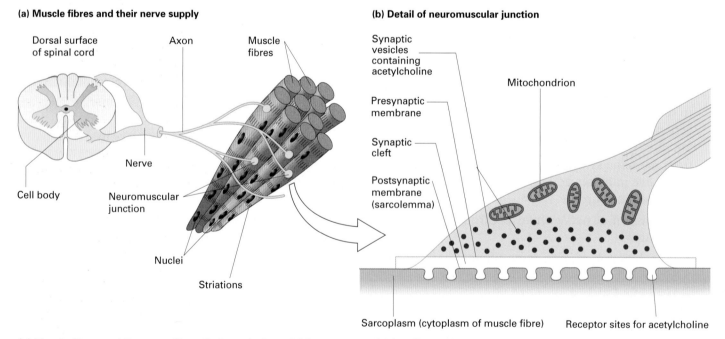

(a) Muscle fibres and the nerve fibres that supply them *(b)* A neuromuscular junction

- Receptor activation causes sodium ion channels to open in the membrane of the muscle fibre.
- An influx of sodium ions into the **sarcoplasm** (the cytoplasm of a muscle fibre) leads to a localized depolarization (a **graded potential** or **end-plate potential**) of the sarcolemma.
- The graded potential does not obey the **all-or-nothing principle**: its amplitude increases with the intensity of the stimulus until the stimulus reaches a threshold level.
- At the threshold level of stimulation, enough acetylcholine is released by the vesicles to generate an action potential across the muscle fibre, causing it to contract.

Immediately after an action potential, acetylcholinesterase breaks down acetylcholine to ensure that the muscle fibre is not overstimulated, and the sarcolemma becomes repolarized. The acetylcholine is then resynthesized and stored in the synaptic vesicles.

If a neuromuscular junction receives a continuous stream of axon action potentials at high frequency, eventually transmission across the junction stops. This is because the neurotransmitter cannot be resynthesized fast enough and it runs out. The synapse becomes **fatigued**.

The effects of drugs on synaptic transmission

A **drug** is any substance that alters the body's actions and its natural chemical environment. Drugs are widely used to prevent, diagnose, and treat disease. The term **medicine** is commonly used to distinguish a therapeutic chemical from substances that are used to alter the state of a healthy body. Drugs that interfere with synaptic transmission include the following.

- **Heroin** and **morphine** are narcotics which mimic (copy) the action of brain neurotransmitters called endorphins by binding onto special receptors (called opiate receptors) and blocking the action of nerve impulses that cause pain. They also act as depressants, reducing the activity of the cardiorespiratory centre. Overdoses can prove fatal if the diaphragm stops contracting or the heart stops beating.
- **Amphetamines** are stimulants that interfere with the storage of noradrenaline (a neurotransmitter in the sympathetic nervous system), causing the noradrenaline to overactivate certain neural pathways in the brain.
- **Nicotine** is a constituent of tobacco that mimics the action of acetylcholine. In addition to being the neurotransmitter between motor neurones and muscles, acetylcholine is an important neurotransmitter in the parasympathetic nervous system.
- **Organophosphates**, used in insecticides and nerve gases, inhibit the action of acetylcholinesterase. The continued presence of acetylcholine causes muscles to contract continuously, in some cases with fatal results.
- **Curare** is a drug that blocks the action of acetylcholine and prevents the contraction of voluntary muscles.
- **Strychnine** is a drug that blocks the action of glycine, an inhibitory neurotransmitter of motor neurones in the spinal cord. In small doses, strychnine acts as a stimulant but in high doses it causes severe muscular spasms and convulsions which can be fatal.

Check your understanding

1 When an action potential reaches a neurone ending, what effect does the release of calcium ions have on the synaptic vesicles?

2 Which neurotransmitter is contained within the synaptic vesicles at a neuromuscular junction?

3 Some drugs and poisons have an antagonistic effect on neurotransmitters, reducing their action, whereas others are agonists, amplifying the effect of a particular neurotransmitter. For example, some nerve gases act by inhibiting the action of cholinesterase at nerve–muscle junctions. Suggest what effect they have on muscle contractions. Are they agonists or antagonists?

4 Benzodiazepines are commonly used tranquillizers, thought to act by increasing the effect of gamma-aminobutyric acid (GABA), a neurotransmitter in the brain. Is GABA an excitatory or inhibitory neurotransmitter?

OBJECTIVES

By the end of this spread you should be able to

- describe the gross and microscopic structure of skeletal muscle

- describe the ultrastructure of a myofibril

Prior knowledge

- neuromuscular junction (8.06)

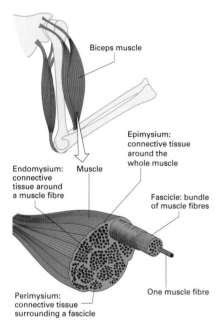

The biceps muscle: a spindle-shaped muscle made up of contractile muscle fibres covered with connective tissue

Gross structure of skeletal muscle

The human body has about 600 skeletal muscles which share a similar internal structure. A typical example is the biceps or upper arm muscle, a spindle-shaped muscle attached to the scapula by two **tendons** and to the radius by one tendon.

The biceps is a large and fleshy muscle covered by a sheath of connective tissue. It is richly supplied with blood vessels. Arteries and veins branch repeatedly into numerous capillaries, forming vast networks in and around the muscle, so that all parts have a good supply of oxygen and nutrients.

Nerves containing both sensory and motor neurones enter the muscle along with the blood vessels. The nerves branch many times to reach all parts of the muscle. **Motor neurones** coming from the central nervous system control the amount of tension in the muscle. **Sensory neurones** carry information from **pain receptors** and **proprioceptors** in the muscle to the central nervous system. Proprioceptors monitor the level of tension in the muscle and provide information about the orientation and movement of body parts.

Microscopic structure of a skeletal muscle

Under a light microscope we can see that the biceps is made up of many thousands of small cells called muscle fibres. Like other cells, a muscle fibre has cytoplasm (called **sarcoplasm**), an internal membrane system (the **sarcoplasmic reticulum**), and a cell surface membrane (the **sarcolemma**). However, each fibre is peculiar in being multinucleated and very long. Three or four blood capillaries surround each muscle fibre in a sedentary person, while up to seven capillaries supply each fibre in a trained athlete.

At this level of magnification, we can see that the axon of each motor neurone branches to supply up to 150 muscle fibres. All the muscle fibres served by the same motor neurone are called a **motor unit** because they work as a unit, all contracting or relaxing at the same time. The motor unit is the basic functional unit of skeletal muscle. Muscles that perform delicate movements may have as few as one muscle fibre per motor unit. Muscles that perform heavy work such as the quadriceps in the thigh may have motor units with more than 150 muscle fibres.

Each branch of an axon terminates at the **neuromuscular junction**.

Ultrastructure of a myofibril

Looking at a single muscle fibre under an electron microscope reveals that it is made up of a bundle of smaller fibres called **myofibrils**. Skeletal muscle appears striated (striped) under a microscope because the combination of different types of myofibrils cause alternate light and dark bands. A myofibril consists of repeating units called **sarcomeres**.

A sarcomere is the region between two dark lines called **Z lines**. The sarcomere is the fundamental unit of action of a muscle fibre. It contains two kinds of filament called thin filaments and thick filaments. A **thin filament** is made up of a double strand of the protein **actin**, along with one strand of a regulatory protein, coiled around each other. Each

thick filament is made up of parallel strands of the protein **myosin**. The thin and thick filaments are arranged within the sarcomere in a way that produces bands of light and dark in electron micrographs. The broad **dark band** (also known as the **A band**) consists of thick filaments interspersed with thin filaments, except in a central region (the **H zone**) where only thick filaments occur. The **light band** (the **I band**) contains only thin filaments along with the proteins in the Z line that connect adjacent thin filaments.

Electron micrograph of a muscle myofibril showing sarcomeres (×6000)

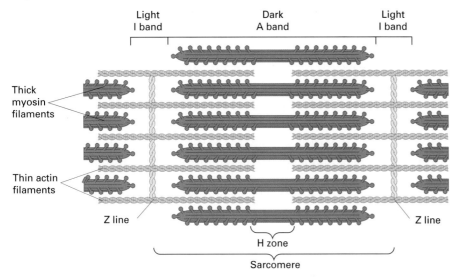

The ultrastructure of a sarcomere

Check your understanding

1 By what structure is a muscle attached to a bone?
2 What major difference would you expect between motor units that perform delicate movements and those that perform powerful movements?
3 What is a sarcomere?

4 Suggest why competitive bodybuilders repeatedly contract and relax their muscles before displaying them.

A bodybuilder repeatedly contracts and relaxes muscles before displaying them.

OBJECTIVES

By the end of this spread you should be able to

- explain the sliding filament theory of muscle contraction
- understand the roles of actin, myosin, calcium ions, and ATP in myofibril contraction

Prior knowledge

- resting and action potentials (8.02–8.03)
- synapses (8.05–8.06)
- skeletal muscle structure (8.07)

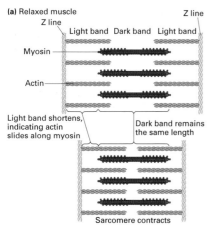

(a) Relaxed muscle

Z line | Z line
Light band | Dark band | Light band

Myosin

Actin

Light band shortens, indicating actin slides along myosin

Dark band remains the same length

Sarcomere contracts

(b) Contracted muscle

A sarcomere in (a) relaxed (b) contracted muscle

A single skeletal muscle is made up of many muscle fibres (muscle cells), each of which is divided into many **myofibrils**. Myofibrils contain **sarcomeres**, repeating units of thick and thin filaments. These filaments are the contractile apparatus of muscles.

How sarcomeres shorten to contract a muscle

In the 1950s, Andrew Fielding Huxley and other researchers used the electron microscope to study sarcomeres of relaxed and contracted muscles. They found that during contraction the Z lines and the thin filaments slide towards the middle of the sarcomere. The sarcomere shortens, but the lengths of the thick and thin filaments do not change.

From these studies, Huxley proposed the **sliding filament theory** of contraction.

According to the theory, a sarcomere shortens when its thin filaments slide along its thick filaments. High-magnification electron micrographs show that the thick **myosin** filaments are rod shaped with a globular end (called the **myosin head**). The head can form a cross-bridge with **actin**, the protein in the thin filaments. When attached to actin, a myosin head can change shape and slide the actin further along the myosin. During a muscle contraction, these actomyosin cross-bridges are formed and broken down repeatedly up to 100 times per second, causing the sarcomere to shorten by a ratchet-like mechanism. The mechanism can only shorten a sarcomere. It cannot actively return the sarcomere to its original length. Muscle elongation is usually brought about by the action of **antagonistic muscles**.

Calcium ions are required for cross-bridges to form, and the breakdown of ATP provides the energy needed by the ratchet mechanism. The combined actions of millions of sarcomeres can contract a whole muscle to about half its resting length. Contraction is started by a nerve impulse which triggers the release of calcium ions and the generation of ATP.

The role of calcium ions

When a muscle is at rest, calcium ions are not present in the **sarcoplasm** because they are stored in the **sarcoplasmic reticulum**, fine membrane-bound channels in the muscle fibres. In the absence of calcium ions in the sarcoplasm, **tropomyosin** (a protein in thin filaments) prevents myosin heads from attaching onto actin by blocking the binding sites.

When a muscle is stimulated sufficiently by nerve impulses, calcium ions are released from the sarcoplasmic reticulum and combine with **troponin** (another protein in thin filaments), causing the tropomyosin to change shape and unblocking the actin binding sites.

Calcium ions are released from the sarcoplasmic reticulum at the end of a sequence of events which begins when an action potential reaches a neuromuscular junction (see spread 8.06). When the graded potential in the sarcolemma (postsynaptic membrane) exceeds the threshold level, an action potential sweeps across the muscle fibre and passes into membranous tubules called **T tubules** or transverse tubules that fold inwards from the sarcolemma. Where the T tubules make contact with the sarcoplasmic reticulum, the action potential causes the sarcoplasmic reticulum to release calcium ions into the sarcoplasm. Calcium ions spread through the sarcoplasm, enabling myosin heads to bind onto actin. Energy from the breakdown of ATP enables the heads to take up a new position.

When action potentials stop arriving, calcium is actively pumped back into the sarcoplasmic reticulum, tropomyosin blocks the myosin head binding sites on the actin, and the muscle relaxes.

In summary, the key components of the sliding filament theory and their functions are

- **actin** and **myosin myofilaments**, the contractile fibrous proteins that are bound together during a muscle contraction by actomyosin cross-bridges and that slide past each other

- **calcium ions**, which are involved in removing the blocking protein tropomyosin from the binding sites on actin, allowing actomyosin cross-bridges to form

- **ATP**, the energy source for myofibril contraction. When hydrolysed it releases the energy required by the myosin heads to detach from actin filaments and to reposition themselves further along the actin. After myofibril contraction, it also provides the energy to actively transport calcium ions back into the sarcoplasmic reticulum so that the sarcomere can assume a relaxed state.

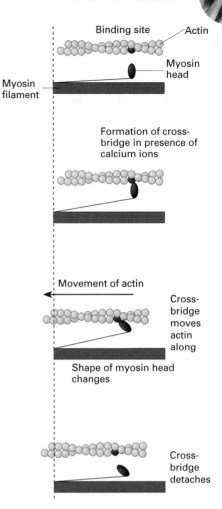

The sliding filament theory. Cross-bridges attach and reattach 50–100 times per second, but only if ATP is available.

Check your understanding

1   During a myofibril contraction, will the H zone of a sarcomere become wider or narrower?

2   According to the sliding filament theory, during a myofibril contraction what is the role of

   a   calcium

   b   acetylcholine

   c   ATP?

- -

3   The sliding filament theory postulates the attachment and detachment of myosin cross-bridges onto actin. The muscles of a dead person become very stiff: a condition known as rigor mortis. What does this suggest about the position of the cross-bridges in the muscles of a dead person, and about the precise role of ATP?

Vertebrate muscles have two main types of muscle fibre: **slow-twitch fibres** and **fast-twitch fibres**. In addition to having different contraction speeds, they have a different appearance: muscles containing mainly slow-twitch fibres are red; those containing mainly fast-twitch fibres are white.

Slow-twitch fibres

Slow-twitch fibres are adapted to function over long periods. They respire aerobically to avoid the build-up of **lactate** which would quickly fatigue them. They have their own metabolic fuel (muscle glycogen) but, because they are aerobic, they can also use the almost limitless supply of fat stores in the body. Their red coloration comes from having a high content of **myoglobin** and a good blood supply. This means that they can obtain sufficient oxygen, and their high density of **mitochondria** ensures that they can use oxygen efficiently to generate large amounts of ATP. However, a disadvantage of **aerobic respiration** is that they cannot generate ATP at a very fast rate, therefore slow-twitch fibres are not very powerful.

Fast-twitch fibres

Fast-twitch fibres are adapted for short bursts of explosive action. They generate ATP quickly and anaerobically from stores of a high-energy compound, phosphocreatine, and by lactate fermentation. When phosphocreatine breaks down it releases energy and phosphate ions which can be used to make ATP for up to 10 seconds of activity. Phosphocreatine is regenerated during aerobic respiration.

Fast-twitch fibres appear white because they have a relatively low myoglobin content and a small number of mitochondria, but they are rich in the enzymes required for **anaerobic respiration**. However, during lactate fermentation, they can only use glycogen as a fuel, and the lactate and hydrogen ions they produce makes them **fatigue** quickly.

Most people have roughly equal numbers of slow- and fast-twitch fibres, but the proportion varies in trained athletes: endurance athletes tend to have more slow-twitch fibres while power athletes tend to have more fast-twitch fibres.

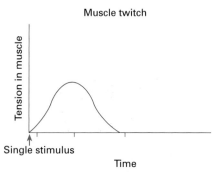
Muscle twitch — Tension in muscle / Time. Single stimulus

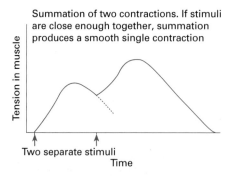
Summation of two contractions. If stimuli are close enough together, summation produces a smooth single contraction. Tension in muscle / Time. Two separate stimuli

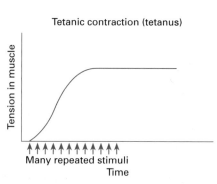
Tetanic contraction (tetanus). Tension in muscle / Time. Many repeated stimuli

Contraction of an isolated muscle induced by electrical stimulation, showing temporal summation

Contraction of a whole muscle

A single **motor unit** (one or more muscle fibres served by the same motor neurone) consists of only fast-twitch or slow-twitch fibres. In humans a fast-twitch motor unit usually contains 300–800 muscle fibres, while a slow-twitch motor unit has only 100–180 muscle fibres. Fast-twitch motor units are therefore much stronger than slow-twitch motor units.

A motor unit obeys the **all-or-nothing principle**: it either contracts completely or not at all. However, a whole muscle can produce graded responses by two mechanisms called temporal summation and muscle fibre recruitment.

Temporal summation

Temporal summation can be demonstrated by stimulating an isolated muscle electrically. If the isolated muscle is given a single electrical stimulus, it will produce a simple **twitch**. If a second stimulus is given before the first twitch is over, the muscle tensions will add together to produce a greater response. If the rate of stimulation is fast enough, the twitches will fuse to produce a smooth sustained contraction called **tetanus**. In normal situations, muscles produce smooth tetanic contractions rather than the jerky movements of muscle twitches.

Muscle fibre recruitment

In **muscle fibre recruitment**, the amount of tension produced in a muscle is altered by changing the number of motor units activated. When a situation requires more force, more muscle fibres are stimulated. Slow-twitch fibres are recruited for small muscular forces and more and more fast-twitch fibres are recruited as the force reaches maximum levels.

Under normal conditions, the nervous system does not recruit all of the available fibres even during maximal voluntary efforts. This prevents damage to muscles and tendons. The potential power of human muscles has been demonstrated many times by people exerting apparently superhuman efforts when confronted with a crisis. For example, apparently weak grandmothers have been known to lift a car to release a trapped grandchild.

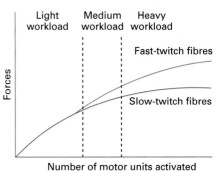

Muscle fibre recruitment

Check your understanding

1 Why do slow-twitch muscle fibres appear redder than fast-twitch muscle fibres?

2 What is the role of phosphocreatine in muscle contraction?

- -

3 Free-swimming fish such as mackerel and tuna have their muscles separated into two distinct masses which have a strikingly different appearance. A deep mass of red muscle is located along the sideline and in towards the vertebral column. The remaining larger mass of muscle is white. Suggest reasons for the relative size and location of each type of muscle.

In the skipjack tuna, the red swimming muscle is located in bands along the sides and in towards the vertebral column. The remaining large mass of muscle is white.

9.01 Homeostasis

OBJECTIVES

By the end of this spread you should be able to

- define homeostasis
- describe the main features of a homeostatic control system
- discuss the importance of maintaining a constant body core temperature and constant pH in relation to enzyme activity
- appreciate the importance of maintaining a constant blood glucose concentration in terms of energy transfer and water potential of blood

Prior knowledge

- factors affecting enzymes (*AS Biology for AQA* spread 3.03)
- water potential and osmosis (*AS Biology for AQA* spread 5.05)

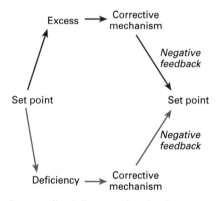

A generalized diagram showing how negative feedback keeps a system in a stable condition. A change from the usual level of a factor (the set point for that factor) triggers a corrective mechanism which restores the factor to its usual level.

A steady state

In order to survive, an organism has to be able to keep its internal environment within tolerable levels. The internal environment of a multicellular organism is the tissue fluid bathing the cells. Keeping the conditions of the tissue fluid such as pH, temperature, and salt content within restricted limits is called **homeostasis** (*homoios* means the same; *statis* means standing).

Artificial homeostatic devices enable humans to go underwater and into space. Our ability to maintain a constant internal environment enables humans to survive in a greater range of habitats than any other animals on Earth.

Negative feedback

Homeostasis is a characteristic of all living things, but the term is used for any system, biological or non-biological, which is in a steady state. In biological systems, homeostasis is usually achieved by a process called **negative feedback**. A change in the level of an internal factor causes **effectors** to restore the internal environment to its original level. For example, an increase in the internal body temperature causes the body to lose more heat; a decrease in body temperature causes the body to generate and conserve more heat. This type of system, in which a change in the level of a factor triggers a corrective mechanism, is called a self-adjusting system.

Control mechanisms

All homeostatic control mechanisms that use negative feedback, whether physical or biological, share similar components. They all have an **output** (for example, an internal factor such as blood temperature) that is controlled, and a **set point** (also called the **norm** or **reference point**).

In a physiological process, the set point is usually determined genetically and is the desired or optimal physiological state for the output. The control mechanism also has **detectors** (sensory receptors) to monitor the actual output. A **comparator** (sometimes called a regulator) compares the actual output with the set point. The comparator produces some sort of error signal which conveys information to the corrective mechanism about the difference between the set point and the actual output. In physiological systems, the error signal is usually in the form of nerve impulses or hormones, and the corrective mechanism may include one or more effectors which restore the output to its set point. In some physiological processes, such as thermoregulation, separate but coordinated mechanisms control deviations in different directions from the set point (for example, rises or falls in body temperature trigger different mechanisms to respond), giving a greater degree of control.

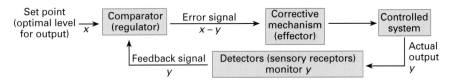

The main components of a simple negative feedback system. The set point is given the value x in the diagram; the actual output is given the value y in the diagram.

Thermostatic control: an example of a homeostatic mechanism

A thermostat is a simple example of a non-biological control system. The output being controlled is room temperature. The desired temperature is set on a control panel. A thermosensor detects the actual room temperature and sends information to a control box. The control box acts as a comparator, comparing the actual room temperature with the set temperature. If the actual temperature deviates from the set point, the control box starts up the appropriate corrective mechanism, switching a heater off if the temperature is above the set point, or switching it on if it is below the set point.

The importance of controlling blood temperature and blood pH

In mammals, the **body core temperature** is the temperature of the brain and vital organs within the chest cavity and abdomen. The ability of mammals to maintain this body core temperature within restricted limits makes them relatively independent of changes to the external temperature and enables them to exploit a wide range of habitats. This is because each organism has enzymes that work most efficiently at an optimum temperature. For example, despite wide variations in environmental temperature, our highly developed internal control mechanisms keep our body temperature close to 37°C, the optimum temperature for most of our enzyme systems. This allows us to remain active in these different environments while other animals such as lizards have limited powers of thermoregulation and cannot function over such a wide range of environmental temperature.

Similarly, each of our enzymes and proteins such as haemoglobin function best at a particular pH. Most of those in the blood function best at a slightly alkaline pH. Our blood pH is therefore maintained at about 7.4. It is kept within restricted limits by excreting urine that varies between pH4 and pH9. When the blood pH becomes too alkaline (for example, after drinking alkaline fluids), we produce urine of a high pH; when it becomes too acidic (such as after intense anaerobic exercise), we produce urine of a low pH.

The importance of controlling blood glucose concentration

Mammalian brain cells can only function aerobically using glucose as an energy substrate. If the oxygen supply is interrupted or if glucose becomes unavailable beyond a critical time, the mammal will lose consciousness. The brain obtains its glucose from blood hence it is vitally important that blood glucose concentrations are maintained above the minimum levels which allow brain cells to function. Too much glucose in the blood decreases its water potential and can have harmful osmotic effects on sensitive cells, such as those in the brain. Consequently, mammals have evolved a complex homeostatic control mechanism to maintain blood glucose levels within restricted limits (see spreads 9.04–5).

By the end of this spread you should be able to

- discuss the contrasting mechanisms of temperature control in an ectothermic reptile and an endothermic mammal
- identify the mechanisms involved in heat production, conservation, and loss

Prior knowledge

- homeostasis (9.01)

(a) Morning

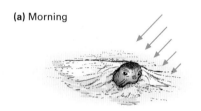

(b) Noon

(c) Afternoon

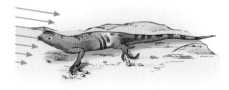

The earless lizard of south-western USA regulates its temperature by its behaviour. (a) The morning sun warms blood in the head protruding from the sand. The rest of the body remains hidden until the lizard is warm enough to be active. (b) At noon the lizard seeks shelter from the hot sun. (c) It emerges in the afternoon when it is cooler, and lies parallel to the sun's rays.

Living in extremes of temperature

Life exists in an amazingly wide range of thermal environments, from sub-zero polar seas to the bubbling waters of hot volcanic springs. Individual species have evolved a diverse array of mechanisms to cope with extreme conditions. However, a particular organism can only withstand a wide range of thermal environments if it can maintain its internal body temperature at a relatively constant level. This is mainly because enzyme-catalysed metabolic reactions work efficiently only within a limited temperature range. If the temperature is too high, enzymes become denatured, disrupting metabolic reactions. If the temperature is too low, metabolic reactions take place too slowly to maintain an active life. At extreme temperatures macromolecules may even change their state: for example, phospholipids become very fluid at very high temperatures and solid at very low temperatures.

Thermoregulation

The process by which an animal regulates its temperature is called **thermoregulation**. Animals that can maintain a stable body temperature are sometimes called **homoiotherms** or **warm blooded**. Animals with a body temperature that is more or less the same as that of the environment are sometimes called **poikilotherms** or **cold blooded**. These terms can be misleading; for example, a hibernating European hedgehog (regarded as warm blooded) maintains its body at a low temperature (about 6°C), whereas some fish and reptiles (commonly regarded as cold blooded) can maintain their bodies at a relatively high temperature (about 30°C) for long periods. To avoid these problems, many biologists use the terms ectotherms and endotherms when describing the thermoregulation of animals.

An **ectotherm** such as a reptile has a body temperature that changes with the environmental temperature. However, ectotherms can use behavioural control mechanisms to regulate their internal body temperature. Earless lizards, for example, can keep their body temperature relatively constant by using external sources of heat gain and loss by moving in and out of the sun.

All mammals and birds are **endotherms**. They control their body temperature independently of the environment using internal physiological control mechanisms as well as behavioural ones. Not all the body of a mammal is kept at a constant temperature, only the **body core**. This consists of the vital organs of the chest and abdomen, and the brain. The skin and tissue close to the body surface are always cooler than the core because it is through these structures that heat is exchanged with the environment.

Generally, endotherms can remain active over a far wider range of environmental temperatures than can ectotherms. Within certain limits, endotherms are free to migrate long distances and maintain high rates of activity in all sorts of weather. This allows them to capture and kill ectothermic prey or escape from ectothermic predators. Maintaining a body temperature different from that of the external environment requires a great deal of energy for metabolism. Consequently, although being endothermic has freed mammals and birds from fluctuations in environmental temperatures, it has made them slaves to their stomachs. They require much more food than ectotherms of equivalent size. A pygmy shrew, for example, consumes about its own mass of food per day whereas a cockroach of the same size can go for days without a meal.

Exchanging heat with the environment

To maintain a constant body temperature, the heat gained by an animal must equal the heat it loses; therefore, heat gained from metabolism and the environment must be balanced against heat lost to the environment. Heat is transferred between an animal and its environment in four main ways: radiation, convection, conduction, and evaporative cooling.

- **Radiation** is the transfer of heat as infrared waves. The amount of heat radiated by a body is proportional to the temperature difference between the body and its surroundings.
- **Convection** is the transfer of heat by fluid molecules (air or water) moving in a current. The faster the rate of fluid movement, the faster the rate of heat transfer.
- **Conduction** is the transfer of heat by physical contact between two bodies.
- **Evaporative cooling** occurs when water changes to water vapour. The evaporation of 1 g of water from the body surface requires the loss of 2.45 kJ from the body.

Check your understanding

1 Why is a lizard described as an ectotherm?

2 Why does a mammal have to eat much more food than a reptile of equivalent size?

3 List the three main ways in which heat can be transferred from the environment to an animal.

- - - - - - - - - - - - - - -

4 Some animals have been called heterotherms because they do not fit neatly into the two main categories of thermoregulation (endotherms and ectotherms). For example, the powerful, streamlined skipjack tuna (*Katsuwonus pelamis*) can maintain its swimming muscle at a temperature as much as 14°C higher than the water in which it swims. Suggest how the fish might achieve this.

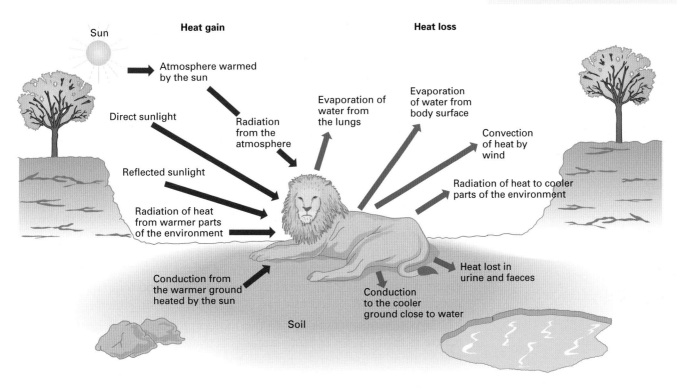

Energy exchanges between a lion and its environment

By the end of this spread you should be able to

- describe the role of the hypothalamus and autonomic nervous system in maintaining a constant body temperature in a mammal

Prior knowledge

- homeostasis (9.01)
- temperature control I (9.02)

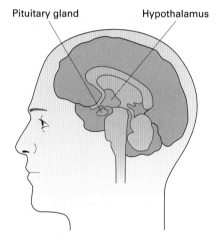

The position of the hypothalamus in the brain

Pituitary gland

Hypothalamus

Two interacting homeostatic systems control the body temperature of mammals: one regulates the temperature of the skin surface; the other regulates the temperature of the **body core** (the vital organs in the abdominal cavity and the chest cavity, and the brain).

Regulation of skin temperature

The system that regulates the skin surface temperature is more obvious to us because we are conscious of most of its main components: the set point, detectors, comparator, and corrective mechanism.

- The set point is the preferred skin temperature, the temperature at which the person feels comfortable.
- The detectors are thermoreceptors in the skin. Although there is disagreement about the nature of these thermoreceptors, most physiologists believe that **heat receptors** detect increases in skin temperature while **cold receptors** detect decreases in skin temperature. These thermoreceptors can only detect changes in skin temperature; they do not give any information about the actual temperature.
- The cortex of the brain (the outer area, responsible for forming our conscious thoughts and feelings) acts as the comparator. If we feel too hot or too cold, we may decide to move to a cooler or warmer area, remove or add clothing, or take some other voluntary action which brings our skin temperature back to its norm.
- The error signals are nerve impulses to voluntary muscles.
- Behavioural responses act as the corrective mechanism.

Regulation of body core temperature

We are not conscious of our second thermoregulatory system. It works mainly by autonomic (involuntary) physiological responses. The set point of this system is the optimal body core temperature, about 37°C. This is genetically determined but can be temporarily shifted. For example, during certain bacterial infections substances known as **pyrogens** raise the set point, causing a fever. Pyrogens may be toxins produced by bacteria, or they may be secreted from white blood cells to raise the body temperature and stimulate the body's defence responses.

The detectors for our autonomic responses are thermoreceptors in the **hypothalamus**, a small part of the brain just above the pituitary gland. The hypothalamus is strategically placed for monitoring the temperature of the blood supplying the nervous tissue of the brain. This tissue is especially sensitive to temperature fluctuations; blood temperatures that are either too high or too low can cause mental derangement.

The hypothalamus contains two thermoreceptor centres. A **heat loss centre** in the anterior hypothalamus is activated by increases in blood temperature. It uses nerve impulses and hormones as the error signals to activate responses that increase heat loss from the body so that the core temperature can be brought back down to its set point. A **heat gain centre** in the posterior hypothalamus is activated by decreases in blood temperature. This uses error signals to initiate a variety of corrective mechanisms which conserve body heat and raise the blood temperature.

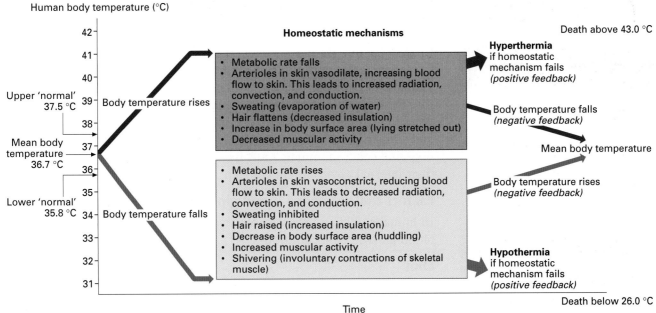

Human body temperature (°C)

Homeostatic mechanisms

Death above 43.0 °C

Hyperthermia
if homeostatic
mechanism fails
(positive feedback)

Body temperature rises

- Metabolic rate falls
- Arterioles in skin vasodilate, increasing blood flow to skin. This leads to increased radiation, convection, and conduction.
- Sweating (evaporation of water)
- Hair flattens (decreased insulation)
- Increase in body surface area (lying stretched out)
- Decreased muscular activity

Body temperature falls
(negative feedback)

Mean body temperature

Upper 'normal' 37.5 °C

Mean body temperature 36.7 °C

Lower 'normal' 35.8 °C

Body temperature falls

- Metabolic rate rises
- Arterioles in skin vasoconstrict, reducing blood flow to skin. This leads to decreased radiation, convection, and conduction.
- Sweating inhibited
- Hair raised (increased insulation)
- Decrease in body surface area (huddling)
- Increased muscular activity
- Shivering (involuntary contractions of skeletal muscle)

Body temperature rises
(negative feedback)

Hypothermia
if homeostatic
mechanism fails
(positive feedback)

Death below 26.0 °C

Time

Homeostatic control of body temperature

Some of the corrective mechanisms involve the skin. The skin dermis contains sweat glands and blood vessels, both of which play an important part in thermoregulation. The dermis also has fat cells which store energy and provide thermal insulation. The watery sweat secreted by sweat glands evaporates from the skin surface, cooling the body in times of heat stress. The capillaries in the dermis are structured so that they can be widened by vasodilation or narrowed by vasoconstriction enabling blood to be shunted to different parts of the skin to help control temperature:

- blood is shunted to the surface so that heat can be radiated away from the body when the body becomes too hot
- blood remains confined to deep layers, insulated by fat, when the body needs to retain heat

Interaction of the two regulatory systems

The two thermoregulatory systems do not work in isolation; they interact. It is thought that the skin receptors pass information to the hypothalamus about changes in the environmental temperature and set up compensatory responses *before* the core temperature starts to change. In most situations, both systems work together to control body temperature.

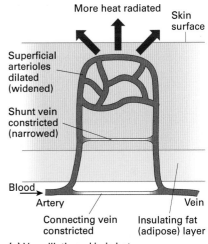

(a) Vasodilation: skin is hot

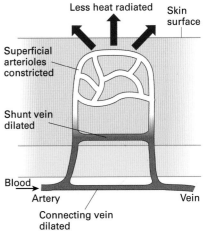

(b) Vasoconstriction: skin is cold

Thermoregulation by the blood vessels of the skin surface

Check your understanding

1 List three structures in the skin that play a part in thermoregulation.
2 Suggest why a person who does not sweat in a hot environment might be in danger.
3 By what means is heat produced in mammals when the body core temperature falls?

OBJECTIVES

By the end of this spread you should be able to

- identify the factors that influence blood glucose concentration

- describe the role of the liver in glycogenesis and gluconeogenesis

Prior knowledge

- carbohydrate structure and function (*AS Biology for AQA* spreads 2.03–2.04)

- homeostasis (9.01)

In 1859 Claude Bernard, a French physiologist, was the first scientist to recognize the importance of homeostasis in mammals after studying variations in glucose concentrations in the blood of dogs. He found that the concentrations remained remarkably stable despite dramatic variations in diet. For example, dogs recently fed meat or sugar-rich food had similar glucose concentrations to a starving dog. He concluded that dogs and other mammals must have a control mechanism that keeps their internal environment constant.

Factors that influence blood glucose concentration

In humans, diet and exercise are two important lifestyle factors affecting blood glucose concentrations.

Diet

The **glycaemic index (GI)** is a system used to rank carbohydrate-containing foods based on their overall effect on blood glucose levels. Slowly absorbed foods have a low GI, whilst foods that are absorbed more quickly have a higher GI. The effect over a 3-hour period of eating 50 g of a digestible carbohydrate is compared with that of eating 50 g of glucose. Glucose is used as a standard reference and is given a GI of 100. New potatoes have a GI of 70 and oranges have a GI of 40.

Eating foods with a high GI causes blood glucose concentrations to rise rapidly. Carbohydrate from foods with a low GI is absorbed into the bloodstream over a relatively long period, helping to maintain stable blood glucose concentrations between meals.

Another measure of the effect of foods on blood glucose concentrations is the **glycaemic load (GL)**. This is calculated using the following formula:

$$GL = \frac{\% \text{ carbohydrate content per portion} \times GI}{100}$$

The same food will have different glycaemic loads depending on how big a portion is eaten.

Exercise

Exercise, particularly **anaerobic exercise**, may make high demands on blood glucose. In most circumstances, the increased metabolism resulting from exercise triggers the conversion of glycogen stores in muscle cells and liver cells to glucose, maintaining the blood glucose concentration relatively constant. However, extremely strenuous exercise may result in a temporary reduction in blood glucose concentration.

Hormones

The most important internal factor that affects blood glucose concentration is the interaction of hormones: **insulin** tends to reduce blood glucose concentration while **glucagon** and **adrenaline** tend to increase it.

The role of the liver

The liver is the largest organ in the mammalian abdomen. It plays a central role in metabolism, regulating the concentrations of a wide range of chemicals in the blood including glucose.

The liver receives, via the hepatic portal vein, all the glucose absorbed into the blood through the intestinal wall. The blood glucose concentration leaving the liver may be kept the same as that entering the liver, it may be reduced, or it may be increased.

Glucose can be added to the blood leaving the liver by

- **glycogenolysis** – 'glycogen splitting'– hydrolysing glycogen to glucose
- **gluconeogenesis** – 'new glucose formation'– converting non-carbohydrate substances such as amino acids and glycerol to glucose

Glucose can be removed from the blood by

- **glycogenesis** – ' glycogen forming'– condensing glucose and storing it as glycogen
- converting glucose to fat
- using glucose as a fuel for cellular respiration (that is, increasing metabolism)

Homeostatic control of blood glucose concentration

Two interacting mechanisms control blood glucose concentration: one compensates for levels that are too high, and the other compensates for levels that are too low. Insulin acts as the error signal or regulating chemical for the first mechanism. Glucagon acts as the main (but not only) error signal or regulating chemical for the second mechanism. The diagram below shows these two mechanisms.

Dual control of the blood glucose concentration means that the body is not over-reliant on one mechanism. Having two mechanisms allows homeostatic control to be more precise and rapid, minimizing fluctuations from the norm or **set point**.

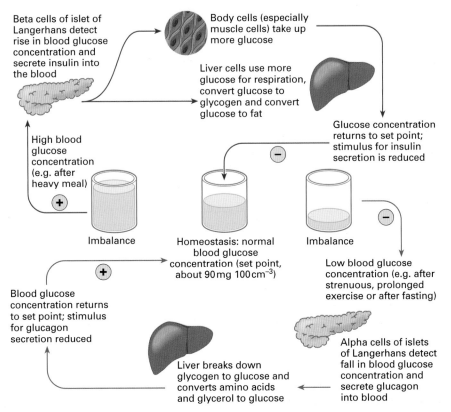

Homeostatic control of blood glucose concentration

Check your understanding

1. What effect does eating a meal with a high glycaemic load have on blood glucose concentration?
2. Distinguish between glycogenesis and gluconeogenesis.

- - - - - - - - - - - - - - - -

3. Endurance training increases the capacity of skeletal muscle to store glycogen. It also reduces muscle glycogen utilization during prolonged, steady, aerobic exercise. Suggest why these effects are important to a marathon runner.

169

OBJECTIVES

By the end of this spread you should be able to

- discuss the role of insulin and glucagon in controlling glucose uptake and the interconversion of glucose and glycogen, and the effect of adrenaline on glycogen breakdown and synthesis

- describe the second messenger model of adrenaline and glucagon action

Prior knowledge

- homeostasis (9.01)
- control of blood glucose I (9.04)

(a) Islet of Langerhans

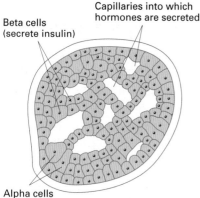

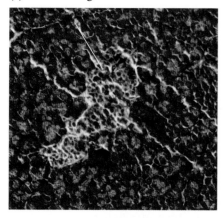

Beta cells (secrete insulin)

Capillaries into which hormones are secreted

Alpha cells (secrete glucagon)

(b) Detail of an islet of Langerhans

Structure of an islet of Langerhans. The pancreas secretes digestive enzymes as well as endocrine hormones.

Insulin

Insulin is a small protein secreted by special cells called **beta cells** in the **islets of Langerhans**, endocrine tissue within the pancreas. When the blood glucose concentration rises above the **set point**, more insulin is secreted from the pancreas.

In order to act on cells, insulin molecules bind to an exposed **glycoprotein** receptor on the cell surface membranes, in much the same way as a substrate binds to an enzyme. The membrane-bound insulin brings about a number of cellular responses which reduce the blood glucose concentration. The responses include changes in both cell surface membrane permeability and enzyme activity which lead to four major effects:

1 an increase in the uptake of glucose and amino acids into cells

2 an increase in the rate of cellular respiration and the use of glucose as a respiratory substrate molecule

3 an increase in the rate of conversion of glucose to fat in adipose (fat-storing) cells

4 an increase in the rate of conversion of glucose to glycogen in liver and muscle cells (**glycogenesis**)

Glucagon

Glucagon is a protein secreted by **alpha cells** in the islets of Langerhans. When the blood glucose concentration falls below the set point, the alpha cells secrete glucagon. Glucagon activates **phosphorylase** (an enzyme in the liver) which catalyses the breakdown of glycogen to glucose (**glycogenolysis**). Glucagon also increases the conversion of amino acids and glycerol into glucose 6-phosphate. This synthesis of glucose from non-carbohydrate sources is called **gluconeogenesis**.

Adrenaline

In addition to glucagon, other hormones can also increase blood glucose concentrations. For example, in times of acute stress or excitement, **adrenaline** is secreted. Adrenaline causes the breakdown of glycogen in the liver, boosting blood glucose concentrations.

The second messenger model

Both adrenaline (epinephrine) and glucagon stimulate glycogenolysis in liver cells. This hydrolysis of glycogen to glucose causes the blood glucose concentration to increase. Neither adrenaline nor glucagon enter liver cells. Instead both hormones act by binding to specific receptors on the surface of the cell surface membrane of liver cells. The binding of the first messenger (adrenaline or glucagon) activates **adenyl cyclase**. This is an enzyme that accelerates the production of **cyclic adenosine monophosphate (cAMP)**. cAMP is a small molecule that acts as the **second messenger** within the cytoplasm. It diffuses through the cytoplasm of the liver cell where it initiates a complex chain reaction that ends with the breakdown of glycogen to glucose phosphate, a source of glucose.

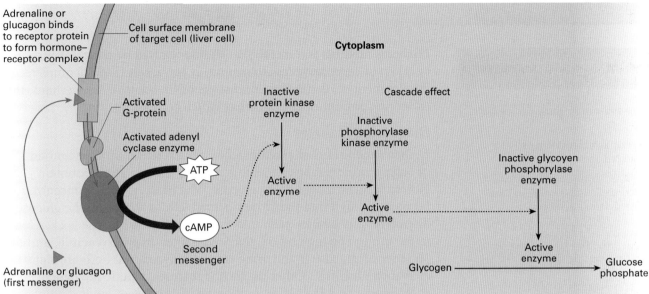

Adrenaline or glucagon binds to receptor protein to form hormone–receptor complex

Cell surface membrane of target cell (liver cell)

Cytoplasm

Activated G-protein

Activated adenyl cyclase enzyme

ATP

cAMP

Second messenger

Adrenaline or glucagon (first messenger)

Inactive protein kinase enzyme

Active enzyme

Cascade effect

Inactive phosphorylase kinase enzyme

Active enzyme

Inactive glycoyen phosphorylase enzyme

Active enzyme

Glycogen

Glucose phosphate

The second messenger model of the action of adrenaline and glucagon on a liver cell

This second messenger pathway involves a complex chain reaction that produces a **cascade effect** which amplifies the response to the hormone. Throughout the chain, each enzyme molecule activates many substrate molecules which become the next enzyme in the chain. This means that a small signal (a few molecules of adrenaline or glucagon on the cell surface) results in a very large response – in this case, a rapid increase in the blood glucose concentration.

Although adrenaline and glucagon use the same second messenger pathway, each hormone binds to its own specific receptor site. However, a maximal production of cAMP by one hormone cannot be increased any further by a second hormone.

The interaction of glucose control mechanisms

The system for controlling blood glucose concentrations is self regulating: the blood glucose concentration itself determines the relative amounts of insulin and glucagon secreted, and these hormones alter the blood glucose concentration so that it remains relatively stable. The two corrective mechanisms regulated by insulin and glucagon act in opposition to each other; they are **antagonistic**. They provide a much more sensitive control system than one that relies on only one set of corrective mechanisms. Cells in the islets of Langerhans are thought to act as detectors and comparators for both sets of corrective mechanisms.

The control systems usually keep blood glucose concentrations close to the norm. However this system, like any other system, can break down. If, for example, the pancreas cannot secrete insulin, or if target cells lose their responsiveness to insulin, the blood glucose concentration can reach dangerously high levels. The resulting condition is called **diabetes mellitus**.

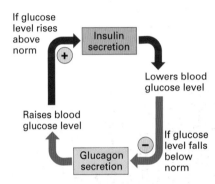

If glucose level rises above norm

Insulin secretion

Lowers blood glucose level

Raises blood glucose level

Glucagon secretion

If glucose level falls below norm

The antagonistic actions of insulin and glucagon

Check your understanding

1 Under what conditions is insulin secretion increased?

2 Why is cAMP referred to as a second messenger?

- - - - - - - - - -

3 Liver glycogen provides the main store of readily available glucose. Muscle glycogen is a second major source of stored glucose. Skeletal muscles lack glucagon receptors. Suggest the significance of this.

OBJECTIVES

By the end of this spread you should be able to

- distinguish between type I and type II diabetes

- explain how diabetes mellitus is controlled by insulin and the manipulation of diet

Prior knowledge

- homeostasis (9.01)

- control of blood glucose (9.04–9.05)

Kris Freeman, a diabetic, is a top cross-country skier. The discovery that insulin can be used to treat diabetes has enabled millions of people with diabetes worldwide to enjoy a full and active life and even to become top-class sportspeople.

Diabetes mellitus is a metabolic disorder caused by a lack of insulin or a loss of responsiveness to insulin. It can result from incomplete development of, damage to, or disease of the islets of Langerhans, the endocrine portion of the pancreas which secretes insulin.

Diagnosing diabetes

Diabetes mellitus is characterized by the excretion of large amounts of sugary urine (diabetes mellitus actually means 'sweet fountain'). Normally, urine contains no glucose. The blood glucose concentration becomes so high that the kidney is unable to reabsorb all the glucose filtered into its tubules back into the blood and so glucose is excreted in the urine. The high blood glucose is called **hyperglycaemia**, and the presence of glucose in the urine is called **glycosuria**; both are signs of diabetes mellitus. Other diagnostic features include the patient complaining of a lack of energy, a craving for sweet foods, and persistent thirst. The main diagnostic test is a **glucose tolerance test** in which the patient swallows a sugar solution and a doctor measures the blood glucose concentration at intervals. Graphs of the results from a diabetic person and a person with normal glucose metabolism are distinctly different, as shown here.

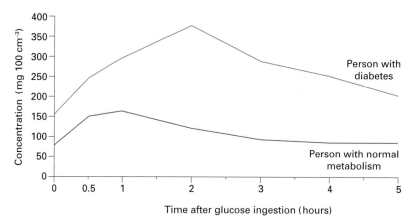

The results of a glucose tolerance test in a person with normal glucose metabolism and a person with diabetes

In addition to disrupting the homeostatic control of blood glucose, diabetes mellitus has other harmful effects. Insulin acts as an anabolic (body-building) hormone, therefore lack of it causes weight loss and muscle wasting. Although the blood of diabetics is rich in glucose, their carbohydrate metabolism is disrupted and their cells are unable to use glucose as a major source of fuel. Consequently, they have to resort to metabolizing fat as a source of energy. In severe cases, the products of this fat metabolism can be harmful.

Types I and II

Type I diabetes (also known as insulin-dependent diabetes or juvenile-onset diabetes) usually occurs suddenly in childhood. It appears to be an autoimmune disease: cells from the immune system attack beta cells in the islets of Langerhans, destroying a person's ability to secrete insulin.

Type II diabetes (also known as insulin-independent diabetes or late-onset diabetes) usually occurs later in life. It is often caused by a gradual loss in the responsiveness of cells to insulin, but it can also be due to an insulin deficiency.

Treating type I diabetes

Most people with type I diabetes are treated with insulin injections. Insulin cannot be taken orally because, being a protein, it would be digested in the alimentary canal. Before insulin became available about 75 years ago, type I diabetes would have meant death by a slow and progressive wasting. Nowadays, with proper use of insulin and careful management of diet and exercise, people with diabetes can lead a normal life.

Insulin used to be obtainable only from non-human sources (mainly pigs and cows), but **recombinant DNA technology** has made human insulin available. In the short term, insulin from non-human sources cannot usually be distinguished from human insulin, but non-human insulin is not absolutely pure and prolonged use may lead to immunological reactions which can damage health. The immune system identifies the non-human insulin as foreign and attacks it with antibodies. The insulin is either destroyed or coated so that it cannot bind to receptor sites on target cells.

Care has to be taken with the levels of insulin injected because an overdose causes too much glucose to be withdrawn from the blood, reducing the blood glucose concentration below the set point (a condition called **hypoglycaemia**). Glucose is the main fuel for brain cells, and a lack of glucose can lead rapidly to unconsciousness and, in extreme circumstances, to coma and even death. This is why a person with diabetes should be given sugar if found unconscious.

All diabetics need to monitor their blood glucose levels. Easy-to-use biosensors and dipsticks are available for this. People with diabetes also need to manage their diet and levels of exercise very carefully. They should avoid eating too much sugar or going for long periods without food.

Treating type II diabetes

Most type II diabetics can control their blood glucose concentration by carefully regulating their diet and exercise. These people tend to become less responsive to insulin as they age. This can be offset if necessary by injecting insulin.

Type II diabetes is less common among communities that have low-fat and high-fibre diets, and that eat complex carbohydrates that are not quickly broken down to glucose in the body. Type II diabetes is linked with high-fat diets and obesity.

As people live longer, type II diabetes is becoming more common and the demand for insulin is increasing annually. The ability to produce human insulin by genetic engineering allows us to treat the millions of diabetic people worldwide.

Check your understanding

1 What is the main cause of type I diabetes?

2 What is the main treatment for type I diabetes?

3 What type of diet is most closely associated with the development of type II diabetes?

- - - - - - - - - - - - - - -

4 The blood glucose pool is the total amount of glucose in the blood at any one time. Although the blood glucose concentration is kept remarkably constant by homeostatic devices, it is important to realize that glucose is always being added to and removed from the blood glucose pool. Suggest why some people become hypoglycaemic after eating a large sugar-rich meal.

By the end of this spread you should be able to

- recall the principles of feedback control
- distinguish between negative feedback and positive feedback
- identify the main phases of the mammalian oestrus cycle

Prior knowledge

- homeostasis (9.01)
- temperature control (9.02–9.03)

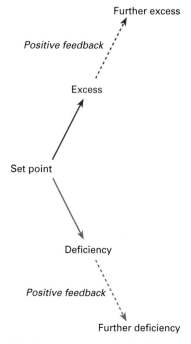

Positive feedback results in deviations from the norm being amplified.

Negative feedback

Homeostatic control systems, such as those that regulate body core temperature in mammals, usually involve **negative feedback** – deviations from the **norm** or **set point** trigger responses that restore the system to its original level. And, as in temperature regulation, there are often separate mechanisms controlling deviations in different directions. Such a **dual system** provides a greater degree of control and means that the body is not over-reliant on one set of responses.

Positive feedback

Sometimes a homeostatic mechanism breaks down and negative feedback does not occur. Deviations from the norm are not corrected. Even worse, the deviations may be made larger. This results in a process called **positive feedback** in which a small change in output causes a further change in the same direction, as shown in the diagram (see left).

Positive feedback is usually harmful because it tends to produce unstable conditions. For example, when the negative feedback mechanism for mammalian temperature regulation breaks down, a rise in body temperature can spiral upwards and threaten death. However, in certain circumstances positive feedback can be useful. It is, for example, an integral part of the control mechanisms of the mammalian oestrus cycle.

The oestrus cycle

Female mammals go through a cycle called the **oestrus cycle** which includes a period of heightened sexual activity. Oestrus refers to the time in the cycle when she becomes sexually receptive and attractive to males. It occurs just before ovulation, when the female is most fertile.

Probably the most familiar oestrus cycle is that of humans who, like most of their primate relatives, are exceptional among mammals: they are sexually receptive throughout the year, and they menstruate.

The uterus lining in all mammals undergoes a similar pattern of thickening during a reproductive cycle. However, if fertilization does not occur, the uterine lining of primates breaks down and is discharged with blood through the vagina, whereas the uterine lining of non-menstruating mammals is reabsorbed and there is no extensive bleeding. The discharge of blood and uterine lining is called **menstruation**.

The reproductive cycle of non-primate mammals is known as the oestrus cycle because oestrus is the most prominent event in it. It lasts from 5 to 60 days, depending on the species. The primate reproductive cycle is also called the **menstrual cycle** because of the prominence of menstruation.

In human females, the menstrual cycle lasts approximately 28 days. A regular sequence of changes is controlled by the interaction of several hormones, the chief of which are **follicle-stimulating hormone (FSH)**, **luteinizing hormone (LH)**, **progesterone**, and **oestrogen**. The cycle has three main phases:

- the **follicular phase**, during which a **follicle** (ball of cells) containing an oocyte (egg cell) develops
- the **ovulatory phase**, during which **ovulation**, the release of an egg cell into the uterus, occurs

- the **luteal phase**, during which an ovarian follicle forms a yellow body called the **corpus luteum**

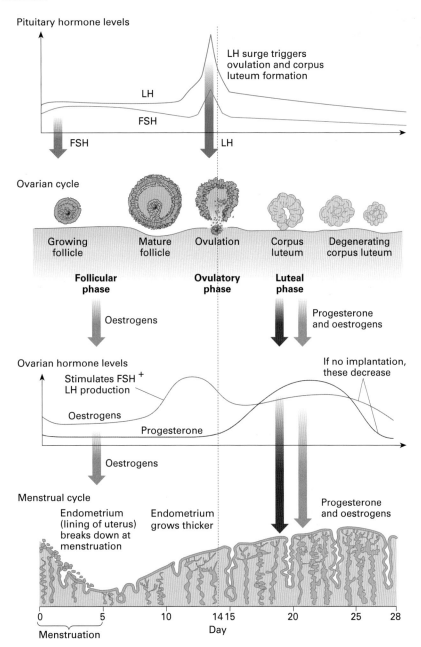

The human menstrual cycle. Note that by convention, day 0 of the menstrual cycle coincides with the start of menstruation.

Check your understanding

1 What is a follicle and what does an ovarian follicle contain?
2 According to the diagram of the human menstrual cycle, on which day does oestrus occur?
3 From what structure does a corpus luteum develop?

- -

4 How does the oestrus cycle of non-primate mammals differ from that of humans?

OBJECTIVES

By the end of this spread you should be able to

- understand how the mammalian oestrus cycle is controlled by the interaction of positive and negative feedback mechanisms involving FSH, LH, progesterone, and oestrogen

- interpret graphs showing the blood concentrations of FSH, LH, progesterone, and oestrogen during a given oestrus cycle

Prior knowledge

- homeostasis (9.01)
- feedback and the oestrus cycle I (9.07)

Hormonal regulation of the oestrus cycle: follicular phase

The menstrual cycle begins when blood is first discharged from the uterus. From day 1 to about day 5, the uterine lining continues to break down and is discharged along with varying amounts of blood. A day before menstruation occurs, the **hypothalamus** produces **gonadotrophin-releasing hormone (GnRH)**. GnRH triggers the secretion of **follicle-stimulating hormone (FSH)** from the anterior **pituitary gland**.

FSH triggers the development of one or more **follicles** in the ovary. This is a ball of cells in the ovary containing an **oocyte** (egg cell). As an ovarian follicle grows in size, it secretes increasing amounts of **oestrogens** (a group of steroid hormones).

Oestrogens stimulate the repair and growth of the uterine lining, and the growth of milk-producing tissue in the mammary glands. During the follicular phase, oestrogens are at low concentrations. They inhibit further production of FSH, so that usually only one follicle matures at a time. This is an example of **negative feedback**: FSH stimulates the production of oestrogens, and oestrogens inhibit the production of FSH.

Oestrogens also stimulate the anterior pituitary gland to secrete **luteinizing hormone (LH)**. Most of the LH is stored in the anterior pituitary gland during the follicular phase, but some LH is released into the blood and stimulates the mature ovarian follicle to produce another hormone, **progesterone**.

Hormonal regulation of the oestrus cycle: ovulatory phase

At about day 13 to day 15 in humans, the level of oestrogens increases rapidly. A slow rise in oestrogen levels during the follicular stage inhibits GnRH production; however, at a critical high level, oestrogen stimulates the hypothalamus to secrete more GnRH. GnRH secretion is also enhanced by a temporary increase in FSH and a sudden release of LH from the pituitary gland. This massive outpouring of LH is called the **ovulatory surge**. The surge of LH and FSH brings about **ovulation**, the release of a secondary oocyte from an ovarian follicle. These effects are a result of positive feedback.

Immediately after ovulation, a woman is fertile and can conceive if she has sexual intercourse, or if sperm are already present in her oviducts.

Hormonal regulation of the oestrus cycle: luteal phase

After ovulation, the remains of the ovarian follicle form the **corpus luteum**. This yellow glandular tissue secretes large amounts of progesterone and smaller amounts of oestrogens. These hormones stimulate further development of the mammary glands and uterus in anticipation of pregnancy.

If the oocyte is not fertilized within about 36 hours of being shed into the oviduct, it dies. Following this, the corpus luteum gets smaller with the result that progesterone and oestrogen secretion is reduced. At day 28, about 14 days after ovulation, lack of progesterone brings about another menstruation and the cycle starts again.

If the oocyte is fertilized, the **zygote** is propelled towards the uterus. During its journey it forms an embryo which becomes implanted in the uterine lining. A **placenta** forms between the embryo and mother; this produces **human chorionic gonadotrophin (HCG)** which stops the degeneration of the corpus luteum, and menstruation is suspended.

In humans, menstruation occurs in females from puberty to **menopause**, after which there are no more fertile follicles so follicular development and ovulation cease.

Chemical contraception and fertility drugs

A knowledge of how hormones regulate the oestrus cycle has been used to control fertility in humans and other mammals. **Chemical contraception** is the use of chemicals which act against conception (the fertilization of an egg cell by a sperm). It usually takes the form of birth control pills which contain combinations of synthetic oestrogen-like and progesterone-like substances. Taking these chemicals stops the secretion of GnRH, FSH, and LH so that neither follicle development (an oestrogen effect) nor ovulation (a progesterone effect) takes place.

Fertility drugs are synthetic chemicals that stimulate ovulation. Some incorporate FSH which stimulates the development of ovarian follicles. Others use chemicals that inhibit the natural production of oestrogens. Lack of oestrogen (the normal FSH inhibitor) results in more FSH being released; this stimulates follicular development.

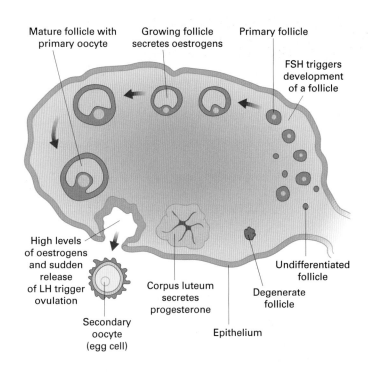

A vertical section through an ovary showing the sequence of events that occurs in a mammalian ovary during one oestrus cycle

Hormone	Site of secretion	Target site	Effects
follicle-stimulating hormone (FSH)	anterior pituitary	ovaries	Stimulates growth of ovarian follicle Stimulates the ovary to produce oestrogen
luteinizing hormone (LH)	anterior pituitary	ovaries	Stimulates growth of ovarian follicle and promotes ovulation Promotes development of corpus luteum
oestrogens	ovarian follicle	anterior pituitary	At low concentrations, inhibit release of LH by negative feedback At a critical high concentration, stimulate release of LH and FSH by positive feedback
		uterine wall	Promote repair and growth of uterine lining
progesterone	corpus luteum	anterior pituitary	At high concentrations, inhibits FSH secretion A sharp decrease along with a decrease in oestrogens triggers menstruation

The main hormones regulating the oestrus cycle

Check your understanding

1 With reference to the changes in hormone levels shown in the graph on spread 9.07, explain how ovulation is triggered.

- - - - - - - - - - - - - - - - -

2 There are alarming reports of high levels of oestrogens (or chemicals with very similar effects) in lakes and rivers used as sources of drinking water. These chemicals are thought to cause male alligators to develop tiny, useless penises, and fish to become sterile. Why do you think levels of oestrogens have increased in drinking water? Suggest possible effects of these high levels of oestrogens on humans.

This spread provides a summary of key points covered in *AS Biology*.

DNA, mRNA, and the genetic code

DNA (deoxyribonucleic acid) and mRNA (messenger ribonucleic acid) are **nucleic acids** consisting of long chains of **nucleotides (polynucleotides)**. The nucleotides are linked together by **phosphodiester bonds** formed as a result of **condensation reactions**.

DNA contains the genetic information inherited by offspring from parents. This information is transcribed into mRNA and used to synthesize polypeptides which make the proteins and enzymes that ultimately determine every inherited physical and behavioural characteristic of an organism.

The polynucleotide chains of DNA and mRNA have four types of nucleotide. DNA has guanine, cytosine, adenine, and thymine, but in mRNA uracil replaces thymine. Other differences between DNA and mRNA are that DNA consists of a double helix (two chains of polynucleotide interconnected by hydrogen bonds), while mRNA consists of only a single strand, and mRNA contains the sugar ribose instead of deoxyribose.

A gene that carries the genetic information for a polypeptide occurs in DNA as a sequence of bases. The information is transcribed into mRNA by complementary base pairing on a DNA template in the nucleus. The transcribed information takes the form of a code. The three most important features of this **genetic code** are that it is:

- **triplet**: each of the 20 amino acids used to make polypeptides is represented by three bases called a codon

- **non-overlapping**: each base is part of only one triplet and is therefore involved in specifying only one amino acid

- **degenerate**: there are more codons than amino acids, therefore some amino acids have more than one codon

After the mRNA is made, and in eukaryotes **introns** have been removed, the single strand moves out into the cytoplasm where the information encoded on the mRNA is translated into a polypeptide. Translation takes place on ribosomes and involves another nucleic acid called tRNA or transfer RNA.

tRNA

tRNA molecules transport specific amino acids to **ribosomes** during polypeptide synthesis.

There are about 20 groups of tRNA. Each is specific for one kind of amino acid (for example, tRNA^{cys}, tRNA^{his}) and acts as an adapter – one end of the molecule combines with its particular amino acid and the other links with the mRNA codon specifying that amino acid.

All tRNAs share the same basic structure:

- each consists of a single polynucleotide strand of RNA, about 80 nucleotides long; the bases are folded to form a clover-leaf arrangement held in place by hydrogen bonds between complementary bases

- one end of tRNA acts as an attachment site for a specific amino acid

Attachment site for phenylalanine

Hydrogen bonds between complementary bases

Site that binds to ribosome

Unpaired bases

Site of association with mRNA

Anticodon for phenylalanine

Schematic diagram of tRNA for phenylalanine. Each type of tRNA has a similar structure.

- a region of each tRNA called the **anticodon** contains three bases which are complementary to the codon for the amino acid it carries

To carry out its function, tRNA must first attach an amino acid to itself and then transfer that amino acid to a ribosome for incorporation into a polypeptide chain. An amino acid is attached to its particular tRNA by a specific enzyme. The energy required for this **endergonic** process is obtained from the hydrolysis of ATP.

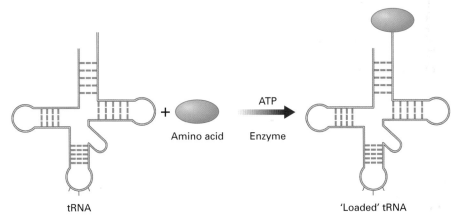

Loading of tRNA with its specific amino acid requires a specific enzyme and ATP.

Check your understanding

1 Explain the meaning of the genetic code being universal.

2 For each of the following codons in mRNA, give the anticodon in tRNA and the base sequence in DNA.

   a AUG   **b** UUU

   c GGU   **d** GGC

   e GCU

3 What does each nucleotide in DNA, mRNA, and tRNA have in common?

- -

4 The genetic code takes the form of base triplets in mRNA. Most base triplets code for amino acids. What do the other base triplets code for?

The nucleic acids compared

Feature	DNA	mRNA	tRNA
chemical make-up	nucleic acid	nucleic acid	nucleic acid
type of chain	double polynucleotide	single polynucleotide	single polynucleotide
origin	replicated from parental DNA	transcribed from DNA templates	transcribed from DNA templates
number of bases	variable number, in the many thousands	variable number, in the hundreds	between 75 and 90
ratio of bases	adenine:thymine = 1 cytosine:guanine = 1	variable	variable
pentose sugar	deoxyribose	ribose	ribose
bases	adenine, guanine, cytosine, and thymine	adenine, guanine, cytosine, and uracil	mostly adenine, guanine, cytosine, and uracil
shape	double helix	singe helix	single strand with 'clover leaf' configuration
location	mainly in the nucleus	made in the nucleus; found throughout the cell	made in the nucleus; found mainly in the cytoplasm
function	store of inherited information	carries genetic information from the nucleus to ribosomes to make particular polypeptides	picks up and carries specific amino acids to ribosomes for polypeptide synthesis

Table summarizing the structure, composition, and function of nucleic acids

OBJECTIVES

By the end of this spread you should be able to

- describe how mRNA is produced by transcription from DNA

- explain the role of RNA polymerase in transcription

- describe how pre-mRNA is spliced to form mRNA in eukaryotic cells

Prior knowledge

- DNA structure and function (*AS Biology for AQA* spreads 11.01–11.02)

- introns and exons (*AS Biology for AQA* spread 11.05)

The blueprint for protein synthesis

DNA carries the inherited instructions (the 'blueprint') for making polypeptides. DNA, however, needs to be kept intact in the nucleus as the permanent store of genetic information. Therefore it is not used directly for protein synthesis. Instead, the information encoded for a particular polypeptide is copied from DNA to messenger RNA (mRNA). The mRNA moves to ribosomes in the cytoplasm where the information is used to make the polypeptide. The copying of information from a DNA molecule to an RNA molecule, whether mRNA or tRNA, is called **transcription**. The conversion of the information in mRNA to make a polypeptide is called **translation**.

The section of DNA that holds the information for one polypeptide chain is called a **cistron** or **gene**. Before transcription can take place, the double helix in the cistron has to unwind and the two polynucleotide chains have to separate ('unzip') in order to expose the nucleotide bases. This is done with the help of an enzyme. Only one of the strands (the transcribing DNA strand) is used as a **template** for the synthesis of mRNA.

Transcribing DNA into RNA

The basic mechanics of transcription is similar in both prokaryotes and eukaryotes:

- The two strands that make up the double helix of DNA are prized apart by an enzyme called **RNA polymerase**. Hydrogen bonds are broken, exposing the bases in the transcribing DNA strand.

- The polymerase attaches to the strand at a particular base sequence, the **promoter**, initiating transcription.

- During transcription, the polymerase moves along the cistron in the 5'–3' direction. It passes over the nucleotides in the transcribing DNA strand one at a time and builds up mRNA by adding complementary nucleotides (U to A, C to G, and so on) as it goes.

- When the enzyme moves on to another region of the transcribing DNA strand, the double helix of the DNA reforms behind it (the 'zip' closes).

- On reaching a special 'stop' sequence called a **terminator**, the enzyme detaches and the mRNA molecule peels away from the DNA.

Transcription: a two-stage process in eukaryotes

In eukaryotes, the nuclear envelope isolates DNA from the ribosomes. This means that transcription and translation have to take place at different times and in different locations. Transcription occurs in the nucleus and translation in the cytoplasm.

The mRNA that leaves a nucleus is much shorter than the length of the DNA that is initially transcribed. For example, the DNA base sequence for a gene that determines a polypeptide chain of 400 amino acids is about 8000 nucleotides long, while the mRNA that leaves the nucleus is only about 1200 nucleotides long. This length difference is because eukaryotic genes contain nucleotide base sequences called **introns** that do not code for polypeptides, as well as base sequences called **exons** that are expressed.

The eukaryotic RNA initially transcribed from DNA is called precursor mRNA or **pre-mRNA**. It includes both exons and introns. The introns are interspersed between the coding segments. To make mRNA, the introns are removed and the exons are spliced together by **spliceosomes**, intracellular structures composed of RNA and protein molecules. Spliceosomes are almost as large as ribosomes.

In the process called **RNA splicing**, a spliceosome interacts with the ends of an intron, cutting it at specific points and then immediately joining the exposed ends of adjacent exons. It is only after this processing that mRNA leaves the nucleus, moves onto ribosomes in the cytoplasm, and is used in translation.

In addition to RNA splicing, before pre-mRNA is converted into mRNA, its ends are modified. The 5' end, made first during transcription, is capped with a modified guanine nucleotide. This **5' cap** helps to protect the mRNA from hydrolysis, and signals the point of attachment when mRNA reaches a ribosome. The 3' end is modified by having 50 to 250 adenine nucleotides incorporated into it. This **poly(A) tail** inhibits degradation, facilitates movement of mRNA into the cytoplasm, and helps ribosomes attach to the mRNA.

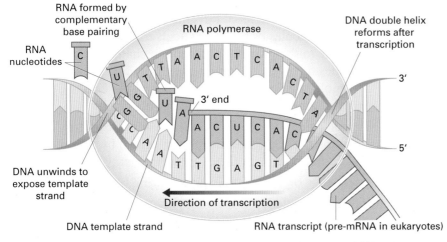

Transcription of DNA to produce a complementary strand of RNA. The RNA polymerase moves along a gene in the 5'–3' direction, assembling an RNA molecule complementary to the template strand of DNA. In prokaryotes the RNA becomes mRNA and is used immediately to direct the process of translation. In most eukaryotes, the initially transcribed RNA is called pre-mRNA and it is processed before leaving the nucleus as mRNA.

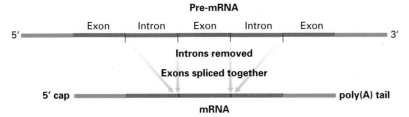

Modification of pre-mRNA to make mRNA

Transcription: a one-stage process in prokaryotes

Prokaryotes have no introns in their DNA and, because they lack a nucleus, the DNA is not isolated from ribosomes. This enables transcription and translation to be coupled. In prokaryotes, ribosomes attach to the leading edge of mRNA as transcription takes place.

Check your understanding

1 Explain why transcription in prokaryotes is a single-stage process while transcription in eukaryotes takes place in two stages.

2 In which direction does RNA polymerase travel when it moves along the DNA template strand, and what happens as each DNA codon is exposed by the enzyme?

3 What are spliceosomes and what is their function?

- - - - - - - - - - - - - - - - -

4 There is much speculation concerning the possible function of introns and RNA splicing. One suggestion is that introns may have a regulatory role, controlling gene activity in some way. For example, the introns may regulate the activity of split genes which can code for more than one kind of polypeptide. Suggest how they might do this.

181

Prior knowledge

- cell ultrastructure (*AS Biology for AQA* spreads 4.05–4.06)
- DNA structure and function (*AS Biology for AQA* spreads 11.01–11.02)
- transcription (10.02)

Even after an organism is fully grown, protein synthesis is essential for survival. Proteins such as digestive enzymes, haemoglobin, and collagen are constantly being broken down and resynthesized. Under normal conditions, the average cell is probably synthesizing several thousand new protein molecules every minute. Proteins consist of one or more polypeptide chains. **Translation** is the process by which information encoded within mRNA is used to make a specific polypeptide chain. It is a complex process which takes place in ribosomes in the cytoplasm.

Ribosomes

Ribsomes are small intracellular organelles. Each consists of two subunits made up of proteins and ribosomal RNA (rRNA):

- the small subunit has a binding site for mRNA
- the large subunit has binding sites for transfer RNA (tRNA)

The process of translation

mRNA carries information in the form of codons which dictate which amino acids are to be used to make a polypeptide. A ribosome moves along a strand of mRNA in the 5'–3' direction and the codons are 'read' sequentially. tRNA molecules bring amino acids to the ribosome. These are added one by one to the growing polypeptide chain. The ribosome allows two molecules of tRNA to combine with the mRNA at any one time. One tRNA molecule holds the growing polypeptide chain; the other carries the next amino acid to be added to the chain.

Before a tRNA molecule moves to a ribosome, a specific enzyme makes sure that it is carrying the correct amino acid specified by its anticodon. The attachment of amino acids to tRNA requires energy from ATP.

The following sequence of events takes place during translation.

- A new amino acid is added to a growing polypeptide chain when the ribosome has reached a part of the mRNA strand containing its specific codon.
- The tRNA molecule with the complementary anticodon and carrying the amino acid attaches onto the mRNA.

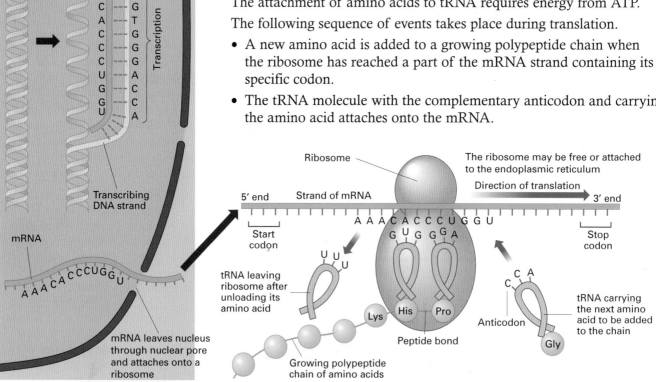

The process of translation: the building up of a polypeptide chain according to the code specified on DNA in the nucleus

- An enzyme (peptidyl transferase) catalyses the formation of a peptide bond between the amino acid on the tRNA and the amino acid at the end of the growing polypeptide chain.
- The tRNA molecule that was holding the amino acid is released from the ribosome, and is free to carry another amino acid molecule.
- The ribosome moves one codon further along the mRNA strand, exposing the next codon so that another amino acid can be added to the chain.

Complementary base pairing between anticodons on tRNA and codons on mRNA ensures that the correct sequence of amino acids is built up in the polypeptide chain, according to the information on the mRNA molecule.

Starting and stopping

Polypeptide synthesis is usually initiated by the codon AUG. This is also the codon for methionine which is therefore the first amino acid in the chain. If the methionine is not needed as a constituent of the final polypeptide chain, it is removed when synthesis is completed. Synthesis ends when the ribosome reaches a 'stop' codon (UAA, UGA, or UAG). The mRNA, ribosome, and tRNA molecules separate, and the polypeptide chain is released.

Processing the polypeptide

On leaving the ribosome, the polypeptide chain is processed according to the final destination of the protein. This may include folding to form the secondary and tertiary structure and, in some proteins, adding chains to form a quaternary structure.

- Proteins used inside the cell, such as haemoglobin, are usually made on free ribosomes and released into the cytoplasm.
- Polypeptides of proteins that are to be exported from the cell, such as digestive enzymes, are made on the rough endoplasmic reticulum. As the polypeptide is made, it is threaded through pores in the endoplasmic reticulum and it builds up in the cisternae. It is transported in vesicles to the Golgi apparatus, where it is modified and packaged. Golgi vesicles then transport the protein to the cell surface membrane from which it is secreted by exocytosis.
- Polypeptides that form membrane proteins follow the same route, but they remain on the cell surface membrane rather than being exported.

Check your understanding

1 What type of bond attaches a new amino acid to a polypeptide chain?

2 By what process does a tRNA molecule attach onto mRNA during translation?

3 What is a polysome?

- -

4 Ribosomes are made of ribosomal RNA. Where is this made, and by what process?

Polysomes: speeding up the process

A single polypeptide chain grows at a rate of about 15 amino acids per second. This is not fast enough for cells that have to produce a great deal of protein quickly. Protein synthesis in these cells is accelerated by the formation of polysomes. A **polysome** consists of 5–50 ribosomes on the same mRNA strand. Each ribosome translates a different part of the mRNA at the same time. When a ribosome reaches the end of the strand, it releases its polypeptide chain and returns to the beginning. Ribosomes can carry out polypeptide synthesis repeatedly on the same mRNA strand, so that a large number of polypeptide chains can be synthesized from one mRNA molecule.

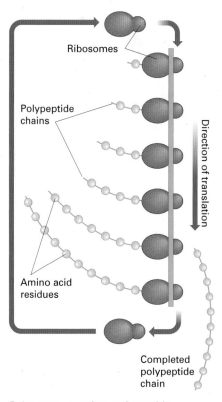

Polysomes speed up polypeptide synthesis.

OBJECTIVES

To answer the questions in this spread you should be able to

- show understanding of how the base sequence of nucleic acids relates to the amino acid sequence of polypeptides

- interpret data from experimental work investigating the role of nucleic acids

- recall and apply relevant information from the *AS Biology* Specification

- use information from the course to answer questions related to unfamiliar topics

Prior knowledge

- amino acids and proteins (*AS Biology for AQA* spreads 2.06–2.07)

- DNA structure and function (*AS Biology for AQA* spreads 11.01–11.08)

- the genetic code (*AS Biology for AQA* spread 11.06)

- nucleic acids (10.01)

- transcription (10.02)

- translation (10.03)

A: The frame-shift experiment

After deducing the structure of DNA with James Watson, Francis Crick and his co-workers studied the genetic code in DNA. They used enzymes to add or delete nucleotide bases in the DNA of a virus that infects bacteria. They found that when one or two bases were added or deleted, the viruses were unable to infect the bacteria. But when three bases were added or deleted, infection occurred. They concluded that adding or removing one or two bases caused a **frame shift** which made the gene dysfunctional, but adding or removing three bases only partially affected the gene. They concluded from this that the genetic code is a **triplet code**.

Questions

1 If a section of DNA has the following base sequence: 5′ TAGGCTTGATCG 3′ what will be the corresponding mRNA base sequence?

2 What are the DNA base triplets in the section of DNA given above?

3 What are the mRNA codons that are transcribed from the DNA?

4 Assuming all the codons are for amino acids, how many amino acids would be in the polypeptide chain translated from mRNA?

5 If guanine is added to the 5′ end of DNA, show what happens to

  a the DNA base triplets   b the mRNA codon sequence.

6 What will happen to the composition of the polypeptide chain after the addition of guanine to the base sequence?

7 What would happen to the amino acid composition of the polypeptide chain if three new bases were added to the DNA base sequence?

8 The results of the frame-shift experiment also showed that the genetic code is non-overlapping. What does this mean in terms of

  a the number of amino acids specified by each base triplet in DNA

  b the number of triplets each base is part of?

9 Explain why replacing one base with another in a single DNA base triplet may have no effect on the amino acid composition of the polypeptide translated from mRNA.

10 Francis Crick and his co-workers reasoned that if the genetic code was a triplet code, adding or deleting one or two nucleotides would make the genetic information in DNA meaningless. A sentence consisting of three-letter words can be used as a simple analogy of the triplet code for an amino acid. The sentence is equivalent to the information in DNA which determines the sequence of amino acids in a polypeptide chain. Devise a sentence of different three-letter words and show what happens to the message when one, two, or three letters in a word are deleted.

B: AZT and HIV

AZT (azidothymidine) was the first **antiretroviral drug** approved for treatment of **HIV** (**human immunodeficiency virus**). HIV uses the enzyme **reverse transcriptase** to make DNA from its RNA, and AZT inhibits this enzyme.

HIV, like other retroviruses, carries its genetic information as RNA rather than DNA. **Reverse transcription** allows the virus to make a complementary DNA copy of its RNA. The viral double-stranded DNA is then spliced into the host DNA where it acts as a gene. The viral gene may either direct the production of more viral particles or it may remain inactive for many cell generations. The integrated viral DNA (called a **provirus**) is passed on to all the host daughter cells. When the viral DNA becomes active, it can cause many new viruses to be produced which eventually burst out of the host cells.

AZT affinity for HIV transcriptase is 100–300 fold greater than its affinity for human nuclear DNA polymerase, accounting for its selective antiviral activity. However, some of its harmful side-effects, such as damage to cardiac and other muscles, have been linked to its action on mitochondrial DNA.

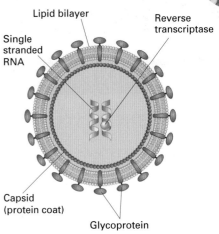

Human immunodeficiency virus (HIV). The virus is cut open to show its two copies of RNA and reverse transcriptase. The protein coat is surrounded by an outer envelope consisting of a lipid blayer (taken from the cell surface membrane of the previous host cell) and protein. The envelope contains glycoproteins which bind the virus to specific receptors on the surface of certain white blood cells called helper T cells.

Questions

1 What is reverse transcription?
2 What happens when a HIV provirus becomes active?
3 Study the figure showing the structure of AZT and a nucleotide containing thymine.
   a What are the three components of the nucleotide?
   b Suggest how AZT might affect the action of reverse transcriptase.
   c What would prevent AZT binding with another nucleotide?
4 Suggest why mitochondrial DNA is more sensitive to AZT than DNA in the human nucleus.

The structure of AZT and a nucleotide containing thymine.

OBJECTIVES

By the end of this spread you should be able to

- recall that gene mutations might arise during replication

- distinguish between deletion and substitution of bases

- appreciate that mutations can arise spontaneously

- explain with reference to the degenerate nature of the genetic code why some mutations result in a different amino acid sequence in the encoded polypeptide while others do not

Prior knowledge

- amino acids and proteins (*AS Biology for AQA* spreads 2.06–2.07)

- DNA structure and function (*AS Biology for AQA* spreads 11.01–11.08)

- mutations (*AS Biology for AQA* spread 18.02)

- nucleic acids (10.01)

- transcription (10.02)

- translation (10.03)

- frame shifts (10.04)

The origin of inherited variation

Mutations are changes in the amount or composition of DNA. They are the ultimate source of all new inherited variation. **Gene mutations** are changes in the nucleotide base sequence in a **cistron**, the section of DNA that constitutes a **gene**. Gene mutations produce different **alleles** of a gene.

Genes determine the **primary structure** of proteins, the sequence of amino acids in each **polypeptide chain**. The primary structure determines how a polypeptide folds up and interacts with other molecules. Therefore a change in the primary structure of a protein may alter its biological properties. The action of a globular protein such as an enzyme depends on its shape. A shape change may bring about a different action which in turn may cause a change in the **phenotype** of the organism, as shown below.

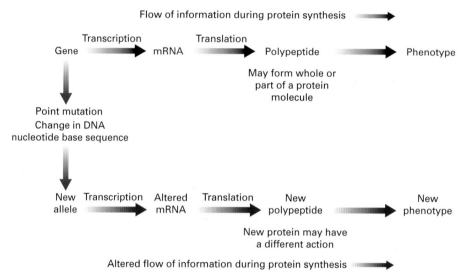

The effects of a point mutation that leads to a change in the amino acid sequence in a polypeptide

How mutations arise

Mutations are random and rare events which might arise spontaneously as a result of errors during the replication of DNA. If the error results in a change of a single nucleotide base, it is called a **point mutation**. If a point mutation occurs in cells that give rise to gametes, or in a gamete itself, it might be inherited by offspring and passed on to successive generations.

Point mutations can be divided into two general types: base-pair deletions or insertions, and base-pair substitutions.

Deletions and insertions

Deletions and insertions are losses or additions of nucleotide base pairs. Unless they occur in multiples of three, they lead to an alteration in all the codons downstream of the deletion or insertion mutation. Such mutations are called **frame-shift mutations**. Unless a frame-shift mutation is near the end of the gene, it will almost certainly produce a non-functional protein.

Substitutions

A substitution is the replacement of one nucleotide pair with another, resulting in the alteration of just one base in one triplet of the DNA template strand. Some substitutions have no effect on the encoded protein because of the **degenerate** nature of the genetic code. That is, there are more codons than amino acids, therefore some amino acids have more than one codon. A change in one base in the coding strand may change one codon into another that is translated into the same amino acid. Such substitutions are sometimes called **silent mutations**. For example, if the mRNA codon UUU is transformed into UCU (by a substitution in the DNA triplet from AAA to AGA), phenylalanine would still be added to the appropriate location in the polypeptide chain.

Other substitutions may result in a change in a single amino acid in a polypeptide chain but have little effect on the action of a protein. The new protein may have almost the same shape as the original protein, or the change may occur in a part of the protein where the exact sequence of amino acids is not important.

However, substitutions can have a significant effect on protein structure and function. One example is sickle cell anaemia.

Sickle cell anaemia

Sickle cell anaemia is an inherited disease caused by a substitution that results in a change in one amino acid in the beta chains of haemoglobin. Haemoglobin is made of four polypeptide chains, two alpha chains and two beta chains, determined by separate genes. The gene responsible for the beta chain has two alleles designated Hb^A (for the normal allele) and Hb^S (for the sickle cell allele). Each person usually has two alleles for a particular gene, so three genotypes are possible:

- Hb^AHb^A
- Hb^AHb^S
- Hb^SHb^S

People with two Hb^S alleles are homozygous for the sickle cell condition and have the full-blown version of sickle cell anaemia. The alteration in the primary structure of their haemoglobin is sufficient to affect its function. At low oxygen partial pressures (for example, in active muscles) a large percentage of red blood cells of people with sickle cell anaemia become sickle shaped and unable to carry oxygen efficiently. The distorted cells tend to form clots and block blood vessels. These effects are connected with the change in the primary structure of the haemoglobin. There is more about sickle cell anaemia in spread 11.07.

False-coloured scanning electron micrograph of red blood cells from a person with sickle cell anaemia. Normal red blood cells (rounded) contrast with elongated sickle-shaped cells (darker). (×1200)

Check your understanding

1 What types of mutation might occur during replication of DNA?

2 Why might deletions or insertions that occur in multiples of three have no effect on the phenotype of an organism?

3 Point mutations may lead to changes in codons that result in an alteration in the amino acid sequence in a polypeptide chain. What other effects might point mutations have?

OBJECTIVES

By the end of this spread you should be able to

- recall that the rate of cell division is controlled by proto-oncogenes that stimulate cell division, and tumour suppressor genes that slow cell division

- explain how a mutated proto-oncogene, called an oncogene, stimulates cells to divide too quickly

- recall that tumour suppressor genes might mutate spontaneously during replication

Prior knowledge

- what is cancer? (*AS Biology for AQA* spread 1.04)

- smoking and lung cancer (*AS Biology for AQA* spread 1.05)

- the cell cycle (*AS Biology for AQA* spread 14.03)

- mitosis and cancer (*AS Biology for AQA* spread 14.05)

- gene mutations (10.05)

Control of cell division

In normal, healthy body tissue, cells divide by **mitosis** so that growth and repair can take place. The regulatory mechanism involves specific genes that carefully control the process. In particular, normal cellular genes, called **proto-oncogenes**, code for proteins that stimulate normal cell growth and division. When growth or repair is completed, other genes called **tumour suppressor genes** inhibit cell division. These genes code for proteins that prevent uncontrolled cell division.

The proteins encoded by tumour suppressor genes may

- repair damaged DNA before it can be replicated

- control cell adhesion and ensure that cells are anchored in their proper place

- inhibit cell division

In a healthy cell, the activities of these proto-oncogenes and tumour suppressor genes are in balance. Problems arise when the genes mutate or other control mechanisms break down so that cells divide uncontrollably.

Uncontrolled cell division: cancer

Proto-oncogenes can mutate into cancer-causing genes called **oncogenes** (*onkos* means tumour). The presence of oncogenes leads to either

- an increase in the amount of proteins that stimulate cell growth and division

or

- an increase in the activity of each growth-stimulating protein molecule – the mutation may lead to an altered protein that is hyperactive or that resists being broken down when it is no longer needed

Either of these effects can lead to abnormal stimulation of the cell cycle and result in the development of a malignant tumour.

Any mutation that decreases the activity of tumour suppressor genes can also lead to malignancy. Such a mutation may allow the rate of cell division to increase through the lack of suppression, or it may lead to cells losing their normal anchorage and invading other parts of the body where they might develop into a malignant tumour.

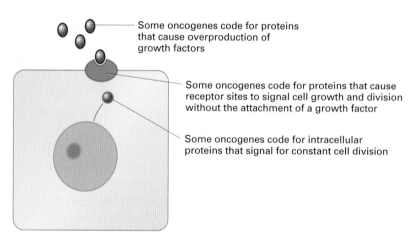

Oncogenes code for proteins that stimulate cell division in several ways.

Causes of mutations

The mutation of either a proto-oncogene into an oncogene or a tumour suppressor gene into one that no longer suppresses cell division may be spontaneous, or it may be triggered by specific cancer-causing agents called **carcinogens**. Spontaneous mutations occur as random events during DNA replication. They may lead to a **point mutation** that alters the nucleotide base sequence in the gene and the protein encoded by it.

Factors that trigger mutations include: chemical carcinogens, such as those in tobacco smoke; physical agents, such as X-rays; and certain viruses. In fact, oncogenes were discovered from research into tumours induced by complex viruses called **retroviruses**. HIV is a retrovirus. Retrovirus capsules contain RNA and **reverse transcriptase**, an enzyme that transcribes DNA from the template formed by the RNA. The resulting viral DNA, which may contain an oncogene, is incorporated into a host's DNA as a **provirus**. The host's **RNA polymerase** transcribes the viral DNA into RNA molecules. This RNA may function as mRNA to make viral proteins, or it may form the RNA contained within the capsules of new virus particles released from the cell.

Most mutated cells are either destroyed by the body's immune system or die, causing no harm. However, a single mutated cell may divide to form a clone of identical cells. Eventually a mass of abnormal cells called a **tumour** is formed. **Benign tumours** such as warts do not spread from their point of origin. Tumours that can spread through the body are called **malignant tumours**. Malignant tumour cells can be carried by the bloodstream or lymphatic system to invade other tissues, causing secondary cancers.

There are many different oncogenes and tumour suppressor genes, some of which act only in specific types of cells at certain times. Also, mutated cells are usually dealt with by the body's immune system before they can proliferate. For example, one function of white blood cells called natural killer cells is to recognize and destroy cancerous cells. Therefore the development of a malignancy is not usually the result of a mutation of a single proto-oncogene or a single tumour suppressor gene. A complex series of events must occur before a normal cell becomes malignant.

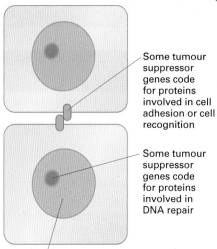

Some tumour suppressor genes code for proteins involved in cell adhesion or cell recognition

Some tumour suppressor genes code for proteins involved in DNA repair

Some tumour suppressor genes code for proteins that stop the cell cycle at G_1, thereby inhibiting cell division

Tumour suppressor genes code for proteins that inhibit cell division and the development of a cancer.

Check your understanding

1 What happens when a proto-oncogene mutates into an oncogene?

2 If the proto-oncogene of a cell mutates into an oncogene, causing the development of a tumour, how might the inactivation of a tumour suppressor gene contribute to the development of a tumour at a secondary site (a secondary cancer)?

- -

3 An inherited type of breast cancer involves a tumour suppressor gene called BRCA1 that has two alleles. About 10 per cent of women who inherit one mutated allele of the gene have a 60 per cent chance of having breast cancer by the age of 50, and an 80 per cent chance of developing it by age 70. Explain why the comparable figures for women who inherit two normal alleles of the gene are much lower (2 per cent and 7 per cent respectively). Why does the risk of developing the breast cancer in both groups of women increase with age?

By the end of this spread you should be able to

- define gene expression

- understand that most of a cell's DNA is not translated

- recall that totipotent cells can mature into any body cell but that during development they become specialized

- recall that in mature plants many cells remain totipotent, and can develop *in vitro* into whole plants

Prior knowledge

- cell specialization (*AS Biology for AQA* spread 14.06)

- plant tissues (*AS Biology for AQA* spread 14.08)

Gene expression

It is one of the most remarkable facts of life that the DNA in each cell of a plant or animal contains all the information required to make a whole organism. When the information in a gene is used to make a functional polypeptide chain by transcription and translation, **gene expression** is said to have taken place. During the life of any particular cell, only a very small proportion of the cell's genes are expressed: most of a cell's DNA is not translated into polypeptides and therefore has no effect on the phenotype of the organism.

Totipotency

The **zygote** (the fertilized egg) of a diploid, sexually reproducing organism has the potential to grow and develop by mitosis and cellular specialization into an adult organism. Cells that retain this potential to form all parts of a mature organism are called **totipotent**. During development, totipotent cells mature and become specialized. A specialized cell translates only the parts of its DNA that are needed to carry out its specialized functions. Most of the DNA is not used.

Totipotency in plants

In specialized cells, DNA that is not needed becomes inactivated. However, in many plant species, this inactivation is reversible and whole new individuals can be grown from differentiated, specialized somatic cells. This was first demonstrated in carrot plants in the 1950s when F. C. Seward and co-workers at Cornell University removed several samples of differentiated cells from the root of a carrot plant and transferred them to separate culture media. Each group of cells grew into a normal adult plant, and each adult plant was genetically identical to the parent tissue. An outline of the experiment is shown on the next page. This ability of differentiated carrot cells to develop *in vitro* into whole plants showed that differentiation does not necessarily result in irreversible changes in the DNA. The carrot root cells remain totipotent.

Since the 1950s, totipotency has been demonstrated in differentiated cells of many plant species and this property has been used in a process called tissue culture, micropropagation, or cloning.

Plant tissue culture

Plant **tissue culture** has the potential to grow hundreds or thousands of genetically identical plants (clones) from a single source in a relatively small amount of space, using few resources, and in a relatively short time. Basically, a piece of a plant (such as a stem tip, node, meristem, embryo, or even a seed) is placed in a sterile nutrient medium where it multiplies. Plant species vary in their response to tissue culture: some are easier to culture than others. For some species, attempts to propagate them by tissue culture have been unsuccessful.

The growth medium

The composition of the growth medium depends on what you are trying to produce. If, for example, you wish to grow hard masses of undifferentiated tissue called **calluses**, you will need a different medium than to produce embryos for 'artificial seeds'. Yet another medium is required for growing whole plantlets.

Special media are available for particular species and circumstances. They can be bought premixed in powder form to mix with sterile water, or made up in test tubes ready to use. No matter what medium is used, it is essential that it is sterile.

The culture environment

In addition to the growth medium being sterile, it is also essential that there are no other sources of contamination by fungal or bacterial spores. Equipment has to be sterilized, work surfaces disinfected, and air contaminants minimized. This is especially important when cutting and transferring the **explant** (the part of the plant that is to be cultured). It is best, for example, to use a special transfer chamber with a highly efficient particulate air filter when transferring explants from one container to another.

In addition to a sterile environment, explants will only grow if the pH and light conditions are suitable. The optimum pH for strawberry explants is about 5.7 and they grow well under quite strong artificial fluorescent light at room temperature (about 20°C). The light may be continuous, but 16 hours of light and 8 hours of darkness is standard.

Transplanting

After about 2–4 weeks of *in vitro* cultivation, plantlets begin to root and are ready to be transplanted to a suitable artificial soil mix in a seedling tray. Initially, because tissue cultured plantlets are more delicate than seedlings, they need to be covered with clear plastic and placed on a lighted shelf or in a shaded greenhouse. (The stomata of tissue cultured plantlets appear to be less responsive to humidity and temperature changes and remain open for longer than normal when conditions are unfavourable.)

After 2–3 weeks the plantlets are uncovered for increasing periods until they slowly adjust to normal humidity and light and they can be left completely uncovered.

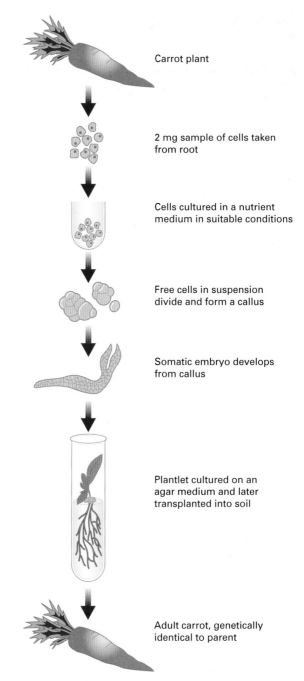

Carrot plant

2 mg sample of cells taken from root

Cells cultured in a nutrient medium in suitable conditions

Free cells in suspension divide and form a callus

Somatic embryo develops from callus

Plantlet cultured on an agar medium and later transplanted into soil

Adult carrot, genetically identical to parent

Cloning carrots

Check your understanding

1 When does gene expression occur?

2 In terms of translation, what happens to most of the DNA in a differentiated cell?

3 What is a totipotent cell?

- -

4 If you clone a strawberry plant, will all the offspring (clones) look alike? Give reasons for your answer.

OBJECTIVES

By the end of this spread you should be able to

- understand that only a few totipotent cells, called stem cells, remain in mature animals

- explain how stem cells can be used to treat some genetic disorders

Prior knowledge

- cells, tissues, and organs (*AS Biology for AQA* spread 14.06)

- stem cells (*AS Biology for AQA* spread 14.07)

- the cell cycle (*AS Biology for AQA* spread 14.03)

Stem cells

Like plant cells, animal cells formed during the early stages of embryonic development are totipotent. These **embryonic stem cells** exist in human embryos for only a few weeks, during which time they can give rise to any type of tissue. When the stem cell divides by mitosis after this initial stage, it produces two daughter cells. One gives rise to differentiated cells which lose their totipotency; the other gives rise to cells called **adult stem cells** that retain some totipotency. Such cells are called **pluripotent**. Adult stem cells normally give rise to only a small range of cell types – for example, those in bone marrow normally produce only blood cells.

Adult stem cells have two important properties:

- they can continually reproduce themselves

- under appropriate conditions, they differentiate into specialized cells of one or more types

Adult stem cells normally have a low level of totipotency, but they can be induced to differentiate into a wider range of cell types if given the appropriate environment. For example, when bone marrow stem cells were transplanted into the brains of mice, the cells proceeded to act like brain stem cells and formed nerve cells. Conversely, brain stem cells transplanted into bone marrow gave rise to blood cells. These experiments indicated that some component of the environment affects what a stem cell does. Subsequent research has shown that specific differentiation factors are involved.

Treating genetic disorders: embryonic stem cells

Theoretically, stem cells can be induced to make any tissue type. Stem cells with normal genes could therefore be used to treat tissues affected by genetic disorders.

Embryonic stem cells, being the most totipotent, are the easiest to induce to form specialized cells. However, genetic differences between the donor and host could cause unforeseen problems in the long term and trigger rejection in the short term.

One possible way of overcoming these problems is to take adult stem cells from the patient, replace the defective allele with a normal allele, implant the genetically manipulated cells into eggs which have had their nuclei removed, and stimulate the eggs to grow into cloned embryos of the patient. The embryo would then be a source of stem cells that could be used to make cells that are genetically the same as those of the patient, except that they will have a normal gene rather than a defective one.

Treating genetic disorders: adult stem cells

Another solution to the problem of incompatibility would be to use adult stem cells from a patient to treat his or her own disorder.

Recent research has shown that **cystic fibrosis** (CF) is one of the disorders that might be treatable using stem cells from bone marrow. CF is caused by the mutation of a gene called **CFTR**. This gene encodes for a substance that channels chloride ions through cell surface membranes and out of the epithelial cells that line the airways. This channeling of chloride ions is essential for keeping airways clear of mucus. People with CF typically have excess mucus in their airways, which becomes an attractive environment for bacteria.

The research showed that bone marrow stem cells can be genetically modified so that the defective gene is replaced by a normal one. Then the stem cells can be coaxed to differentiate into airway epithelial cells. *In vitro*, in epithelial cells made from the genetically modified stem cells, the CFTR genes are expressed in the normal way, and the cells function in the same way as those lining the airways of healthy people.

A similar technique has been used to treat **sickle cell anaemia** in mice. Cells obtained from the tail were converted into **induced pluripotent stem cells (iPSCs)** by inserting special genes into the cells using **retroviruses**. iPSCs have a similar totipotency to embryonic stem cells. The defective sickle cell gene was corrected and the iPSCs were changed to a specialized stem cell that produces various blood cells. These were injected back into the mice they originally came from where they significantly reduced the effects of sickle cell anaemia. However, the treatment had serious side effects, including increasing the risk of cancers. These will have to be overcome before the technique can be used to treat sickle cell anaemia in humans.

Another source of adult stem cells is the patient's own defective tissue. Stem cells have been found in a variety of tissues including the brain and pancreas. If these cells can be located and extracted, they could be genetically modified so that the normal allele replaces the defective allele. Then the cells could be transplanted back into the target tissue. If the factors responsible for inducing them to make the specialized cells can be identified, it might be possible to stimulate them to proliferate and differentiate into normally functioning cells.

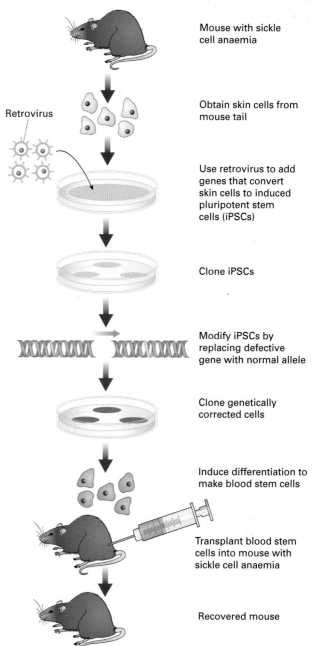

Mouse with sickle cell anaemia

Retrovirus

Obtain skin cells from mouse tail

Use retrovirus to add genes that convert skin cells to induced pluripotent stem cells (iPSCs)

Clone iPSCs

Modify iPSCs by replacing defective gene with normal allele

Clone genetically corrected cells

Induce differentiation to make blood stem cells

Transplant blood stem cells into mouse with sickle cell anaemia

Recovered mouse

Stem cell therapy of a mouse with sickle cell anaemia

Check your understanding

1 Which cells have the greater totipotency: embryonic stem cells or adult stem cells?

2 Explain the significance of research into the use of stem cells to treat cystic fibrosis being limited to *in vitro* investigations.

3 Suggest a possible cause of the cancer-increasing side effects of using iPSCs to treat sickle cell anaemia. Suggest a possible solution.

By the end of this spread you should be able to

- interpret data relating to the tissue culture of plants from samples of totipotent cells

Prior knowledge

- gene expression: totipotent cells (10.07)

African violet plants such as this are commonly produced by tissue culture.

Plant tissue culture of the African violet

Because tissue culture has the potential to produce large numbers of plants that are genetically identical with each other and the parent, it has important commercial and research applications. One of the easiest plants to culture from tissues is the African violet (*Saintpaulia ionantha*).

Plant tissue cultures can be initiated from almost any part of a parent plant that is healthy and free from obvious signs of disease or decay. In the African violet, leaf sections are a common source of tissue containing totipotent cells that are used to start a culture. This tissue is called the explant.

Sections are taken of the leaf, resulting in the extracted cells having the characteristics of **callus** cells. (These appear on cut surfaces of a plant and gradually cover and seal the damaged area.) Callus cells have a smaller vacuole than normal and lack chloroplasts and the photosynthetic pathways generally found in most plant cells.

After extraction, an explant is grown on a sterile nutrient medium, the composition of which controls the development of the cells. By varying the concentration of plant growth substances, particularly auxins and cytokinins, the cells can be kept as an undifferentiated mass, or encouraged to differentiate into plant organs or a whole plant. An undifferentiated mass of cells can be maintained indefinitely as long as it is subcultured regularly into a fresh growth medium.

For each species, the precise conditions required to culture a whole plant from an explant is determined through carefully designed and observed experiments that involve varying the growing conditions.

A tissue culture experiment

In the following experiment, explants from the African violet were cultured in different media in sterile working areas using aseptic techniques. The media were designed to show the effects of auxin, cytokinin, sucrose, and mineral salts on development.

Squares with 1 cm sides were sectioned from the leaves of healthy adult plants. The youngest and oldest leaves were avoided. Before sectioning, any dust was washed off the leaves using distilled, sterilized water. The leaf was then immersed in 70% alcohol for 30 seconds. Explants were extracted using sterilized scissors and forceps and then placed in 10% hypochlorite (bleach) for 5 minutes, after which the bleach was removed by gently washing the explants in sterile, distilled water.

Explants were placed in Petri dishes containing different growth media. All media were mixed with 10 dm^3 of agar and sterilized by autoclaving, then dispensed into 9 cm Petri dishes. The upper epidermis of each explant was pressed gently against the solidified agar surface. The Petri dishes were sealed with a plastic film, incubated at 25°C, and exposed to fluorescent light in a cycle of 16 hours of light/8 hours of darkness.

The tissue cultures were inspected at weekly intervals for 8 weeks after which the observations were recorded. Only cultures free of any obvious signs of bacterial or fungal colonies were included in the results.

Culture number	Mineral salt concentration (g dm^{-3})	Auxin source (mg dm^{-3})	Cytokinin source (mg dm^{-3})	Sucrose (%)	Observations
1	5	0.5	0.5	0.0	Very few shoots developed on the explant No roots and no callus
2	5	0.5	0.5	1.0	Shoots developed on edges of upper surface Some root development
3	5	0.5	0.5	3.0	Explant completely covered with green shoots Roots developing on the lower surface
4	5	0.5	0.5	5.0	Explant completely covered with green shoots Roots developing on the lower surface
5	5	0.5	0.0	3.0	Profuse development of roots and a reduced number of shoots Undifferentiated callus developing at the edges of the explant
6	5	0.5	0.1	3.0	More balanced development of roots and shoots Undifferentiated callus developing at the edges of the explant
7	5	0.5	5.0	3.0	Reduced development of roots Shoots developed around edge of explant Callus initiated
8	5	0	0.5	3.0	Poor shoot development and no roots
9	5	0.1	0.5	3.0	Poor shoot development and no roots Extensive callus generation
10	5	0.5	0.5	3.0	Profuse development of roots over the explant upper and lower surface, and few shoots Some callus
11	0	0.5	0.5	3.0	No sign of regeneration from explant at all
12	0.5	0.5	0.5	3.0	Some root growth but reduced development of shoots
13	1.0	0.5	0.5	3.0	Shoots developed on upper surface of explant No roots or callus

Results of tissue culture investigation on the African violet. In cultures 1–4 the sucrose concentration was varied; in 5–7 the cytokinin concentration was varied; in 8–10 the auxin concentration was varied; and in 11–13 the mineral salt concentration was varied.

Check your understanding

1 What is a totipotent cell?

2 How do callus cells differ from typical plant cells?

3 Suggest one advantage to a grower of being able to maintain a culture as an undifferentiated mass of cells for an extended period.

4 Why is the surface of the explant sterilized before extracting cells to set up a culture?

5 Why were the Petri dishes sealed with a plastic film?

6 Why are cultures with any signs of fungal or bacterial colonies not recorded in the results table?

7 From the observations, what is the composition of the medium or media that produced healthy plantlets? (A healthy plantlet cultured from an explant should have many small lateral buds that lead to many small shoots on the upper surface where the leaf is not in contact with the medium.)

8 What is the effect on the development of plantlets of different concentrations of auxin?

9 What is the effect of low or zero levels of mineral salts on the development of a plantlet?

OBJECTIVES

By the end of this spread you should be able to

- evaluate the use of stem cells in treating human disorders

Prior knowledge

- diabetes mellitus (9.06)
- gene expression: totipotent cells (10.07)
- gene expression: stem cells (10.08)

Using stem cells to treat type I diabetes

Theoretically, stem cells can be induced to make any tissue type and this newly made tissue could be used to treat a human disorder. For example, it should be possible to treat type I diabetes by making healthy insulin-secreting beta cells to replace those destroyed by a patient's own immune system.

Type I diabetes affects about 300 000 people in Britain. The number of British children under the age of five with type I diabetes had increased five fold in the 20 years between 1987 and 2007. Type I diabetics have to regularly inject themselves, some more than four times a day, with the hormone insulin to control their blood sugar levels.

Type II diabetes tends to affect people later in life. Generally, it is linked to lifestyle factors such as obesity. Type II diabetics have beta cells but their body cells do not respond appropriately to the insulin they produce. There are about 2 million type II diabetics in Britain. Most control their blood sugar levels by lifestyle changes (diet and exercise) without the need for daily insulin injections.

Although embryonic stem cells are totipotent, adult stem cells have the advantage that they could be obtained from a person with diabetes to treat his or her own disorder without the problem of genetic incompatibility.

In 2007, a report was published in the *Journal of the American Medical Association* of a three-year clinical trial of such a procedure carried out on a group of 15 people with type I diabetes by a team of US and Brazilian scientists. The subjects were aged between 14 and 31 years. Each was given a powerful drug to suppress their immune system in order to eliminate the white blood cells attacking their beta cells. They were then injected with a chemical that loosened stem cells from the bone marrow. The adult stem cells were then filtered out, collected, and injected back into the patient's blood.

According to the report, 14 of the 15 subjects overcame their dependence on insulin injections for variable periods of time. One volunteer was eliminated from the study because of complications. One was free of insulin dependency for nearly 3 years. The toxicity of the treatment was low, with no subjects dying.

The therapy, known as **autologous haematopoietic stem cell transplantation**, has been shown to have benefits to people with a range of autoimmune diseases such as rheumatoid arthritis and Crohn's disease.

Although stem cells have the potential to treat a wide range of human disorders, there is still a long way to go before they can be widely used to treat patients. Doctors need to be able to produce stem cells that

- make sufficient cells of the desired cell types
- are accepted by and fully integrated into the recipient's tissue
- function well for the whole life of the recipient
- cause no harm to the recipient

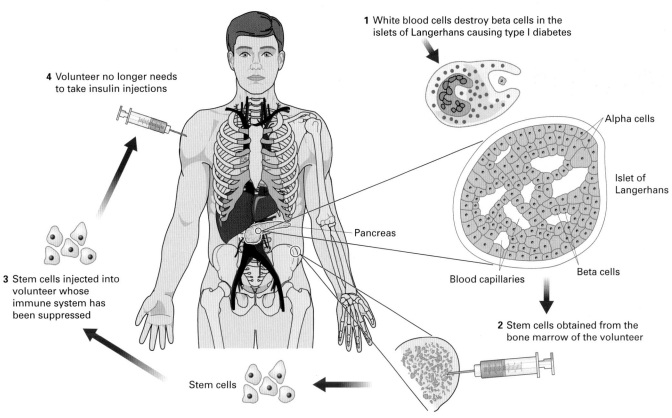

1 White blood cells destroy beta cells in the islets of Langerhans causing type I diabetes

4 Volunteer no longer needs to take insulin injections

Alpha cells

Islet of Langerhans

Pancreas

Beta cells

3 Stem cells injected into volunteer whose immune system has been suppressed

Blood capillaries

2 Stem cells obtained from the bone marrow of the volunteer

Stem cells

Summary of the clinical trials into the use of stem cells to treat type I diabetes

Check your understanding

1 Where are human beta cells normally found?

2 What is the function of the beta cells?

3 What is thought to have happened to beta cells in a person with type I diabetes?

4 Why is stem cell therapy not generally considered for the treatment of type II diabetes?

5 **a** Give one advantage of using adult stem cells rather than embryonic stem cells in stem cell therapy.

  b Give one disadvantage.

6 What did the researcher do to stop white blood cells attacking the beta cells?

7 What might have caused the loss of insulin dependence in 14 of the type I diabetics who were injected with bone marrow stem cells?

8 One of the volunteers went 12 months without needing insulin injections but then relapsed after suffering a viral infection, and started daily injections again. Suggest how a viral infection might have caused the relapse.

9 List the limitations of the research into the effects of bone marrow stem cells on the diabetic volunteers.

10 Do the research results show conclusively that bone marrow stem cells

  a proliferate sufficient beta cells

  b are accepted by and fully integrated into the recipient's tissue

  c function well for the whole life of the recipient

  d cause no harm to recipients?

11 In 2008, an international team of researchers reported that they had located stem cells in the pancreas of mice. This was the first conclusive evidence that there are stem cells in any mammalian pancreas. If these stem cells exist in humans, what would be needed to convert them to beta cells?

By the end of this spread you should be able to

- explain how the transcription of target genes is regulated by specific transcriptional factors that move from the cytoplasm into the nucleus
- describe the effect of oestrogen on gene transcription
- describe small interfering RNA (siRNA) as a short double strand of RNA that interferes with the expression of a specific gene

Prior knowledge

- nucleic acid structure and function (10.01)
- transcription (10.02)
- translation (10.03)

Every cell in the body of a eukaryotic organism has chromosomal DNA that contains tens of thousands of genes. Typically, only a small percentage of a cell's genes are expressed as a polypeptide at any one time. Many genes will never be expressed. Those that are expressed will be continually switched on and off in response to signals from the cell's external and internal environments. When the appropriate signals are received, polymerases and other enzymes that transcribe DNA have to be mobilized into action and locate the correct gene at the right time.

After a gene has been transcribed, a complex series of processes has to take place before the gene is properly expressed. If anything goes wrong with the signalling, transcription, or translation processes, the cell may stop functioning properly. This can lead to a variety of disorders and diseases, including cancers. An understanding of how genes are expressed is therefore of paramount importance for medicine as well as for biology in general.

Opportunities for regulating gene expression

Gene expression is regulated in prokaryotes at the transcriptional stage by special switches. For example, the bacterium *Escherichia coli* has a gene switch for three enzymes that catalyse the absorption and breakdown of the sugar lactose, one of its most important energy sources. The switch consists of **structural genes**, a **regulator gene**, and an **operator gene**. The structural genes encode the three enzymes. The regulator gene produces a **repressor molecule** which, in the absence of lactose, stops the structural genes from being expressed so that the enzymes are not produced when there is no lactose to break down. The repressor molecule acts on the structural genes indirectly, through another gene next to it – this is the operator gene. The operator and structural genes are called the **lac operon**. This gene switch ensures that the enzymes are produced only when lactose is present.

Similar gene switches occur in eukaryotes, but the barrier of the nuclear membrane enables more regulation to take place between transcription and translation, for example, at the RNA splicing stage, rather than at transcription.

The regulation of gene expression can occur before transcription in eukaryotes, and an example of this involves oestrogen. An example of post-transcriptional gene regulation involves a special type of RNA, called small interfering RNA (siRNA).

The effect of oestrogen on gene transcription

Oestrogen, like other hormones, is transported in the bloodstream and can reach almost any cell in the body. It acts as a signal molecule initiating specific cellular activities, but only in cells that have special oestrogen receptors onto which it can bind.

The binding of oestrogen to its receptor molecule is similar to that between a substrate and an enzyme: both depend on complementary shapes. Being a **steroid hormone**, oestrogen is soluble in lipids and can pass through the hydrophobic phospholipid part of plasma membranes easily. The oestrogen moves through the cytoplasm and into the nucleus where it binds to its receptor protein molecule, activating it. With the oestrogen attached, the active form of the receptor protein becomes a **transcriptional factor**. A transcriptional factor is defined as a regulatory protein that binds to DNA and stimulates transcription of specific genes into mRNA, which directs the synthesis of particular polypeptides. In the case of oestrogen, the oestrogen–receptor complex is the transcriptional factor for a number of different genes, including those responsible for the development of female secondary sexual characteristics and those involved in repairing the uterine wall during the oestrus cycle. In this way, oestrogen can have multiple effects on the body.

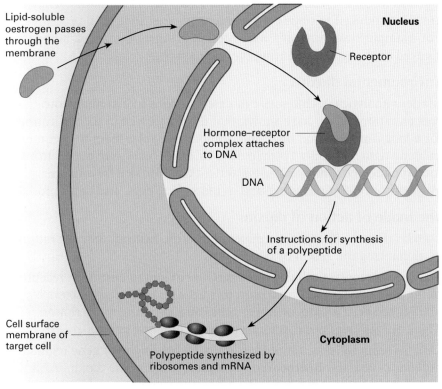

The control of gene expression by oestrogen. Oestrogen acts as a signal which triggers transcription and the synthesis of specific polypeptides in cells that have oestrogen receptors.

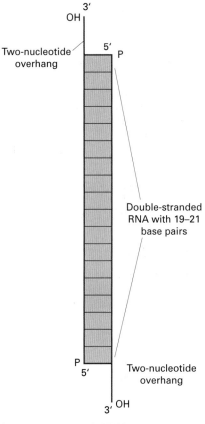

Outline structure of siRNA.
P = phosphate; OH = hydroxyl group

The effect of siRNA on gene expression

Small interfering RNA (siRNA) consists of pieces of double-stranded RNA, usually between 19 and 21 nucleotides long, with overhangs of two nucleotides at each 3' end. They are key to a process called **RNA interference**, the inhibition of gene expression at the translation stage. The targets for RNA interference by naturally occurring siRNAs include RNA viruses.

siRNA acts by having nucleotide sequences complementary to the targeted mRNA. Specific proteins guide siRNA to the target and the siRNA unwinds so that one strand (called the **guide strand**) can combine with the targeted mRNA by complementary base pairing. After combining, the siRNA interferes with the translation of the mRNA into a polypeptide thereby stopping expression of the gene.

siRNAs with a tailor-made nucleotide sequence can be synthesized. Their ability to selectively silence specific genes makes them valuable tools in both research and medicine.

Check your understanding

1 What is a transcriptional factor?

2 In terms of size and structure, how does siRNA differ from mRNA?

3 How does siRNA interfere with the action of mRNA?

- - - - - - - - - - - - - - - -

4 With reference to transcription regulation, suggest how *E. coli* uses glucose preferentially when lactose and glucose are present together.

Investigating the effects of dioxins

Dioxins are a group of lipid-soluble substances with similar chemical structures. One compound, 2,3,7,8-tetrachlorodibenzo-p-dioxin, referred to as TCDD, has been studied in most detail. It is considered the most potent. The potency of other dioxins is expressed as fractions of TCDD potency, called toxic equivalents (TEQs).

Dioxins are formed during most forms of combustion, both natural (such as forest fires) and industrial (for example, incinerators). They are also produced as trace contaminants in other industrial processes. For example, trace amounts were present in Agent Orange, a herbicide sprayed onto vegetation during the Vietnam War.

Environmental dioxins

Modern analytical techniques can detect dioxins in extremely low concentrations. They have shown that dioxins persist in the environment for a long time and that their lipid solubility enables them to accumulate in living tissue. For most people, 90% of dioxin exposure comes from food produced on land and using water contaminated by dioxins. Highest concentrations are found in meat, fish, eggs, and dairy products.

The mode of action of dioxins

Dioxins do not affect genetic material directly. Instead, they act indirectly by mimicking the action of oestrogens.

Oestrogens regulate cellular processes by binding to oestrogen receptors in the nucleus. The activated receptors act as transcription factors by binding to specific parts of target genes and initiating transcription. There are different oestrogen receptors for different genes.

Dioxins seem to bring about their effect by binding with gene regulatory proteins called **aryl hydrocarbon receptors** (AhR). The dioxin–AhR complex activates some types of oestrogen receptors, causing them to become transcription factors for what are normally oestrogen-targeted genes. One study identified 38 oestrogen-targeted genes that can be activated by TCDD in human cells.

Dioxin toxicity: animal studies

Animal tests have suggested that dioxin can be lethal and can cause birth defects, cancers, liver and thymus damage, and immune system suppression. The toxic effects are dose related but they vary from species to species: for example, a hamster is not affected by a dose that can kill a guinea pig.

The most consistent and sensitive effects of dioxins were observed when they were given to pregnant animals. In particular, the reproductive system of male embryos did not develop normally when exposed to dioxins in the womb. This resulted in decreased fertility because of changes in sperm production and quality.

Dioxin toxicity: human studies

Although dioxin toxicity has been demonstrated in animal studies, its effects on humans are less clear. The fact that a substance is toxic to other mammals does not necessarily mean that it is toxic to humans.

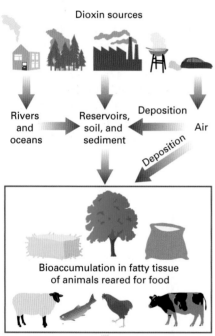

Dioxin sources

Rivers and oceans

Reservoirs, soil, and sediment

Deposition

Air

Deposition

Bioaccumulation in fatty tissue of animals reared for food

Animal products

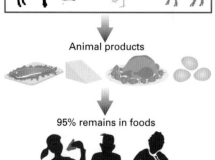

95% remains in foods

Sources of dioxin in the human diet

Evidence that dioxins are harmful comes from studies of people exposed to different levels of dioxins. A major breakthrough came in 1995 when the *American Journal of Epidemiology* reported a study into the effect of dioxins and related compounds in 1189 workers from a factory in Hamburg that had been producing herbicides for 32 years. The estimated level of exposure of the workers to dioxin was based on actual measurements of blood dioxin levels in 190 workers. The study included all who worked at the plant for more than three months from 1952 until the factory closed down in 1984. The workers were medically tracked to the end of 1992.

A control group consisted of 2528 workers at a nearby gas supply company who were not exposed to dioxins in their employment. Both groups had the same proportion of smokers. The results showed a close relationship between exposure to dioxin and the incidence of cancer. Employees with the highest dioxin exposures were three times more likely to have cancer compared with the control group.

In another study, the number of people developing cancer in a group of 250 people who were exposed to high levels of dioxin was 8. This compared with 2.56 expected from national statistics for the general population. Statistical analysis of the data showed that the increased risk of cancer in the high dioxin group was statistically significant at the probability (p) = 0.01 level.

Tolerable daily intake

According to studies conducted so far, the effects of dioxin are not significant, or they are reversed, when intakes are below a certain point (the **threshold**). It is therefore possible to establish a **tolerable daily intake (TDI)** for dioxins. It is assumed that intakes below this level are safe.

The TDI is the amount of a substance that can be ingested daily over a lifetime without appreciable health risk to the most vulnerable in the population. It is expressed in relation to body mass in order to allow for different body sizes. The TDI for dioxins has been set in the UK by the Committee on Toxicity (a Government agency) at 2 picograms (pg) per kilogram of body mass per day ($1 g = 10^{12}$ pg). Other toxic substances such as tobacco smoke can damage genetic material directly and may cause cancers by their effects on a single cell. For these substances, a precautionary approach assumes there is no threshold and therefore no safe intake.

Check your understanding

1 Dioxins appear to mimic the action of oestrogens. Explain how this might help explain the variety of the effect of dioxins on different body systems and in different animals.

2 What property enables dioxins to move easily through cell membranes?

3 Is it possible to have a zero risk of dioxin toxicity? Explain your answer.

4 Using the levels set by the Committee on Toxicity, what is the tolerable daily intake of dioxins for a person weighing 60 kg?

5 Why is there no tolerable daily intake for smoking?

6 What proportion of the herbicide workers in the Hamburg study had their dioxin blood levels measured?

7 Why was it important that the control group had the same proportion of smokers as the herbicide workers?

8 For the study in which 8 out of 250 people exposed to high levels of dioxin reported cancers compared to 2.56 expected from data for the general population,

a Express as a ratio the risk of cancer in those exposed to dioxin against the risk of cancer in the general population.

b State the expected risk ratio if there was no difference between the two groups.

9 The risk ratio was statistically significantly elevated at the probability (p) = 0.01 level. What does this mean? (Hint: see the chi-squared test, spread 1.06.)

10 Although human data indicates a link between high exposures to dioxins and certain illnesses such as cancers, it is not possible to be certain that a particular individual became ill as a result of exposure to dioxin. Why not?

OBJECTIVES

By the end of this spread you should be able to

• understand information about the use of oncogenes and tumour suppressor genes in the prevention, treatment, and cure of cancer

Prior knowledge

• gene mutations (10.05)

• cell division: stimulation and suppression (10.06)

Preventing cancer

Detailed studies of the development of colon cancer, one of the most common forms of cancer, show that it does not result from a single mutation, but only develops after several somatic mutations have taken place.

Colon cancer usually develops gradually. The first sign is the growth of a polyp (a small benign tumour) in the colon wall. The polyp cells appear normal but they divide at an abnormally fast rate. The tumour grows, but only becomes malignant if an accumulation of mutations occurs. The multi-step path development model for colon cancer has been confirmed for many other types of cancers that develop gradually.

The gene mutations that must occur for a cell to become fully cancerous usually include the transformation of at least one proto-oncogene into an oncogene, and the mutation or deactivation of several tumour suppressor genes. Because mutated tumour suppressor alleles are usually recessive, the gene becomes non-functional only if both alleles are mutated. In contrast, it only takes one mutation to convert a proto-oncogene into an oncogene.

Even if proto-oncogenes and tumour suppressor genes mutate, cancers usually only develop after other enzymes which normally regulate cell division are affected.

The fact that several gene mutations are required to produce a cancer cell helps explain why some people are more predisposed to cancer than others. An individual inheriting an oncogene or a mutated allele for a suppressor gene will be one mutation closer to developing a cancer.

Geneticists have identified several of the specific alleles which when mutated can contribute to the development of cancers. The ability to detect these genes by genetic screening provides information about an individual's predisposition to the cancers. Screening of cells from those found to have a higher than normal predisposition to a cancer is a very important prevention factor. The screening enables a doctor to detect a precancerous tumour and to remove it before it can develop into a cancer.

Although an inherited predisposition increases the risk of developing colon cancer, for most people the most important risk factors are linked to lifestyle. In fact, the links between diet, weight, exercise, and colon cancer risk are among the strongest for any type of cancer. While diets high in red meats and processed meats increase the risk, diets high in vegetables and fruits have been linked with a decreased risk. A high alcohol intake, physical inactivity, obesity, and smoking all increase the risk.

In addition to these lifestyle factors, any exposure of the colon cells to environmental **carcinogens** will increase the rate at which proto-oncogenes and tumour suppressor cells mutate, thereby increasing the risk of colon cancer.

Although the risk of cancer can be reduced, there is nothing that anyone can do to guarantee protection. There is always a chance, albeit a small one, that a person without an inherited predisposition and who adopts a healthy lifestyle in a healthy environment will develop a cancer.

Treatments and cures

Surgery is the main form of treatment for those with cancer. Despite the removal of a visible tumour, cancers may recur. Chemotherapy and radiotherapy used to destroy cancerous cells can improve the success rates, but these vary depending on the type of cancer.

One of the great difficulties of treating cancers is that they are caused by many different things and they respond differently to different treatments. Therefore no one treatment strategy can be used on all cancers.

Despite the difficulties, many cancers can be cured these days. According to Cancer Research UK statistics, seven out of ten children are cured of cancer. Testicular cancer, Hodgkin's disease, and many forms of leukaemia can all be cured in adults by chemotherapy. Many types of cancer, including breast cancer, can be cured if detected early enough (in the case of breast cancer, this is when the first cancerous cells appear and before **metastasis** has taken place).

One new treatment strategy involves using **gene therapy** on a mutant tumour suppressor allele called p53 which occurs in most human cancers. Genetic engineers have been working on a way to replace the mutated allele with a normal one. In the laboratory, they have used viruses as vectors to carry the new allele into the cancer cells. If used in people, the carrier viruses would need to be weakened so that they do not cause disease. However, weakened viruses might be killed by the body's immune system. Researchers are investigating how to stop the virus from causing disease while still allowing it to be strong enough to get past the immune system without being killed. This technical difficulty will have to be solved before the treatment can be developed any further.

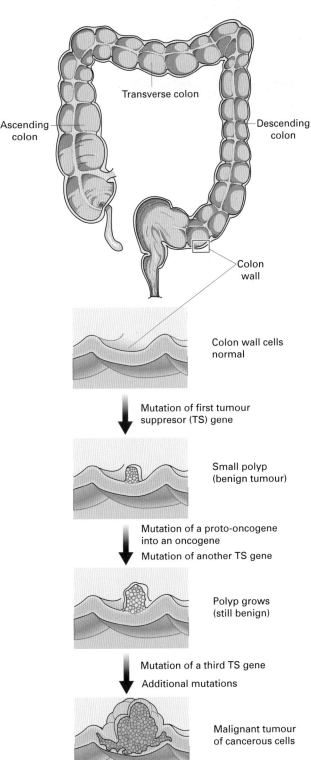

Colon wall

Colon wall cells normal

Mutation of first tumour suppresor (TS) gene

Small polyp (benign tumour)

Mutation of a proto-oncogene into an oncogene

Mutation of another TS gene

Polyp grows (still benign)

Mutation of a third TS gene

Additional mutations

Malignant tumour of cancerous cells

Colon cancer develops in many stages.

Check your understanding

1 Explain why the incidence of cancer increases greatly with age.

2 How can genetic screening help prevent colon cancer?

3 Why is it not possible to use one treatment strategy for all cancers?

4 How might gene therapy of tumour suppressor genes be used to cure cancer?

Inherited and acquired mutations: what are they?

• An **inherited mutation** is a genetic change present in gametes and passed on from parent to offspring.

• An **acquired mutation** occurs in somatic cells, not gametes, and so cannot be transmitted to offspring. Acquired mutations *are* passed on to all cells descended from the mutant cell, giving rise to a colony of cells marked by the mutation.

Most mutations are harmful and, if expressed, cause disorders. However, the chance of a person having a disorder caused by a mutation depends on an interplay between genes and the environment. In this context, the environment includes the conditions in which the person lives and lifestyle factors such as diet and exercise. Which component, genes or environment, has the greater effect will depend on the particular disorder and other circumstances.

Many disorders caused by inherited mutations, such as cystic fibrosis, sickle cell anaemia, and haemophilia, are present from birth (and even before). Others, such as Huntington's disease, do not appear until later in life, possibly after a person has had children. No matter when they appear, the inherited mutation will be passed on to offspring.

Not all inherited mutations are expressed as a disorder. Those with a single recessive mutant allele will be carriers for the disorder although they may not show any signs of it themselves. People with a disorder caused by a mutant dominant allele will express that disorder whether they are heterozygous or homozygous for it. Nevertheless, the extent to which the condition is expressed (that is, how serious the condition is) again depends on an interplay of genetic and environmental factors.

Although disorders caused by acquired mutations are not inherited, some are present from birth. They are usually caused by **mutagens** (environmental factors that increase the rate of spontaneous mutations) to which the developing fetus has been exposed. Thalidomide, for example, is a drug that was taken by pregnant women for morning sickness and found to cause fetal abnormalities such as deformed limbs. (Thalidomide is no longer used as an anti-morning sickness drug.) Other substances that can cross the placenta and cause fetal malformations include alcohol and retroviruses.

Genetic predispositions to disorders

Most non-inherited disorders that develop after birth have many causes. Epidemiological research has shown that some disorders tend to occur more frequently among members of the same family. In the case of certain cancers, individuals become predisposed to the disorder because they have inherited a mutant allele for a tumour suppressor gene. However, the cancer only manifests itself if acquired mutations are added to the inherited mutation.

Gene tests

The whole of the human genome, the sequences of the approximately 3 billion base pairs and 20 000–25 000 genes that make up human DNA, has been identified by the **Human Genome Project**. Medical scientists are now discovering how mutant alleles of specific genes are linked to particular disorders. This forms the basis of **gene tests** carried out to help clarify a diagnosis and direct a doctor toward an appropriate treatment.

One type of gene test compares the sequence of DNA bases in a person's gene with the sequence in a version of the gene known to be normal. When testing for an inherited mutation, it is possible to take a DNA sample from any tissue including blood. However, when testing for an acquired mutation, the DNA sample has to be taken from cells in which the mutation has occurred or is suspected to have occurred. For some cancers this might involve surgically removing a polyp and examining the DNA in its cells.

A couple expecting a baby are given genetic counselling.

Genetic counselling

Gene tests should be followed by genetic counselling, the giving of advice and information about the risks of a genetic disorder and its outcome. Counselling is a very challenging task. Counsellors must have a good understanding of medical genetics and need to be well trained in sympathetic counselling techniques. They must give information that helps clients come to their own decision rather than imposing their own views. Clients should be made aware of the limitations of gene tests. Although a result of a gene test is positive, in many cases it only gives the probability of developing a disorder. For example, the identification of a single inherited mutant allele can inform a healthy person with a strong family history of Alzheimer's disease of their chances of developing the disease. However, some people who carry the disease-linked mutant allele never develop the disease.

Gene therapy

With the ability to identify harmful mutations comes the potential to treat disorders by **gene therapy**. Gene therapy involves correcting a mutant allele by either repairing it, swapping it with a normal allele, or by regulating its activity so that it no longer produces harmful effects.

Acquired mutations are limited to the originally mutated cell and the **cell line** produced from it. As long as the cells remain together as a single colony, the effect of the mutant alleles will be localized and gene therapy can be targeted at the colony of cells.

Inherited mutations occur in every cell of the body but they are not always expressed. For example, a mutant allele linked with cystic fibrosis is only expressed in the epithelial tissue of the lungs. This enables gene therapy to be targeted at the affected tissues and organs rather than the whole body.

Check your understanding

1 With reference to somatic (body) cells and germ cells (gametes), distinguish between an inherited mutation and an acquired mutation.

2 Why is gene therapy unlikely to be successful in the later stages of cancer?

3 Suggest why some people who carry the inherited disease-linked mutant allele for Alzheimer's disease never develop the disease.

- - - - - - - - - - - - - -

4 Gene tests, like all medical tests, are susceptible to laboratory errors.

a Suggest two ways in which errors might occur.

b How might errors of diagnosis be minimized?

OBJECTIVES

By the end of this spread you should be able to

- describe how fragments of DNA can be produced by converting mRNA to cDNA using reverse transcriptase

- explain the role of restriction endonucleases and ligases

Prior knowledge

- DNA structure and function (*AS Biology for AQA* spreads 11.01–11.05)

- the genetic code (*AS Biology for AQA* spread 11.06)

Gene cloning refers to the making of multiple, identical copies of a fragment of DNA that includes one or more genes. A piece of DNA coding for a particular polypeptide can be made artificially if its base sequence is known. The base sequence can be deduced from the amino acid sequence of the protein. The process of gene cloning is most useful for small proteins that have a correspondingly short DNA base sequence.

Larger fragments of DNA can be made by converting mRNA to DNA. This process uses special enzymes that act as molecular tools. Three of the most important groups of enzymes are reverse transcriptases, restriction endonucleases, and ligases.

Reverse transcriptase: converting mRNA to cDNA

Reverse transcriptase is an RNA-dependent DNA polymerase. It is used to convert mRNA to its **complementary DNA (cDNA)** which, for example, contains the gene for a particular polypeptide. Cells that produce large amounts of a desired polypeptide will have large amounts of mRNA that encode it. If this mRNA can be isolated, its cDNA can be synthesized by **reverse transcription**, a process catalysed by the reverse transcriptases. Reverse transcription, as its name implies, is the reverse of normal transcription: in normal transcription DNA acts as the template for mRNA; in reverse transcription mRNA acts as the template for DNA. In both types of transcription, free DNA nucleotides are required as well as the appropriate enzymes.

After a strand of mRNA has been copied into its cDNA, the mRNA is removed and a second strand of DNA is made by adding the enzyme **DNA polymerase** and more DNA nucleotides. The result is a double-stranded DNA molecule identical with the original DNA molecule.

Reverse transcriptases were first obtained from retroviruses that contain RNA. These viruses include human immunodeficiency virus (HIV, the cause of AIDS). They use reverse transcriptases to copy their RNA into the DNA in the host cell.

Restriction endonucleases: molecular scissors

Restriction endonucleases are enzymes which function as 'molecular scissors', cutting DNA molecules into fragments. There are different restriction endonucleases that identify where the DNA is to be cut by specific base sequences called **recognition sequences**. If the appropriate recognition sequence is present, the DNA will be cut at a location called the **cleavage site** near the recognition sequence or overlapping it.

Cleavage (separation) of the two strands may occur at sites directly opposite each other. This results in DNA fragments with **blunt ends**. Alternatively, the cleavage sites may be offset slightly, resulting in fragments with '**sticky ends**', so called because their exposed bases readily form hydrogen bonds with the complementary bases in the 'sticky ends' of other DNA molecules cut by the same restriction endonuclease.

One widely used restriction endonuclease has the recognition sequence GAATTC. The sequence is termed a palindrome because it is the same as its reverse complement of bases (CTTAAG). By cleaving the G–A bond on both strands, the restriction enzyme creates fragments with 'sticky ends'.

Restriction endonucleases occur naturally in bacteria. The bacteria use the enzymes to defend themselves against **bacteriophages** (viruses that infect bacteria). Restriction endonucleases restrict the growth of invading viruses by chopping bacteriophage DNA into smaller non-infectious fragments. The bacteria protect their own DNA by incorporating a methyl (CH_3) group in one or more bases in or near the restriction sequence wherever it occurs in the bacterial DNA. This blocks the restriction endonuclease and prevents it from working on bacterial DNA.

Only a few bases are modified in each DNA strand so that methylation does not interfere with the normal functioning of DNA, such as base pairing and transcription.

Ligases: sticking DNA fragments together

DNA ligases are enzymes that can link together DNA strands that have been broken. The normal function of ligases is to join the 'sticky ends' of two strands of DNA together during replication and repair. The DNA ligase catalyses the formation of a covalent phosphodiester bond between the 3' hydroxyl end of a nucleotide in one strand with the 5' phosphate end of a nucleotide in another strand. ATP is required for the reaction.

A DNA ligase will join together two blunt ends of DNA, but it does this with reduced efficiency: joining blunt ends requires a higher enzyme concentration and different reaction conditions from when DNA ligases join 'sticky ends' together.

Restriction endonucleases and DNA ligases complement each other. By using the two enzymes, molecular biologists can cut and rejoin DNA fragments from any source to produce *in vitro* **recombinant DNA** (that is, DNA from more than one source).

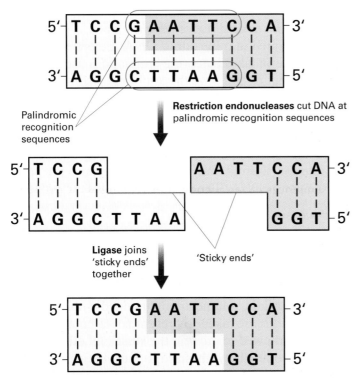

A piece of double-stranded DNA is cut by a restriction endonuclease which recognizes a palindromic sequence of bases. Two fragments are created with overlapping 'sticky ends' that can be joined together using ligase.

Check your understanding

1 Which enzyme catalyses the conversion of mRNA into cDNA?

2 What are 'sticky ends' of DNA?

3 Which enzyme is involved in sticking the ends of DNA together?

4 Using your knowledge of how enzymes work and how retroviruses reproduce, suggest how a drug might prevent a retrovirus such as HIV from reproducing.

OBJECTIVES

By the end of this spread you should be able to

- describe *in vivo* gene cloning

- explain how plasmids can be used to transfer a gene from one organism to another

- describe the use of recombinant DNA technology to produce transformed organisms

Prior knowledge

- gene cloning and transfer I (11.01)

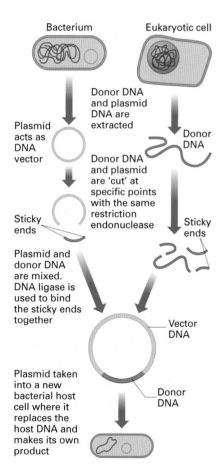

Bacterium — Eukaryotic cell

Plasmid acts as DNA vector

Donor DNA and plasmid DNA are extracted

Donor DNA

Donor DNA and plasmid are 'cut' at specific points with the same restriction endonuclease

Sticky ends — Sticky ends

Plasmid and donor DNA are mixed. DNA ligase is used to bind the sticky ends together

Vector DNA

Donor DNA

Plasmid taken into a new bacterial host cell where it replaces the host DNA and makes its own product

Recombinant DNA: the transfer of DNA from a eukaryotic cell to a bacterial cell using a plasmid. Note that by using the same restriction endonuclease on the plasmid DNA and donor DNA, the sticky ends will match (they will have complementary bases).

In vivo gene cloning

In vivo gene cloning involves inserting a DNA fragment into a vector, which can transfer the DNA into host cells that can clone the gene. The most commonly used host cells are bacteria, and the most commonly used vectors are plasmids.

Plasmids are small rings of DNA contained within some bacteria. They are quite separate from the main bacterial DNA. They replicate independently and can be transferred from one bacterial cell to another, and even from bacteria to plant cells. Plasmid replication and transfer occurs naturally, but it can also be encouraged artificially.

Recombinant DNA

Using restriction endonucleases and DNA ligases, a plasmid can be cut open so that short stretches of donor DNA can be inserted. The donor DNA may be from any other living organism: microorganism, plant, animal, or fungus.

The new combination of DNA created by joining together fragments of DNA from different organisms is called **recombinant DNA**. The techniques used to construct and manipulate the DNA form part of **recombinant DNA technology**.

After inserting the donor DNA, the modified plasmid can be transferred to the host cell. The donor DNA is cloned every time the plasmid containing it replicates. A newly inserted plasmid replicates several times inside a bacterium both before and after the bacterium divides. Therefore the potential for gene cloning by using recombinant DNA within the plasmids of bacteria is immense.

To ensure that only the desired donor gene is cloned, a gene for antibiotic resistance is incorporated into the donor DNA. Before any large-scale culturing of the bacteria takes place, the antibiotic is used to kill all the unwanted bacteria that do not have the antibiotic-resistant gene and therefore do not have the desired donor gene.

Because the genetic code is universal, the bacteria can recognize and use the donor DNA even though it comes from another organism or species. If appropriate fragments of DNA (called **control sequences**) are also incorporated into the plasmid, the bacteria will produce the polypeptide coded for by the donor DNA. The figure here shows the process.

Industrial processes

Industrial fermenters are used to clone genes and culture bacteria on a large scale. The fermenters are large vessels containing a liquid culture medium in which the bacteria are kept in conditions that optimize the production of the gene product. Environmental factors such as temperature, pH, and fluid volume are maintained within very strict limits. Also, precautions are taken to prevent contamination. Only the desired organism is allowed to grow in the vessel; all others are excluded. After the required product is synthesized, it must be separated from the host cells and then purified.

One problem with using bacteria to manufacture a valuable product is that the product itself often inhibits bacterial growth at concentrations that are too low for the product to be extracted and purified. In some cases, the problem can be overcome by separating the growth phase from the production phase:

- the bacteria are allowed to multiply rapidly at first without producing the required product (the growth phase)
- when there are enough bacteria, the donor gene is 'switched on' by an appropriate transcription factor
- the product is then made by the bacteria (the production phase)

Another problem with industrial-scale fermentation is that in large populations, the microorganisms may be damaged by the heat they generate. Cooling the fermenter requires energy and costs money. Genetic engineers sometimes overcome this problem by using a **thermophilic organism** to synthesize the desired product, or by inserting the gene for temperature tolerance into the host DNA.

Viruses and non-biological vectors

Other biological vectors include viruses, which transfer DNA into cells naturally. For example, a human gene can be inserted into a herpes virus and transferred by infection into a human host cell, where it is expressed.

Non-biological methods of gene transfer include:

- **ballistic impregnation**: a DNA gun fires tungsten or gold particles coated with DNA into cells; this has been used to modify several important crop plants, including wheat
- **electroporation**: bursts of electricity create temporary pores in cells, allowing donor DNA to enter
- **micro-injection**: a very fine pipette is used to inject the DNA into a host cell nucleus
- **liposome transfer**: a bubble of fatty substance called a liposome enables a gene to be carried through the cell surface membrane and into the host cell, where the gene may then replace the normal one beneficial

In vivo gene cloning and transfer has been used to make many beneficial products. Two examples are alpha-1-antitrypsin and insulin.

The production of alpha-1-antitrypsin

Alpha-1-antitrypsin is a protein that helps protect the lungs from damage during an infection. Microinjection has been used to introduce the human gene for the protein into the fertilized eggs of sheep. The gene is preprogrammed to work in the mammary gland when the sheep lactates. The protein can be separated and purified from the milk constituents after milking. The gene for the protein is inherited by the sheep's offspring, so they will also produce this valuable protein.

Human insulin

Human insulin was the first human protein to be manufactured and used by humans as a result of gene cloning, using the process shown in the diagram above. It is now produced on a very large scale and is used by thousands of diabetics. Human insulin made by recombinant DNA technology produces fewer side effects than insulin prepared from cow or pig pancreatic extracts, previously the main source of insulin.

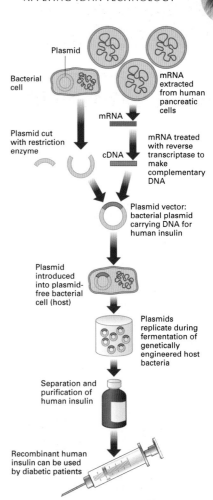

The manufacture of recombinant human insulin

Check your understanding

1 What is a gene clone?

2 What is the function of a plasmid in gene cloning?

3 How do genetic engineers ensure that the recombinant DNA to be cloned in bacteria contains the desired gene?

- - - - - - - - - - - - - - -

4 Suggest how, unless appropriate precautions are taken, the use of antibiotic-resistant genes in recombinant DNA technology might affect our ability to use certain drugs to treat an infection.

OBJECTIVES

By the end of this spread you should be able to

- describe *in vitro* gene cloning

- explain how the polymerase chain reaction is used to form multiple DNA fragments

- discuss the relative advantages of *in vivo* and *in vitro* cloning

Prior knowledge

- gene cloning and transfer I and II (11.01–11.02)

A thermocycler goes through a regular cycle of temperature changes to bring about PCR.

In vitro gene cloning

In 1983, Kary Mullis devised a method of cloning genes in a test tube without involving bacteria. This *in vitro* method of cloning DNA is called the **polymerase chain reaction (PCR)**. It uses special DNA polymerases to make multiple copies of a short strand of DNA. The polymerases must be able to function at high temperatures, so the enzymes are usually obtained from thermophilic bacteria such as *Thermus aquaticus*.

The piece of DNA to be cloned is mixed and incubated with the DNA polymerase in a solution containing nucleotide triphosphates as building units for new DNA. Specific, artificially synthesized **primers** (short nucleotide sequences) which act as signals for replication are added to the solution. Two primers mark the ends of the fragment of DNA to be copied. Typically, the DNA between the two primers is copied during a precisely controlled three-step cycle of temperature changes.

- Heating to 94–96°C causes double-stranded DNA to become separated into single strands.

- Cooling to 50–65°C allows the specific primers to bind to DNA, bracketing the DNA region to be copied.

- Raising the temperature to about 73°C enables the thermostable polymerase to move along each single strand of DNA, read its code, and replicate a complementary strand using the DNA nucleotide triphosphates.

In PCR machines, these steps are automated. Each time the thermal cycle is repeated, DNA is replicated. The first replication produces two DNA molecules; the second produces four molecules; the third produces eight molecules, and so on. Typically the process is repeated 30 times so that millions of copies of the DNA can be produced in about an hour.

The primers are usually 15–20 bases long. They are an essential ingredient in PCR. They flag the beginning and end of the DNA stretch to be copied. Therefore, to carry out PCR, the base sequences at the ends of each strand of the target DNA have to be known so that the

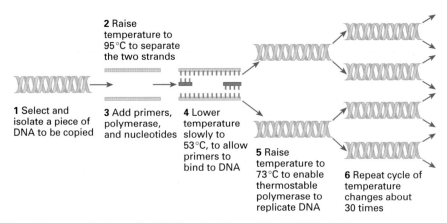

2 Raise temperature to 95°C to separate the two strands

1 Select and isolate a piece of DNA to be copied

3 Add primers, polymerase, and nucleotides

4 Lower temperature slowly to 53°C, to allow primers to bind to DNA

5 Raise temperature to 73°C to enable thermostable polymerase to replicate DNA

6 Repeat cycle of temperature changes about 30 times

The polymerase chain reaction (PCR): an in vitro *method of DNA cloning*

complementary primers can be made *in vitro*. Usually, there will be only two regions in the DNA of an organism's genome to which specific primers of this length will bind. This specificity of the primer is a key feature of PCR.

Analysing DNA fragments

Originally, PCR was performed in a test tube and when the reaction was completed the amplified DNA fragments were analysed and visualized by **gel electrophoresis**. However, recent advances in instrumentation enable the products to be analysed as they are produced by a technique called quantitative PCR (Q-PCR). This is achieved by using special fluorescent dyes which react with specific base sequences in the amplified product and which can be measured by the Q-PCR instrument. In terms of base sequences, the instrument can tell a researcher both what is present and how much is present, allowing samples from different sources to be compared quantitatively.

Relative advantages of *in vitro* and *in vivo* cloning

The major advantages of *in vitro* cloning using PCR include the following.

- Speed and ease of use: DNA cloning using PCR can be completed in a few hours while *in vivo* cloning may take several weeks. PCR is easy to carry out – it requires a test tube, a few relatively simple reagents and a means of controlling the temperature of the reagents.

- Sensitivity: only one molecule of DNA is required for the PCR to generate billions of copies. However, this extreme sensitivity means that great care has to be taken to prevent contamination.

- Robustness: PCR can be used on a broad range of samples, from pure DNA from fresh living sources to partly degraded DNA from ancient sources. This makes it particularly useful for palaeontological and anthropological studies of, for example, frozen or mummified tissues and archaeological remains.

There are also some major limitations of PCR as a method of gene cloning.

- Specific primers need to be used, and this requires prior knowledge of at least some of the nucleotide sequence in the DNA.

- Although theoretically *in vitro* cloning is limitless and yields of DNA rise exponentially, in practice, the amount of DNA cloned is much less because of gradual losses of efficiency. Typically relatively small amounts of product are obtained compared with *in vivo* gene cloning using cells.

- PCR is limited to making copies of relatively short sections of DNA; it becomes increasingly difficult to clone DNA as the size of the fragment increases. Unlike *in vivo* cloning which can clone DNA sequences approaching 2 Mb, DNA sequences cloned by PCR are typically only 0.15 kb in size. (A **mega base pair**, **Mb** or Mbp = a million base pairs; a **kilo base pair**, **kb** or kbp = a thousand base pairs.)

So *in vivo* gene cloning in cells is the best way to produce large amounts of a particular gene. However, when the target DNA is small or impure, PCR is quicker and more selective.

Uses of PCR

Since it was devised, PCR has had a major impact on biology. Any piece of DNA can be quickly copied many times (amplified). The sample may be from fresh tissue such as blood, or from mummified tissue thousands of years old. PCR can even be used on a sample containing only one DNA molecule, producing billions of copies of the DNA in a few hours.

PCR is a common and indispensable technique used in biological and medical laboratories throughout the world. It has been used in basic biological research to determine evolutionary relationships between organisms, in medicine to detect and diagnose infectious diseases, hereditary diseases, and malignancies, and in forensic science to identify genetic fingerprints.

Check your understanding

1 What is the function of the DNA primers in PCR?

2 Why is the temperature raised to about 95°C during PCR?

3 There are many types of DNA polymerase. They all share the ability of the one used in PCR to move along segments of DNA, read its code, and assemble a copy using nucleotide phosphate building blocks. What makes the DNA polymerase used in PCR particularly special?

4 Which type of cloning is better if multiple copies of DNA are required from a very small sample, *in vitro* or *in vivo* sampling?

5 Suggest why the amount of DNA cloned at each thermal cycle eventually levels off.

OBJECTIVES

By the end of this spread you should be able to

- interpret information related to the use of recombinant DNA technology

- evaluate ethical, moral, and social issues associated with the use of recombinant DNA technology in agriculture, in industry, and in medicine

- balance the humanitarian aspects of recombinant DNA technology with the opposition from environmentalists and anti-globalization activists

Prior knowledge

- ethical implications of research into human genetic diversity (*AS Biology for AQA* spread 12.01)

- gene cloning and transfer (11.01–11.03)

rDNA research controversies

Recombinant DNA technology (rDNA technology) provides the tools to genetically manipulate viruses and any cellular organisms, often to develop specific products. When DNA was first extracted and genetic engineers started to cut and splice bits of DNA, scientists and the general public began to debate the benefits and potential dangers of rDNA research. Many fears were expressed. They mainly hinged around the possibility of creating new kinds of plagues, of irreversibly altering human evolution, or of irreparably damaging the environment. These fears resulted in scientists worldwide halting experiments using rDNA technology until 1975.

The Asilomar Conference: resolving the controversy

In February 1975, a group of 140 experts (mainly biologists but also including lawyers and medical scientists and practitioners) participated in a conference at Asilomar State Beach in the USA. The participants identified and evaluated potential biohazards of rDNA research, and discussed ways to mitigate the perceived risks of rDNA. The **Asilomar Conference** established a set of guidelines for carrying out rDNA research safely. Using these guidelines, scientists were able to restart their research. Since the Asilomar Conference, the development of rDNA technology has flourished to the extent that it dominates much of modern biology. The isolation and manipulation of genes from any organism on Earth, alive or dead, is now routine. rDNA technology has also greatly influenced the direction of research in other fields as diverse as anthropology, forensics, information theory, and computer science. In biology and all of these fields, it has altered the way questions are formulated and the way solutions are sought.

Benefits of applying rDNA technology

The first commercial application of rDNA technology was in 1982 with the production of recombinant human insulin for the treatment of diabetes. Since then, the technology has been used in agriculture, industry, and medicine.

Ethical, moral, and social issues

Recombinant DNA technology has many potential benefits to agriculture, industry, and medicine, but there are many ethical, moral, and social issues associated with its application.

Both ethical and moral issues relate to the rights and wrongs of doing or not doing something. However, doing something ethically means doing it in accordance with principles of conduct considered correct by a community of people, while doing something morally means doing it in accordance to an individual's beliefs, traditions, and values. Therefore, an ethical code of conduct is shared by a group, but a moral code of conduct is personal.

Some of the ethical and moral issues associated with the application of rDNA technology are the same as the fears about rDNA research and pose questions such as

- Is it safe to tamper with genes?
- Will the application and development of rDNA technology, such as whole organism cloning, lead to its use in humans? If so, is this good or bad?
- What effect will the worldwide use of rDNA technology have on the environment and biodiversity?

Some people regard rDNA technology as inherently bad. They argue that the potential risks outweigh the benefits so much that there should be a blanket ban on its use. The potential risks they cite include

- harm to humans, for example, through the transfer of antibiotic resistance markers, or the creation of new pathogenic organisms
- potential damage to the environment, for example, through the unintended transfer of genes such as those for herbicide resistance to natural plant populations through cross-pollination, or unknown effects on soil microorganisms which could disrupt nutrient cycles and lead to a loss of biodiversity
- loss of freedom of choice for consumers, for example, if genetically modified organisms (GMOs) are not labelled clearly, or if GMOs and traditional organisms are mixed

The opponents to rDNA technology are also concerned about the potential effects on society. Because the development of rDNA technology is so expensive, it tends to be in the hands of a few companies in rich countries. There is a danger, for example, that one or two companies which use rDNA technology for food production could dominate the world market and that the least developed societies will become more dependent on highly industrialized ones.

Balancing benefits and concerns

Humanitarian aspects of rDNA technology relate to its use for the whole of humankind, not just for a privileged few. Those who share a humanitarian view believe that everyone in the world has the right to sufficient food and basic healthcare.

Many experts believe that the only way that we will be able to feed our increasing global population and to administer adequate healthcare will be to increase our use of rDNA technology to improve food production and medicines.

The great challenge that faces humanitarian organizations, such as the Food and Agriculture Organization (FAO) of the United Nations, is to ensure that

- new advances in rDNA technology are not skewed to the interests of rich countries
- benefits in food production and healthcare are shared globally
- the advances in rDNA technology take place without harmful effects to individuals, societies, and environments

One way of working towards these aims might be to emulate the work done at the Asilomar Conference and for each new application of rDNA technology to develop guidelines that govern and regulate it to the benefit of all. The guidelines will need to balance the expectation of companies to profit from the advances they make in rDNA technology, with the rights of all people to share in the benefits.

Some uses of rDNA technology

Agricultural uses

- new strains of crops and breeds of animals to improve food production
- transgenic crops with increased resistance to diseases, herbicides, and pesticides
- improved animal health through development of new tools to diagnose and treat diseases and disorders

Industrial uses

- manufacture of new chemicals such as adhesives
- improvement to the shelf-life, taste, and quality of foods
- development of new security equipment based, for example, on DNA fingerprinting
- improved waste management using transgenic organisms

Medical uses

- new tools to detect and diagnose disorders
- new treatments such as stem cell therapy to treat genetic disorders
- new pharmaceuticals to make new vaccines

Check your understanding

1 Distinguish between an ethical and a moral code of conduct.

2 What was the main outcome of the Asilomar Conference?

3 Why is the development of rDNA technology for food production dominated by a few companies?

What is gene therapy?

Gene therapy is a technique for correcting mutated genes that cause disease. There are several correction methods:

- a normal gene may be inserted in a non-specific location into the genome to replace a mutated gene
- a mutated gene could be exchanged for a normal gene at its normal gene locus by recombination
- a mutated gene could be repaired by selective reverse mutation to make it normal
- the regulation of a faulty gene could be altered to switch it off or on so that it no longer has harmful effects

Most gene therapy involves **gene replacement** (replacing a mutated allele with a normal one) or **gene supplementation** (adding one or more copies of a normal allele to a cell without removing any of the pre-existing ones) using recombinant DNA technology. In gene supplementation, the added alleles are dominant and mask the effects of recessive alleles. In both cases, a vector is needed to deliver the normal allele into the target cells. Currently, viruses are the most common vectors. The genetic material of a virus is altered so that it includes the therapeutic gene. Then the target cells (for example, the lung cells of a patient with a pulmonary disorder) are infected with the viral vectors, which unload their altered genetic material into the cells. In theory, when the therapeutic genes are expressed, they should produce a functional polypeptide which restores the target cells to normality.

Which cells can gene therapy treat?

Gene therapy may be either **germ-line gene therapy** or **somatic-cell gene therapy**, depending on which cells are treated.

In germ-line gene therapy, sperm or egg cells, or the cells that form them, are treated so that every cell of the offspring contains the inserted, normal gene and can function normally. The alteration of the genome would be passed on from one generation to the next. Because of this, germ-line gene therapy is regarded as unethical and its use to treat humans is prohibited.

In somatic-cell gene therapy, the body cells are altered, but not the sperm or eggs, or the cells that give rise to them. The alteration is therefore not passed on to future generations.

Evaluating the effectiveness of gene therapy

Laboratory research and animal studies have demonstrated that gene therapy has the potential to treat a variety of disorders. However, clinical trials have been conducted on humans for only a relatively short time. Some have been halted because of health or other problems. For example, in 2002, two out of ten children given gene therapy for severe combined immune deficiency developed leukaemia-like symptoms and the trial was temporarily halted and reviewed. The leukaemia might have been induced by the altered gene being integrated in the wrong place in the genome. Other problems include

A physiotherapist treating a cystic fibrosis patient. The mutant allele causes mucus to accumulate in the lungs, and this has to be removed by aggressive massage.

- the viral vector being rejected before the new gene has been inserted
- the immune response being heightened so that the target cells with the altered genome are rejected
- target cells with the altered genome not dividing rapidly enough so that patients have to undergo multiple rounds of gene therapy

To be judged effective, it should be shown that a particular gene therapy has the desired effect on the patient's genome and provides a permanent cure for the targeted disorder without causing significant harm. Although the results from several clinical trials of gene therapy are encouraging, it is too early to judge their long-term effectiveness.

Gene therapy for cystic fibrosis

Cystic fibrosis (CF) is due to a mutant gene located on chromosome 7. The gene is large and many different mutations have been linked with CF. Some mutations give rise to mild symptoms; others have much more severe effects. Although every cell in the body of a CF patient contains a mutant gene, the most harmful effects are in the epithelial tissue of the lungs. Gene therapy therefore targets the epithelial tissue lining the lungs. The aim is to put normal alleles of the gene into the cells. In the UK, clinical trials are underway using liposomes (spheres made of lipid) designed to carry the gene that codes for the functional polypeptide that is missing in CF patients. If all goes well, this normal gene will be fully integrated into the epithelial cells so that it produces the functional protein, and the cells will divide normally to achieve a permanent cure.

- -

Check your understanding

1 Suggest why cystic fibrosis is a better candidate for gene therapy than heart disease.

2 a What is the vector used in the UK trial of gene therapy for cystic fibrosis?

b What advantage might this have over a viral vector?

3 Is the gene therapy described for cystic fibrosis an example of

a gene replacement or gene supplementation

b germ-line gene therapy or somatic-cell gene therapy?

4 In Britain, the Warnock Committee concluded that germ-line gene therapy is unethical. Suggest arguments for and against germ-line gene therapy.

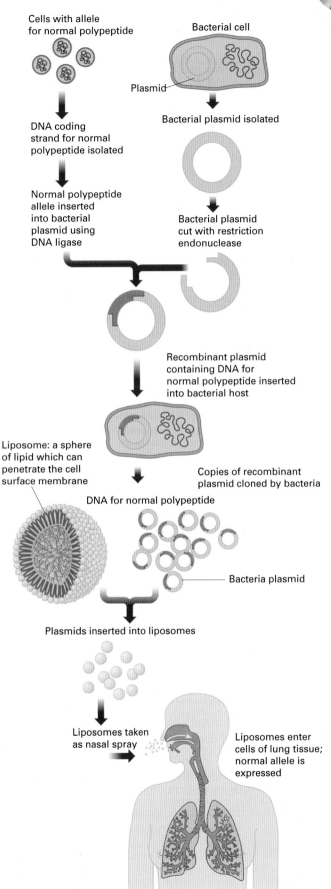

An outline of gene therapy for cystic fibrosis

OBJECTIVES

By the end of this spread you should be able to

- discuss the use of labelled DNA probes and DNA hybridization to locate specific genes

- explain how the base sequence of a gene can be determined by restriction mapping

Prior knowledge

- classification and DNA (AS Biology for AQA spread 17.02)

- DNA–DNA hybridization (AS Biology for AQA spread 17.03)

Recombinant DNA technology has been used to locate specific genes as well as to isolate and modify them. One technique uses **DNA probes** and **DNA hybridization**; another uses **restriction mapping**.

DNA probe

A DNA probe consists of a single strand of DNA that contains a nucleotide base sequence complementary to the known sequence of bases in the mutant gene. Therefore, to apply this technique, at least part of the DNA base sequence in the required gene must be known.

The following is an outline of the procedure which is carried out *in vitro* under controlled conditions designed to prevent contamination. The DNA probe is used to find out whether a particular gene is present in a sample of test material, such as DNA taken from cells in blood.

- Restriction endonucleases are added to the test material. The enzymes cause the DNA to be released and double strands to separate into single strands.

- The specific DNA probe is added. If, for example, DNA in the mutant gene is known to include the base sequence CTAAGTCCCA, the probe will be made with the complementary base sequence GATTCAGGGT. The probe is labelled with a radioactive or fluorescent marker.

- If the test material contains the mutant gene, the DNA probe hybridizes with DNA in the test material by complementary base pairing.

- The label enables the DNA probe to be tracked and the mutant gene to be located.

DNA probes and DNA hybridization have been used to locate mutant genes associated with specific medical disorders. For example, the mutant gene for cystic fibrosis was found to be located at a particular locus on chromosome 7. They are also used in gene tests to detect whether or not a person has any copies of a mutated gene.

Restriction mapping

Restriction endonucleases are used to cut a DNA molecule at specific **restriction sites** into several fragments. The number and size of the fragments can be determined using gel electrophoresis. The smaller the fragment, the further it moves along the gel. By using DNA fragments of known size to calibrate the gel, it is possible to estimate the number of base pairs in each test fragment. A fragment may have, for example, 12 kb (12 kilobase pairs, 12 000 base pairs.)

If the ends of the test DNA are marked with a fluorescent dye, and the DNA is completely digested with two or more different restriction endonucleases both singly and together, enough information is gained to make a **restriction map**. This is so called because it locates the restriction sites on a DNA molecule at which the specific restriction base sequence occurs. The diagram on the next page shows how the restriction map is pieced together.

The scientist needs to take into account the following points when compiling a restriction map.

- DNA can be circular or linear.
- Specific restriction endonucleases cleave (cut) the original DNA at specific restriction sites.
- Partial digestion of the DNA will produce larger fragments than total digestion. If the DNA is incubated with the enzymes for long enough to achieve total digestion, the larger fragments will be cut by the enzymes into smaller ones.
- With linear DNA, the total number of fragments is always one more than the number of sites; with circular DNA, the number of fragments equals the number of sites.
- Similarly sized fragments can mask each other – although they have different base sequences, they can appear as one band on a gel because of their similar sizes.
- The combined length of the DNA fragments should equal the original length of the undigested starter DNA. If it does not, one or more DNA fragments might be masked.
- When two restriction endonucleases have been used, it is helpful to look for fragments in one enzyme's pattern that appear to be cut by the other enzyme. It is often good to start with the frequent cutter's fragments and look for those that are cut again by the less frequent cutter enzyme.
- When linear DNA is digested by two enzymes, the ends of fragments may be the free ends (F) of the starter DNA, exposed ends (E1) of the internal parts cut by enzyme 1, or exposed ends (E2) cut by enzyme 2. So the ends of a DNA fragment might be any of the following:

  F_____E1

  F_____E2

  E1_____E1

  E2_____E2

  E1_____E2

  With circular DNA there will of course be no free ends.
- Determining what kinds of ends are on each fragment provides valuable information about where the restriction sites are for restriction mapping. For example, the free ends could be labelled with a fluorescent dye.

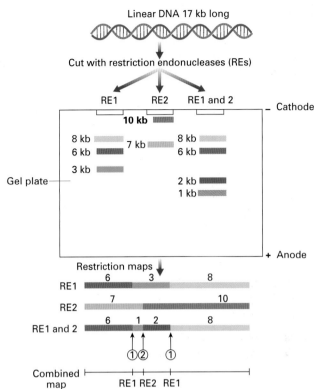

Restriction mapping. A sample of linear DNA has been digested by two different restriction endonucleases, RE1 and RE2, singly and together. The fragments of DNA have been separated by gel electrophoresis. All the DNA fragments had the same negative charge and moved towards the anode. Fluorescent markers showed which fragments were at the ends of the original DNA. By using all the information, a restriction map was compiled showing the locations of the restriction sites 1 and 2, which correspond to the locations on the DNA cut by RE1 and RE2, respectively.

Check your understanding

1 What is a DNA probe?

2 What is a restriction map?

3 By what process are DNA fragments separated?

OBJECTIVES

By the end of this spread you should be able to

- explain how DNA sequencing of genomes is carried out

- discuss how gene tests can be used to diagnose disorders caused by mutant genes and the role of genetic counselling in interpreting the test results

Prior knowledge

- DNA sequencing by the chain-termination method (*AS Biology for AQA* spread 17.02)

- oncogenes (10.13)

- acquired and inherited mutations (10.14)

- gene cloning and transfer (11.01–11.03)

- applying recombinant DNA technology (11.04)

- gene therapy (11.05)

DNA sequencing of the human genome has required the use of many high-powered computers programmed with sophisticated software and operated by specially trained computer analysts to analyse millions of DNA base sequences.

The way in which many disorders are diagnosed is changing, following advances in recombinant DNA technology and our increased knowledge and understanding about mutant genes resulting from the **DNA sequencing** of the human genome. Doctors are now able to carry out genetic tests for a wide range of disorders. DNA probes can be used to detect mutant alleles associated with a number of acquired disorders (for example, cancer of the colon) and inherited disorders (for example, cystic fibrosis and sickle cell anaemia).

Genetic tests often involve cloning a small sample of DNA from a person using the **polymerase chain reaction** (**PCR**). The tests are designed to identify gene mutations, chromosome abnormalities, or the production of specific proteins associated with particular disorders.

Positive gene test results

If an apparently healthy person gives a positive gene test result, this might show that each of their cells has one of the following:

- a genetic mutation that is expressed only at a later date (for example, Huntington's disease, an inherited genetic disorder caused by a dominant allele, which usually manifests itself between the ages of 35 and 45 years)

- a single recessive allele (that is, they are heterozygous) which makes them predisposed towards a disease such as cancer of the colon, although they are healthy at the time of the test

- a single recessive allele which makes them a symptomless carrier of a disorder such as cystic fibrosis

A positive test may also confirm diagnosis of a particular gene-based disorder in a person showing symptoms, such as sickle cell trait or sickle cell anaemia.

Mutant alleles, sickle cells, and malaria

The mutant allele associated with sickle cells is particularly interesting because it is expressed differently in the heterozygous and homozygous condition. In the homozygous condition, the two mutant alleles are expressed in the production of abnormal haemoglobin which causes red blood cells to become sickle shaped at low oxygen partial pressures. This causes full-blown sickle cell anaemia, an inherited disorder characterized by tiredness and breathlessness (similar to other forms of anaemia). Sickle-shaped blood cells in the limbs may cause acute pain and, although those affected can have normal lifespans, the disorder may also lead to kidney failure, collapse, and even death.

In the heterozygous condition, the single mutant allele causes sickle cell trait. The normal allele is expressed as normal haemoglobin, and the mutant allele as abnormal haemoglobin. People with the trait have blood cells with a lower than normal binding capacity for oxygen. This sometimes leads to mild anaemia, particularly when oxygen partial pressures in the blood are very low (for example, after heavy exercise or at high altitudes), but usually those with sickle cell trait have no adverse symptoms.

People with sickle cell trait have an advantage, in that they also have a higher than normal resistance to malaria. Because the malarial parasite has a demanding aerobic metabolism requiring a plentiful supply of oxygen, it cannot grow and reproduce within the red blood cells of those with sickle cell trait. This probably explains why sickle cell anaemia is very common in parts of Africa, the Middle East, and India, where malaria is most prevalent.

Negative gene test results

If a person has a negative gene test result this should indicate one of the following:

- they are not affected by a particular disorder
- they are not a carrier of a specific genetic mutation
- they do not have an increased risk of developing a certain disease

However, it is possible that a negative test result may be false, either because it was carried out incorrectly, or because it was not sensitive enough and missed a disease-causing mutation that was actually present.

There is also the possibility of positive test results being false (for example, due to contamination or wrong labelling), so further testing may be required to confirm the result.

Interpreting gene tests

Whatever the result, interpreting gene tests is rarely straightforward. It often requires the expertise of a trained genetic counsellor who will consider a patient's own medical history, family medical history, and the specific test that was carried out. For example, because blood relatives are bound to have some genetic material in common, a positive test result may also have implications for other members of the family. Also, for inherited genetic disorders such as cystic fibrosis, although heterozygous carriers show no symptoms, there is a risk that their children will have the disease. Genetic counsellors will explain these risks and provide parents with sound information which they can use to make their own decisions about having or not having children.

Genetic counsellors also explain that even if a test result is positive, it is often impossible to determine with certainty if a person without symptoms will develop them in the future, or how severe the symptoms will be. A positive result from a test for cancer of the colon, for example, would only show that the person has an increased risk of developing the cancer. It does not show that the person *will* develop it. Nevertheless, with the knowledge of a positive result, the person could be screened more frequently for any early signs of the disease and adopt a lifestyle that reduces the risks of a cancer developing in the future.

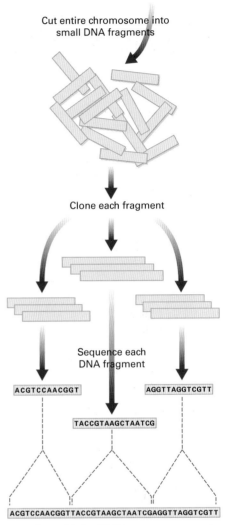

Chromosome

Cut entire chromosome into small DNA fragments

Clone each fragment

Sequence each DNA fragment

ACGTCCAACGGT   AGGTTAGGTCGTT

TACCGTAAGCTAATCG

ACGTCCAACGGTTACCGTAAGCTAATCGAGGTTAGGTCGTT

Assemble the DNA base sequence for the whole chromosome

DNA sequencing *refers to techniques, such as the* **chain-termination method**, *used to determine the base sequence in a piece of DNA. Although it's relatively easy to do this for small fragments of DNA, it has been a monumental challenge to DNA sequence complete genomes. Methods include the following:*

- *cutting DNA of entire chromosomes into small fragments using specific restriction endonucleases*
- *cloning the fragments in vivo (for example, in plasmids) or in vitro using PCR*
- *sequencing each fragment*
- *analysing the separate sequences and assembling them into an overall sequence*

Genome projects have been successful for a number of species including humans by combining highly automated recombinant DNA technology with sophisticated computer technology that analyses and assembles partial sequences.

Check your understanding

1 What is the role of PCR in DNA sequencing?

2 What is meant by a false negative gene test result?

3 In the future, gene therapy may be able to eliminate the mutant sickle cell anaemia allele. Suggest why it may not be advisable to eliminate this allele completely from the world population.

OBJECTIVES

By the end of this spread you should be able to

- describe the technique of genetic fingerprinting and discuss its uses

Prior knowledge

- gene cloning and transfer (11.01–11.03)
- applying rDNA technology (11.04)

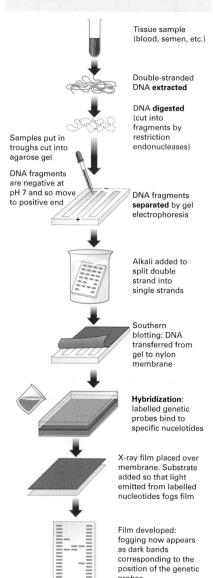

Tissue sample (blood, semen, etc.)

Double-stranded DNA **extracted**

DNA **digested** (cut into fragments by restriction endonucleases)

Samples put in troughs cut into agarose gel

DNA fragments are negative at pH 7 and so move to positive end

DNA fragments **separated** by gel electrophoresis

Alkali added to split double strand into single strands

Southern blotting: DNA transferred from gel to nylon membrane

Hybridization: labelled genetic probes bind to specific nucelotides

X-ray film placed over membrane. Substrate added so that light emitted from labelled nucleotides fogs film

Film developed: fogging now appears as dark bands corresponding to the position of the genetic probes

The main stages of genetic fingerprinting

Each chromosome in a eukaryotic cell contains a single molecule of DNA, elaborately coiled and intertwined with proteins. In humans, one DNA molecule may contain up to 4000 genes. These are located at different positions (loci) along the chromosome. However, about 95 per cent of human DNA does not code for polypeptides. Stretches of non-coding DNA may lie between different genes, and the genes themselves contain non-coding portions called introns.

It might be expected that the nucleotide sequences in the coding part of DNA could be used to identify individuals from their DNA. After all, it is the coding nucleotides that determine each person's characteristics. It therefore came as a surprise when, in 1984, Alec Jeffreys at Leicester University devised a method of identifying individuals using the non-coding regions of DNA. The method was called **DNA profiling**, but it became popularly known as **genetic fingerprinting** because it is used, like traditional fingerprinting, to help police with investigations. Forensic scientists can use genetic fingerprinting to compare DNA from hair, saliva, blood, or semen found at the scene of the crime with that of suspects. The same type of procedure is used to compare genetic material from different organisms.

The principle

Non-coding DNA used in genetic fingerprinting contains **hypervariable regions**. These have repeating base sequences called **core nucleotide sequences**. The number and length of the repeats vary between individuals, but are similar in related individuals: the closer the relationship, the greater the similarities. The four main steps involved in genetic fingerprinting are extraction, digestion, separation, and hybridization.

Extraction

Extraction of the DNA is normally carried out by mixing a tissue sample with a solvent (for example, a mixture of water-saturated phenol or chloroform) which dissolves DNA but precipitates protein.

Digestion

Restriction endonucleases are then used to digest the DNA into fragments. The enzymes cut the DNA at specific points and produce fragments of varying size, some of which contain hypervariable regions.

Separation

The DNA fragments are separated by gel electrophoresis.

- The fragment mixture is placed in a trough cut into agarose gel. The gel is immersed in buffer at pH 7, and a current is applied to the gel.
- All the DNA fragments move towards the positive electrode because they are negatively charged at pH 7. However, the smaller fragments move faster through the gel than the larger ones, so gel electrophoresis separates fragments according to their size.
- Markers (DNA fragments of known size) are also separated in the gel.

Up until this point, the DNA is double stranded. After electrophoresis, it is converted into single strands by immersing the gel in an alkali. The DNA fragments are blotted onto a nylon membrane using a technique called **Southern blotting**:

- A thin nylon membrane is laid over the gel and several sheets of blotting paper or filter paper are laid on top.
- The buffer containing the DNA is drawn up through the filter paper by capillarity and some of the DNA is deposited on the nylon membrane. The positions of the DNA fragments on the membrane correspond with their positions on the gel.
- The DNA is fixed on the membrane by exposing it to short-wavelength ultraviolet light.

DNA hybridization

The separated single-stranded DNA is mixed with DNA probes which contain nucleotide sequences that are complementary to the core nucleotide sequences known to occur in hypervariable regions. The probes are labelled with the enzyme **alkaline phosphatase**. When the nylon membrane is incubated at the correct temperature, pH, and ionic strength, complementary base pairing occurs between the probes and the DNA from the sample.

After hybridization, the membrane is covered with a phosphate-containing substrate and placed on an X-ray film in the dark. The alkaline phosphatase removes the phosphate; and this causes the substrate to fluoresce, fogging the X-ray film.

When developed, the film shows dark bands where the probes were bound to the DNA in the hypervariable regions. The positions of these bands (the genetic profile or fingerprint) can be used to identify individuals, and are used to work out the genetic relationships between different individuals.

Uses of genetic fingerprinting

DNA profiling is used in many thousands of criminal investigations each year. With the exception of identical twins, the probability of two people having the same genetic fingerprint is very small. However, small samples of degraded DNA may give only a few bands, giving a less reliable fingerprint. This problem is largely overcome by PCR which can clone millions of copies of DNA from a very small sample. PCR enables DNA profiling to be used successfully on very small samples of even partly degraded DNA from ancient, long-dead sources.

Genetic fingerprinting is also used to resolve paternity disputes, check immigration applications, confirm animal pedigrees, and establish the genetic relationships between ancient humans preserved in peat bogs and ice graves and people alive today. It is also used extensively in biological research to establish evolutionary relationships between organisms.

Check your understanding

1 What are core nucleotide sequences, and in which region of DNA are they found?
2 Why is DNA profiling commonly referred to as genetic fingerprinting?
3 In DNA profiling, what is the function of alkaline phosphatase?

---

4 The genetic fingerprints in the diagram were made from a blood sample from a rape victim, a specimen of semen taken from the victim, and the blood of three suspects.
   a On the basis of the genetic fingerprints, which suspect was the rapist?
   b How reliable do you think this evidence would be?
   c Suggest how the reliability could be improved.

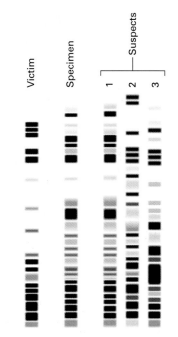

Diagram of genetic fingerprints from a victim's blood

OBJECTIVES

By the end of this spread you should be aware

- of the key biological principles that you are expected to understand for Unit 5

- that you will be expected to apply that understanding to unfamiliar contexts

Prior knowledge

- regulation of the internal environment (9.01)

- stimulus perception (7.01 and 7.04–7.05)

- chemical and electrical coordination in animals (7.06–7.07)

- response by effectors (7.02–7.03)

- plant growth factors (7.08)

- the genetic code and polypeptide determination (10.01–10.03)

- regulation of gene expression (10.07–10.08 and 10.11)

- artificial manipulation of gene expression (10.13 and 11.01–11.08)

The key biological principles that you have studied in Unit 5 are summarized below.

- Organisms regulate their internal environment and so maintain optimum conditions for their metabolism.

- Animals respond to their internal and external environment as a result of stimulus perception, chemical and electrical coordination, and a response by effectors.

- Plants respond to their external environment as a result of specific growth factors that regulate cell growth.

- The genetic code is held in the base sequence of nucleic acids and determines the amino acid sequence of polypeptides produced by a cell.

- Regulating gene expression enables a cell to control its own activities and development. Scientists are able to manipulate gene expression for many agricultural, industrial, and medical purposes.

These biological principles have been the subjects of chapters 7–11 and are commented on below. You are expected not only to understand the biological principles in the specific contexts described in the chapters, but also to be able to apply those principles to unfamiliar contexts.

Regulation of the internal environment

Homeostasis, the ability of an organism to maintain the physical and chemical conditions of its internal environment in a steady state, is one of the distinguishing characteristics of living organisms. Living organisms therefore tend to be self maintaining and self equilibrating. The internal environment refers either to the cytoplasm of a unicellular organism, or to the tissue fluids surrounding the cells of a multicellular organism. In A2 Biology, you are expected to understand that homeostasis usually involves negative and positive feedback, and that these mechanisms play a key part in the regulation of body temperature, the oestrus cycle, and blood glucose in mammals, and many other processes.

Stimulus perception

A stimulus refers to a change in the environment, for example, an increase or decrease in light intensity. Perception refers to the mental processes by which the brain interprets and gives meaning to the information it receives from the sense organs. The information is processed much more thoroughly in the complex and highly developed brains of mammals than in the simple brains of invertebrates. Even in mammals, some behavioural responses, such as a simple reflex, have the minimum of internal processing in the central nervous system. Other behavioural responses, for example those relying on receptors in the retina of the eye, involve processing in the synaptic connections between the sensory cells and the optic nerve, and in the occipital region of the brain.

Chemical and electrical coordination in animals

Animals have two main types of coordination: chemical coordination, which is generally slow acting and which can affect many different parts of the body at the same time; and nervous coordination, which is usually fast acting and has more short-lived and localized effects. There are similarities between the two systems – both involve chemical transmitters (hormones for chemical coordination and neurotransmitters for nervous coordination) and both usually involve receptor molecules that have shapes complementary to those of the transmitters.

Response by effectors

The effectors of an animal include any gland, muscle, or organ that brings about an action as a result of the stimulus it receives. The coordinated actions of several effectors are usually required to bring about behavioural responses, such as kineses and taxes in woodlice. The particular effectors that you are expected to know about include striated muscle and those involved in the regulation of blood glucose, body temperature, heart rate, and the oestrus cycle.

Plant growth factors

Plant coordination is almost exclusively chemical. Plant growth factors are chemicals produced by glandular cells that bring about movement responses through their effects on cell division and growth. One factor that features strongly in the A2 Specification is indole-3-acetic acid (IAA). You are expected to understand how this chemical brings about a positive phototropism in aerial parts of plants (for example, the coleoptiles of maize), and also how plant growth factors can be applied in horticulture and agriculture.

Genetic code and polypeptide determination

The genetic code is mentioned in the AS Specification, but in A2 Biology you develop your understanding of its importance. In particular, you are expected to know how the genetic code on mRNA is translated on ribosomes into specific polypeptides. The universal nature of the genetic code is the basis of the many modern aspects of biology which involve recombinant DNA technology.

Regulation of gene expression

Gene expression is said to take place when the information in a gene is transcribed into mRNA and then translated into a specific polypeptide. You should be aware that most genes in any cell of any organism are not active and that activation involves special factors called transcription factors. Many disorders linked to an inherited or acquired mutation result from a breakdown in gene expression: in some, genes that should be active are inactive; in others, genes that should be inactive are active. Effective treatment of the disorder often depends on a detailed understanding of the mechanisms regulating gene expression.

Artificial manipulation of gene expression

A detailed understanding of the regulation of gene expression is enabling molecular biologists to manipulate it artificially using recombinant DNA technology. You are expected to be able to explain how this technology is being used in agriculture (for example, to increase crop productivity), in medicine (for example, in cancer treatments), and in industry (for example, to make particular chemicals).

Meerkat (Suricatta suricatta) standing together outside their burrows in the Kalahari Desert, South Africa. Meerkats use behavioural as well as physiological mechanisms to regulate their body core temperatures. Around midday during the hot season, they usually withdraw into their extensive network of burrows to protect themselves from the heat of the desert sun. At night when desert temperatures can plummet, they sleep huddled together in one chamber of the burrow to combat heat loss. On cold clear days of the dry season, they can often be seen sunbathing, absorbing heat through the dark patch of sparsely furred skin on their stomach which acts like a solar panel.

Check your understanding

1 Define homeostasis.
2 Distinguish between a hormone and a neurotransmitter.
3 What is indole-3-acetic acid?
4 When does gene expression take place?

Question 1: Sensitivity – the retina

Study the figure on the left showing connections between cells in the retina.

a Explain how the connection of several rods to a single bipolar cell influences
  i visual acuity
  ii sensitivity [4]

b Study the graph below showing the distribution of rod cells and cone cells across a human retina.
  i Describe the relative distribution of rods and cones in the retina. [4]
  ii Explain why an image formed at 0 degrees from the centre of the retina is perceived in more detail than an image formed at 40 degrees either side of the centre. [4]

c Describe how cone cells allow us to see different colours. [3]

[Total marks = 15]

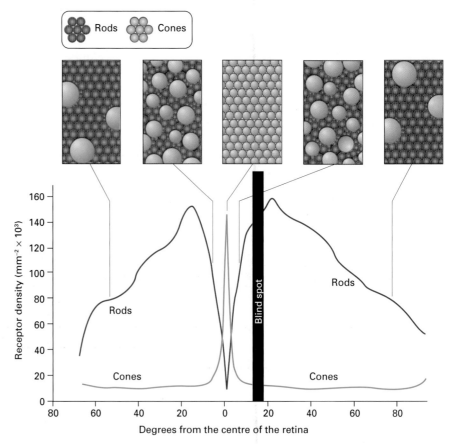

Graph showing the distribution of rod cells and cone cells across a human retina

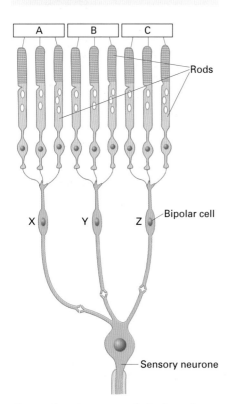

Connections between cells in the retina

Question 2: Sensitivity – plant growth substances

Coleoptiles are protective sheaths around the germinating shoots of grasses and cereal crops such as wheat and maize. Coleoptile tips produce a plant growth substance which can be collected by placing cut coleoptiles on to agar blocks.

a Name one common plant growth substance involved in the response of plants to light. [1]

b Suggest how you could demonstrate that the plant growth substance involved in phototropisms is produced at the tip of the coleoptile and not further down where the coleoptile bends. [2]

c Study the figure and the table of results below of an experiment carried out to investigate the movement of the plant growth substance through coleoptiles. In the experiment, the tip of a coleoptile was removed and a block of agar (labelled A) containing plant growth substance was placed on the cut surface. Another block of agar (labelled B) was placed under the lower cut end surface of the coleoptile. A small section was cut out of the coleoptile so that a third block of agar (labelled C) could be placed on the left side of the coleoptile. The concentration of plant growth substance in each block was estimated by carrying out a bioassay. The results are shown in the table.

i Describe what the results of this investigation show about the movement of the plant growth substance through the coleoptile. [4]

ii This plant growth substance stimulates elongation of cells in the shoot and coleoptile and is involved in phototropism. Explain how the movement of the plant growth substance brings about a positive phototropism of the coleoptile. [2]

iii Distinguish between a positive phototropism and a positive phototaxis. [2]

d Suggest two ways the plant growth substance involved in phototropism of coleoptiles could be used in horticulture and agriculture. [4]

[Total marks = 15]

A barley seedling coleoptile

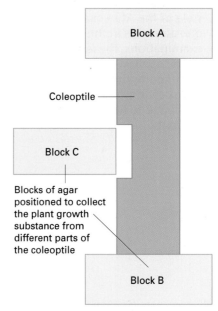

Collecting a plant growth substance on agar blocks

Agar block	Amount of plant growth substance (arbitrary units)	
	Dark	Unidirectional light (from right side)
block A	100	100
block B	75	62
block C	25	38

Question 1: Nerve impulses

a Define the term **threshold stimulus**. [2]

b The resting potential of a motor neurone is about −70 millivolts. Describe how a resting potential is produced. [4]

c Study the figure of the action potential in a motor neurone. Describe what happens during

i the depolarization phase and

ii the repolarization phase

in terms of membrane permeability and the movement of ions across the membrane. [4]

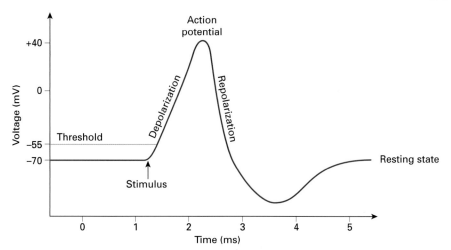

An action potential

d Study the figure of the synaptic connections between neurones A, B, and C.

i Name the structures which contain neurotransmitters. [1]

ii Explain how neurones A and B might have an excitatory effect or an inhibitory effect on neurone C. [4]

[Total marks = 15]

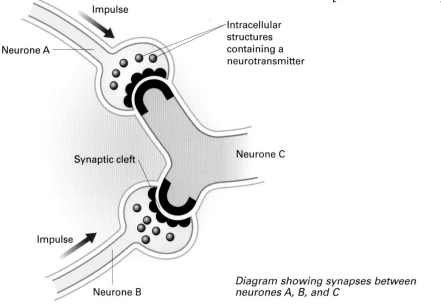

Diagram showing synapses between neurones A, B, and C

Question 2: Muscle structure and function

Study the figure of an electron micrograph of part of a myofibril of a skeletal muscle. It shows the line along which the myofibril was cut at two different stages of muscle contraction. The accompanying figures show the ends of the protein filaments at these two different stages of contraction.

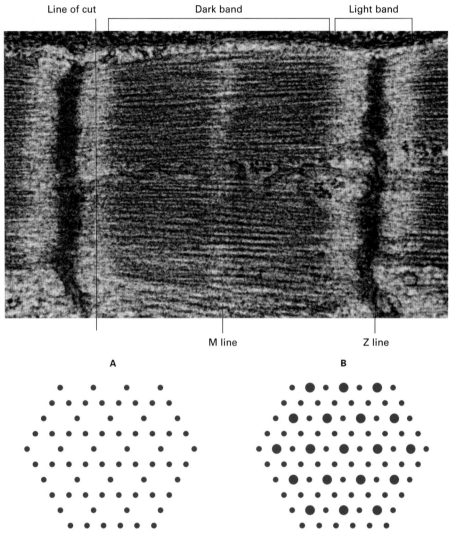

An electron micrograph of part of a myofibril of a skeletal muscle. It shows the line along which the myofibril was cut at two different stages of muscle contraction. Figures A and B show the ends of the protein filaments at these two different stages of contraction.

a Name the protein filaments and explain the differences between them.

[5]

b Distinguish between acetylcholine and acetylcholinesterase.   [2]

c Mammalian skeletal muscle has two main types of fibres. Name the two types and describe how they differ.   [5]

d Marathon runners may suffer from muscle fatigue during which the contraction mechanism is disrupted. One factor thought to contribute to muscle fatigue is a decrease in the availability of calcium ions within muscle fibres. Explain how a decrease in the availability of calcium ions could disrupt the contraction mechanism in muscles.   [3]

[Total marks = 15]

The harp seal, Phoca groenlandica

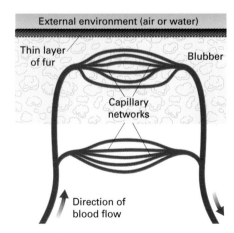

Blood vessels in the skin surface of the seal

Question 1: Temperature control

Harp seals (*Phoca groenlandica*) live in the Arctic. Like most other marine mammals, they have a core body temperature of about 37°C and, like all other mammals, their skin has hair. In seals, this forms a layer of fur on the body surface. In addition, seals have a layer of blubber under the skin that is about 50 mm thick.

a Study the figure below of the seal's temperature at different distances from the skin surface when the seal is in water at 0°C, and the figure below left of a section through a seal's body.

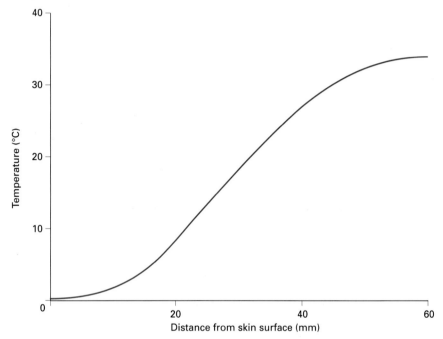

The temperature of a seal at different distances inside the body from the skin surface

  i Explain the meaning of the phrase 'core body temperature'. [2]

  ii What is the temperature of the seal at its skin surface? [1]

  iii By how much does the temperature of the seal at 50 mm from the body surface differ from its core temperature? Show your workings. [2]

  iv What does the graph indicate about the function of the blubber? [1]

  v Name two other functions of blubber in seals. [2]

b Study the figure opposite that shows the arrangement of blood vessels that lead from the body tissues to the skin surface of the seal. Explain how this arrangement helps the seal maintain a constant body core temperature. [4]

c The flippers of seals have very little fat and no thick layer of blubber. Study the arrangement of blood vessels in the flippers of a seal opposite. Explain how this arrangement minimizes heat loss. [3]

[Total marks = 15]

Question 2: Insulin and the control of blood glucose

a Study the graph showing the effects of an intake of 100 g of glucose on a person who has been fasting for 12 hours.

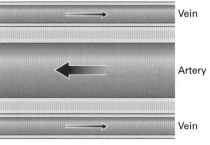

→ Direction of blood flow

Blood vessels in the flippers of the seal

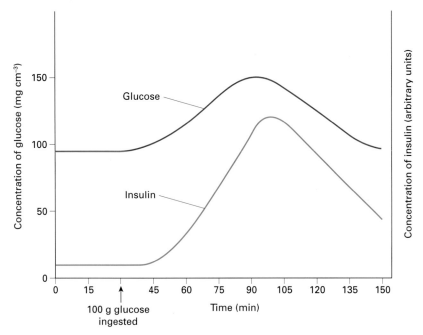

Blood glucose and insulin concentrations after an intake of glucose

  i Describe the relationship between the concentrations of insulin and glucose in the blood after swallowing glucose. [1]

  ii With reference to the graph, explain negative feedback. [2]

  iii Explain why the concentration of glucagon in the blood rises during exercise while that of insulin falls. [2]

b Diabetes mellitus is a complex disorder that has two main types: type I, insulin-dependent or early onset diabetes; and type II, non-insulin-dependent diabetes. People with type I diabetes require regular injections of insulin because the cells that produce the insulin appear to have been destroyed by the body's own immune system. Type II diabetes can develop at any age, but its frequency of occurrence increases with age. Its development is often associated with obesity. Many type II diabetics have a lower sensitivity to insulin, possibly as a result of a decrease in the numbers of insulin receptors.

  i Which cells are destroyed in type I diabetes? [1]

  ii Outline the actions of insulin which enable it to control bl od glucose levels. [3]

  iii Why is it not possible for diabetics to take insulin orally? [1]

  iv What sort of molecules are insulin receptors and where do they occur? [2]

  v Suggest how diet and exercise can be used to control type II diabetes. [3]

[Total marks = 15]

Question 1: The genetic code and protein synthesis

a Copy and complete the table showing three amino acids and the DNA base sequences and RNA base sequences that code for them. [2]

Amino acid	mRNA base sequence	DNA base sequence
serine	UCU	
proline		GGC
tyrosine	UAU	

b Sketch the diagram of a tRNA molecule. Indicate where an amino acid is attached and where the anticodon would be. [2]

c Explain the role of the anticodon in protein synthesis. [2]

d Give two ways in which the structure of a molecule of tRNA differs from the structure of a molecule of mRNA. [2]

e The mRNA coding for a protein has 372 bases. The protein is secreted in an inactive form. It is activated when an enzyme removes three amino acids from one end. Calculate how many amino acids there are in the activated protein. Show your working. [2]

f Study the diagram outlining polypeptide synthesis. Indicate which of the processes A and B is 'transcription' and which is 'translation'. Explain why these processes are necessary for an organism's survival. [5]

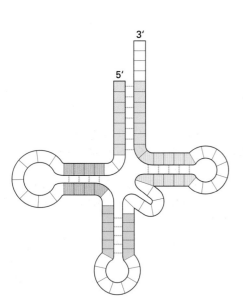

A tRNA molecule

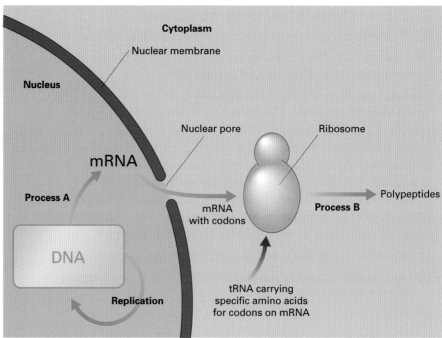

Outline of polypeptide synthesis

[Total marks = 15]

Question 2: Cancer and chemotherapy

Read the following passage carefully.

Cell division is a complex process regulated by a number of factors. If the regulation breaks down, cells may divide uncontrollably to produce a tumour. Some tumours are benign; others are malignant and lead to cancers. Breakdown of the regulation of cell division is caused by mutations. These mutations are often destroyed before they have any damaging effect. Even when a mutated cell survives, it is very rare for one mutation to lead to a cancer. At least three separate mutations are required for most cancers, and in some cases a cancer will develop only after 20 mutations. It is for this reason that the incidence of cancer is age related, and why it occurs more commonly in people exposed to mutagens.

Cancer cells can often be distinguished from normal cells by the proteins they produce. They may also differ in appearance, especially after being stained with suitable dyes. Early diagnosis of cancers is usually the key to their effective treatment. Diagnosis can take many forms including DNA probes, the use of analytical enzymes to test blood and body fluid samples, and microscopical examination of cells.

Chemotherapy is one of the methods used to treat cancerous growths. Drugs are administered which kill dividing cells, both normal and tumour. However, normal cells are less sensitive to the drug than are tumour cells, so a suitable dose will kill proportionately more tumour cells than normal cells. A chemotherapy programme typically consists of administering the same dose of drug at regular intervals (for example, every 21 days) until all the tumour cells are destroyed.

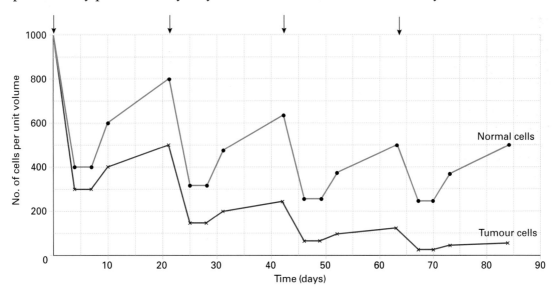

The results of a chemotherapy programme

Use information from the passage and your own knowledge to answer the following questions.

a Name two different mutagens that can lead to cancer. [2]

b Explain why cancer is age related. [2]

c State two ways in which cancer cells differ from normal cells. [2]

d What makes enzymes suitable analytical reagents for use in the diagnosis of cancers? [2]

e With reference to the graph, describe the way in which both types of cell respond during the first 30 days of treatment. [2]

f Tumour cells could be more effectively destroyed by increasing the dosage of chemotherapy drug or by decreasing the time interval between administering the doses. Suggest why the amount of drug is not usually increased on each occasion and the time interval is not shortened. [2]

g Explain why early diagnosis of cancer is so important. [3]

[Total marks = 15]

OBJECTIVES

By the end of this spread you should have answered questions on

- cloning and applying rDNA technology

Prior knowledge

- cloning and applying rDNA technology (11.01–11.08)

Stretch and challenge

These long structured questions on cloning and applying rDNA technology are posed in the style of the AQA examination questions. As in the AQA examinations, these questions are designed with an incline of difficulty such that the later subquestions offer a genuine challenge to the most able students.

Question 1: Cloning

a An egg cell from a sheep was fertilized *in vitro* and allowed to develop into an eight-celled embryo. This was split into four pairs of cells. Each pair of cells developed into a new embryo which was later transferred into different surrogate sheep.

  i Why are the new embryos referred to as clones? [2]

  ii Suggest why it was not possible to produce clones by splitting embryos that number more than eight cells. [1]

  iii Explain why all the sheep cloned from the same embryo are not identical in all respects. [2]

  iv Give one advantage and one disadvantage to a farmer of producing cloned sheep. [2]

b The polymerase chain reaction (PCR) is an *in vitro* form of cloning which can be used to produce large quantities of DNA.

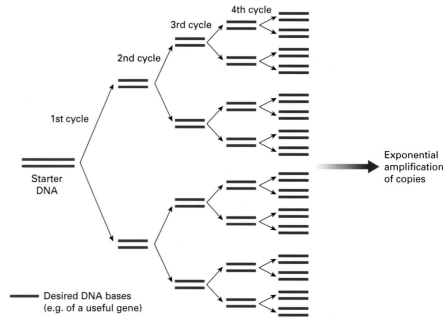

1st cycle
2nd cycle
3rd cycle
4th cycle
Starter DNA
Exponential amplification of copies

— Desired DNA bases (e.g. of a useful gene)

An outline of PCR

  i Distinguish between *in vitro* and *in vivo* cloning. [2]

  ii Describe how the PCR is carried out. [6]

[Total marks = 15]

Question 2: Restriction endonucleases and gene therapy

a Study the figure which shows the position of four different restriction sites. The distance between these sites is measured in kb. The circular DNA was cut using only two restriction endonucleases. The resulting fragments were separated by gel electrophoresis.

i There are many different restriction endonucleases which are used to cut DNA molecules into small fragments. Some cut the DNA at palindromic restriction sites. What are palindromic restriction sites?

[2]

ii What does kb mean? [1]

iii With reference to the gel electrophoresis of the DNA fragments cut by the two enzymes, which of the restriction sites A, B, C, or D were cut? Give reasons for your answer. [2]

iv State where restriction endonucleases occur naturally and give their function. [2]

b With reference to germ-line gene therapy, somatic-cell gene therapy, gene replacement and gene supplementation, explain what gene therapy is. [8]

[Total marks = 15]

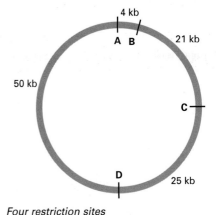

Four restriction sites

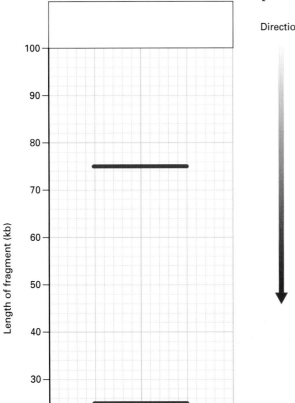

Gel electrophoresis of the DNA fragments

Question 1

a The oestrus cycle occurs in all female mammals. The key event during the cycle is ovulation. Describe what happens during ovulation. [2]

b Distinguish between the oestrus cycle and the menstrual cycle. [1]

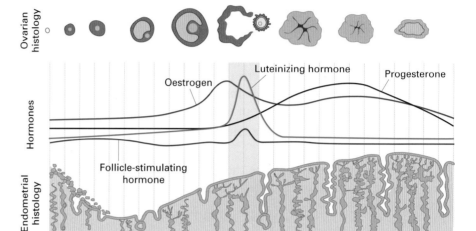

The menstrual cycle

c What makes hormonal control better than nervous control for regulating the oestrus cycle? [2]

d Recombinant follicle-stimulating hormone (rFSH) and clomiphene are two drugs used to treat women who are infertile because they do not ovulate.

  i Suggest why rFSH would not be effective if it were taken orally. [1]

  ii Explain how taking rFSH by injection might help a woman who was infertile. [2]

  iii Clomiphene inhibits the action of oestrogen. Suggest how it might do this. [2]

  iv How might taking clomiphene help a woman who was infertile? [2]

  v By using calculations based on the data in the table, which shows the results of a trial comparing rFSH and clomiphene, find which of the two drugs is more effective in treating infertility. Show your workings. [3]

	Drug used	
	rFSH	**Clomiphene**
Number of women treated with the drug	50	375
Number of pregnancies following drug administration	14	16
Total cost of the drug in the trial (£)	7000	12 000

[Total marks = 15]

Question 2

Nitrogen has two isotopes: ^{15}N is a heavier isotope than the normal isotope ^{14}N.

Meselson and Stahl carried out a classic investigation of DNA replication by culturing bacteria in media containing the different isotopes. First, bacteria (generation 0) were cultured in a medium containing only ^{15}N. The bacteria were then transferred to a medium containing only ^{14}N and allowed to divide once (generation 1). A sample of these generation 1 bacteria was removed and the DNA in the bacteria was extracted and spun in an ultracentrifuge. The bacteria were allowed to divide in the medium containing only ^{14}N one more time, to form generation 2. The DNA was also extracted from these generation 2 bacteria and spun in an ultracentrifuge.

Study the diagram showing the results of this investigation.

a **i** Which part of the DNA molecule contains nitrogen? [1]

   ii Explain why the DNA from generation 1 is in the position shown in the ultracentrifuge tube. [2]

   iii Copy the third tube and show the results for generation 2. [2]

b In 1950, Erwin Chargaff studied the base composition of DNA from different organisms. Study the table showing some of his results.

   i Copy and complete the table to show the figures for human DNA. [2]

   ii What does the table indicate about the relative proportions of pyrimidines and purines? [2]

   iii The DNA found in prokaryotes and eukaryotes is double stranded. Using evidence from the table, explain why the structure of the viral DNA molecule is probably different. [2]

c Chargaff could only measure the base composition of DNA in terms of percentages. Name the methods that you could use to

   i obtain a fragment of DNA

   ii find out what size it is

   iii clone the fragment *in vitro*

   iv find the sequence of bases occuring in the fragment [4]

[Total marks = 15]

| Bacteria (generation 0) cultured in medium containing only ^{15}N | Bacteria transferred to medium containing only ^{14}N and allowed to divide once (generation 1) | Bacteria allowed to divide in medium containing only ^{14}N one more time to form generation 2 |

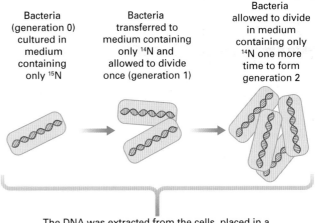

The DNA was extracted from the cells, placed in a special salt solution, and spun in an ultracentrifuge

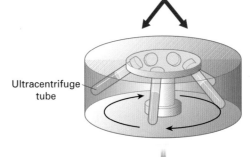

Ultracentrifuge tube

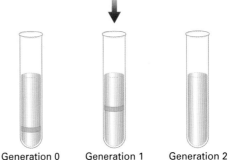

Generation 0   Generation 1   Generation 2

Meselson and Stahl's investigation

Organism	Base composition (%)			
	Adenine	**Cytosine**	**Guanine**	**Thymine**
virus	24	26	18	32
bacterium	30.8	19.0	21.0	29.2
wheat	27.3	22.8	22.7	27.1
sheep	29.3	21.0	21.4	28.3
human	40.0			

The percentage of different bases in the DNA of different organisms

Question 1

In 2005, carbon dioxide was at a globally averaged concentration of approximately 380 ppm by volume in the Earth's atmosphere, while in 1960 it was at a globally averaged concentration of approximately 320 ppm. However, the actual concentration varies from place to place, and varies in one place from one time of the day to another and from one season to the next.

Carbon dioxide is an important greenhouse gas. It makes a major contribution to global warming.

a For the globally averaged carbon dioxide concentration calculate

   i the absolute increase from 1960 to 2005 [1]

   ii the average yearly increase between 1960 and 2005 [1]

   iii the percentage increase between 1960 and 2005 [1]

   In all cases, show your workings. [2]

b i Suggest three reasons why the globally averaged carbon dioxide concentration is increasing in the atmosphere. [3]

   ii Explain how carbon dioxide in the atmosphere acts as a greenhouse gas. [2]

c The table shows the effect of increasing carbon dioxide concentration on the rate of photosynthesis in maize.

Carbon dioxide concentration (ppm)	Rate of photosynthesis (arbitrary units)
100	30
150	40
230	50
300	60
400	65
500	64

   i Using information in the table, describe and explain the effect of increasing carbon dioxide concentration on the rate of photosynthesis. [3]

   ii Suggest why increasing carbon dioxide concentration to 1% may cause a significant decrease in the rate of photosynthesis. [2]

[Total marks = 15]

Question 2

As part of a medical screening procedure, samples of blood were taken from three people and used for blood grouping, gene tests, and mitochondrial DNA analysis.

a The blood groups were determined using anti-A and anti-B antibodies and the results are shown here.

	Person 1	Person 2	Person 3
Anti-A			
Anti-B			
Blood group			
Key	Agglutination		No agglutination

The blood group test results of three people

  i What is agglutination and how does it occur? [2]

  ii Copy the table and complete it to show the blood groups of the three people. [3]

b Gene tests involved using a DNA probe.

  i Explain what a DNA probe is. [2]

  ii Where in the blood sample would a DNA probe be applied? Give a reason for your answer. [2]

c Gene mutations in mitochondria are important in a number of disorders.

  i With reference to the origin of mitochondria in eukaryotic cells, suggest why mitochondrial DNA is likely to differ from nuclear DNA. [2]

  ii Suggest a type of disorder for which a mitochondrial gene mutation might be important. Give a reason for your answer. [2]

  iii As well as being used to detect certain genetic disorders, genetic fingerprints of mitochondrial DNA (mDNA) are also used in anthropological studies to study familial relationships. One advantage of human mDNA is that it is inherited along the maternal line. Suggest why it is inherited along the maternal line. [2]

[Total marks = 15]

By the end of this spread you should

- understand the criteria of assessment for the essay question in Unit 5

Prior knowledge

- all AS Biology subject content
- all A2 Biology subject content

The essay is an important part of the synoptic assessment. It is the last part in the Unit 5 assessment and takes the form of an 'either... or...' question with a choice of two titles. For example,

Write an essay on one of the following topics.
EITHER

a Globally averaged carbon dioxide levels are rising. Describe and explain the physiological and ecological effects that increases in carbon dioxide might have on living organisms.

OR

b Define the term recombinant DNA technology (rDNA technology). Discuss the uses of rDNA technology in agriculture, medicine, and industry.

The titles are designed to give you the opportunity to demonstrate your

- knowledge and understanding of the biological principles and concepts that you have studied in the different units
- ability to express ideas relevant to the essay title, coherently and logically, using appropriate specialist vocabulary

In addition to being an important part of the synoptic assessment, the essay also forms part of the 'stretch and challenge' requirement of the AQA Specification. It provides an opportunity for you to demonstrate your reading round the subject. You can gain marks by including material that goes beyond that expected of a good A level Biology candidate.

Much of A level assessment requires short, pithy answers, but the essay is an extended piece of writing that should be written in continuous prose. You are expected to spend about 40 minutes planning and writing the essay. It is a misconception to think that you have to write extraordinarily quickly to provide enough information to gain high marks. If the essay meets all the criteria, it is possible to gain the full 25 marks with the amount that a good student can write within the expected time.

The components that make up the mark allocation are given below, along with what you have to do to gain full marks for each component. (Note that the marks are allocated independently of each other; therefore it is possible to gain full marks in one category without gaining many marks in another.)

Scientific content

This is the most important component of the essay. It is allocated 64% of the marks. The maximum mark of 16 can be obtained by writing an essay containing accurate scientific material at a high standard throughout. The essay should reflect a sound understanding of biological principles related to the essay and be free from fundamental errors. The essay should also reflect a comprehensive knowledge and understanding of the relevant key biological principles that you are expected to understand from Units 4 and 5; these are listed in spreads 6.01 and 12.01

respectively. In addition, to gain full marks, you will need to include significant references, at least two paragraphs, to scientific material that demonstrates a depth of study that goes beyond that expected in the subject content of the A level specifications. This is where your reading round the subject can be demonstrated.

Breadth of knowledge

To obtain the maximum 3 marks for breadth, the essay should be a balanced account of the title topic and make reference to most of the relevant areas that are covered in the A level Biology specifications. For example, for the essay question 'Explain what recombinant DNA technology is and discuss its uses in agriculture, medicine, and industry', you would be expected to refer to all three fields of use, but not necessarily in equal depth, to gain the full marks for breadth of knowledge.

Relevance

To gain the full 3 marks for relevance, the material you include in your essay should all be clearly relevant to the title, although allowance is made for the use of marginally relevant material in the introduction. You must also write sufficient material to demonstrate this ability; if you write a very short essay (under a page in length) you might only be able to gain one or two marks even if all the material is relevant.

Quality of written communication

To obtain the full 3 marks, you will need to write an essay of sufficient length to demonstrate your ability to present material logically in clear English and in an appropriate scientific style. You will also be expected to use technical terms effectively and accurately throughout.

A useful essay-writing strategy

Although the essay question is at the end of the Unit 5 assessment, it is a good idea to look at the title choices *before* you start answering the questions on the rest of the paper. By doing this, you are giving the subconscious parts of your brain an opportunity to retrieve information about the titles before writing the essay. Some students find it useful to copy the titles down on a piece of rough paper and then jot down ideas as they come to mind.

When writing the essay, you should always be mindful of the precise wording of the title. This will indicate the topics that you are expected to cover. You should also understand the meaning of instructions in the title such as 'describe' and 'explain' (see the Glossary of terms used in questions on the next page).

Your revision should include analysing essay titles so that you know precisely what you are expected to write about. During the pressure of formal assessments, some students make the basic mistake of focusing on only one or two words, missing key elements in the title, and including irrelevant material.

Check your understanding

Analyse each of the following essay titles:

1 Globally averaged carbon dioxide levels are rising. Describe and explain the physiological and ecological effects that increases in carbon dioxide might have on living organisms.

2 Define the term recombinant DNA technology (rDNA technology). Discuss the uses of rDNA technology in agriculture, medicine, and industry.

Identify and define the instructional terms. Break down the title to identify the topics to be covered.

Glossary of terms used in questions

The following terms are generally used in the question papers set by the AQA for A level Biology assessments. Other terms may also be used if they make the meaning of the question clearer.

Calculate: indicates that only a numerical answer is required (given, where appropriate, with correct units).

Define: you need to give a formal statement using appropriate scientific terminology which gives the meaning
of a word or biological term.

Describe how you should…: instructs you to give an account of how *you* specifically should do something within the context of the A level Specification, for example, 'Describe how you should take random samples during a piece of fieldwork.' It is worth remembering that you are expected to include only appropriate and safe procedures in such a description and you will not be given credit for anything inappropriate or unsafe.

Describe: this tells you to 'give a description of' something. This might involve the translation of one form of information to another (for example, describe the shape of a graph), or it might involve giving an account of a process, such as an experiment, investigation, or clinical trial. For example, 'Describe a field experiment' means give an account of how a particular piece of field experiment should be carried out.

Evaluate: here you need to make a judgement about the worth of something; you will be expected to do more than merely list advantages and disadvantages.

Explain: you need to give a reason for something or an interpretation of something. Explain and describe are often confused – whereas explain answers the question 'Why?', describe only answers the question 'What?'. For example, when presented with a question asking you to explain the curve on a graph, you are expected to give biological reasons for any changes of direction or pattern that can be seen on the graph. Answers to 'Explain…' questions often start with the word 'Because …'.

Give: this is used as an instruction when a statement or an account of the similarities and/or differences between two or more items is required, for example, 'Give **two** differences between a prokaryotic cell and a eukaryotic cell.'

Give the evidence from…: you need to use information contained in the relevant material provided as evidence for your answer. Answers that do not use this material may not be given credit, even if they are biologically correct.

List: this requires you to give a number of features or points, often as single words or simple phrases, with no further elaboration or detail.

Name: this usually requires a technical term or its equivalent.

Show your working: this is usually used in conjunction with 'calculate' to indicate that details or methods of the calculation need to be shown.

Sketch: usually used in relation to graphs, this instructs you to make a drawing which can be on ordinary lined paper of a simple estimate of a curve. Although not drawn on graph paper, you are expected to include correctly labelled axes.

Suggest: this requires you to attempt to answer a question for which there may be no single correct answer, or a question on a topic outside the A level Biology Specification. Although your answer may be speculative, to gain credit it must be biologically plausible and based on an understanding of the general biological principles you are expected to have covered in A level Biology.

Using examples from…: this instructs you to use information contained in the relevant material provided as examples in your answer. Answers that do not use this material may not be given credit, even if they are biologically correct.

What is meant by…?: this shows that a less formal statement that gives the meaning of a word or biological term will be acceptable.

What is the name of…?: this usually requires a technical term or its equivalent; often no more than one or two words are expected.

Where? What? and Why?: these preface direct questions and retain their common everyday meaning.

Glossary

Terms first introduced in *AS Biology for AQA* are presented in blue text.

A band (dark band): Part of a muscle sarcomere that comprises several thick myosin filaments held between thin actin filaments.

abiotic factor: A non-living factor that forms a non-biological part of the environment.

abiotic phase: In relation to nutrient cycling, the non-living parts of a cycle through which nutrients pass.

ABO blood group: A classification of blood into four groups A, AB, B, and O. It involves three alleles represented by I^A, I^B, and I^O of which only two occur in any individual; I^A and I^B are codominant and both of these alleles are dominant to I^O.

abscisic acid: A plant growth substance that acts as a powerful growth inhibitor and that promotes dormancy.

absolute refractory period: Brief period after excitation when a neurone or muscle cell becomes completely unexcitable.

absorption: The uptake of matter or energy into a system e.g. nutrients and water from the gut into the blood stream.

absorption spectrum: The wavelengths of light that a particular pigment absorbs.

abundance scale: A qualitative measure of how many organisms occur within a locality, e.g. the ACFOR scale.

accessory pigment: A pigment that does not participate directly in photosynthesis but which absorbs and conveys light energy to chlorophyll; the light energy is then used for photosynthesis.

acetyl coenzyme A: A two-carbon compound that acts as a link between glycolysis and the Krebs cycle in aerobic respiration.

acetylcholine: A chemical that acts as a neurotransmitter between motor nerve fibres and skeletal muscles, and which also acts as the neurotransmitter of some synapses in the peripheral nervous system.

acetylcholinesterase: An enzyme that breaks down the neurotransmitter acetylcholine.

ACFOR scale: An acronym for an abundance scale: A = abundant, C = common; F = frequent, O = occasional, and R = rare.

Acquired immune deficiency syndrome (AIDS): A disease that reduces the ability of the immune system to fight other diseases such as tuberculosis. It is caused by a blood-borne virus called human immunodeficiency virus or HIV.

acquired mutation: A change in the genetic make-up of a somatic cell that is not present in gametes and not passed on from parent to offspring.

actin: A contractile protein that forms the thin filaments in sarcomeres of muscle fibres.

action potential: The transient change in potential difference across the membrane of a neurone or muscle cell when the cell is stimulated. The resting potential, typically about −70 mV, is reversed, reaching about +30 mV, due to an influx of sodium ions.

action spectrum: The wavelengths of light that bring about a particular process such as photosynthesis.

activation energy: The energy needed to start a chemical reaction.

active immunity: Resistance to disease brought about by antibodies produced during exposure to an antigen.

active site: A particular area of an enzyme which combines with a substrate to form a temporary enzyme-substrate complex.

active transport: Movement of ions or molecules across a cell membrane against a concentration gradient (i.e. from a region of lower concentration to a region of high concentration); it requires the expenditure of energy.

adaptive radiation: An evolutionary process in which organisms sharing a common ancestry multiply and diverge to occupy and exploit different habitats and modes of life.

adenine: An organic base that forms part of ATP.

adenosine: A nitrogen-containing compound made up of the base adenine attached to the sugar ribose.

adenosine diphosphate (ADP): A nucleotide consisting of a base, sugar, and two phosphate groups.

adenosine triphosphate (ATP): A nucleotide consisting of a base, sugar, and three phosphate groups.

adenyl cyclase: A membrane-associated enzyme that is involved in the production of cyclic AMP from AMP.

ADEPT: Antibody-directed enzyme producing therapy, a way to target a drug to a specific tissue; used in cancer treatment.

adhesion: The force of attraction between molecules of different substances, e.g. between water molecules and the walls of xylem vessels.

ADP (adenosine diphpsphate): A nucleotide consisting of a base, sugar, and two phosphate groups.

adrenaline (epinephrine): A hormone secreted by the adrenal medulla (the central portion of the adrenal gland) that prepares the body for action, e.g. by increasing the heart rate. It also causes the breakdown of glycogen in the liver, boosting blood glucose concentrations.

adult stem cell: An undifferentiated cell in an adult that normally has the potential to give rise to a limited number of specialized cell types.

adventitious root: Small lateral root growing out of a stem.

aerobic exercise: Physical activity, usually of long duration and relatively low intensity, that uses aerobic respiration as the main energy source.

aerobic respiration: A series of biochemical reactions requiring energy in which energy-rich compounds are broken down to produce ATP.

agglutinins: Substances, such a specific type of antibody, that cause cells to clump together.

AhR (aryl hydrocarbon receptor): A gene regulatory protein which, when it forms a complex with the toxin dioxin, activates some types of oestrogen receptors.

AIDS (acquired immune deficiency syndrome): A disease that reduces the ability of the immune system to fight other diseases such as tuberculosis. It is caused by a blood-borne virus called human immunodeficiency virus or HIV.

air sacs: In insects, dilations of the tracheae (air tubes) important in respiration.

alcoholic fermentation: A process that takes place during glycolysis and anaerobic respiration in yeast and plant cells in which carbon dioxide generated by the breakdown of pyruvate and hydrogen released from NADH is used to make alcohol. Alcholic fermentation enables NAD to be regenerated.

algal bloom: A sudden rapid increase in the population of algae in an aquatic ecosystem.

alkaline phosphatase: An enzyme used to label DNA probes.

alkaloids: Nitrogenous organic bases found in plants that have toxic or medicinal properties, e.g. nicotine, caffeine, and strychnine.

allele: One of the forms of a gene which in eukaryotes occupies a specific locus on a chromosome.

allele frequency: The proportion, often expressed as a percentage, of a particular allele of a gene in a population, relative to other alleles of the same gene.

allergy: A type of medical condition in which a person is more than normally sensitive (hypersensitive) to one or more substances that are harmless to most other people.

allopatric speciation: The process by which two or more species evolve from geographically isolated populations of one original species.

all-or-nothing principle: The principle that a single neurone or muscle cell will respond to a stimulus, irrespective of its intensity, either completely or not at all.

alpha carbon: In amino acids, the carbon atom to which the various functional and R-groups are attached.

alpha cell: Type of cell in the islets of Langerhans in the pancreas; it secretes glucagon.

alpha glucose: The form of glucose molecule that combines to form starch in plants and glycogen in animals.

alpha helix: In proteins, a type of secondary structure of a polypeptide chain, or part of a chain, that takes the form of a spiral.

alpha-1-antitrypsin: A protein that helps protect the lungs from damage during an infection.

altitude acclimatization: Reversible physiological adaptations, such as an increase in red blood cell count, to high altitudes.

alveoli (sing. alveolus): Microscopic spherical sacs at the ends of bronchioles in the lungs which provide a very large surface area for gaseous exchange. They have walls one cell thick and are covered in small capillaries.

amino acid: A nitrogen-containing organic molecule with an acidic carboxyl (—COOH) group and a basic amino group (—NH_2). Amino acids are the monomers of proteins.

ammonification: The formation of ammonia from dead organisms by decomposers.

amoeba: A microscopic, single-celled eukaryotic organism characterized by the ability to change shape and the possession of pseudopodia ('false feet'), temporary extensions of the cytoplasm formed during locomotion.

amphetamine: A drug that acts as a stimulant.

amphoteric: A term applied to chemicals, such as amino acids, that have both acidic and basic properties.

amylase: An enzyme that catalyses the breakdown of starch and glycogen to the sugars maltose, dextrin, or glucose.

amyloplast: A colourless, membrane-bound starch-storing organelle in plants. In roots, amyloplasts are thought to have a gravity-sensing function.

anabolism: Type of metabolism in which complex chemicals are synthesized from simpler ones.

anaerobic exercise: Physical activity, usually of short duration and high intensity, that uses the anaerobic respiration of carbohydrates as the main energy source.

anaerobic respiration: Cellular respiration that takes place in the absence of oxygen.

analogous structure: Part of the body in one species that performs a similar function and appears superficially similar to one in an unrelated species, but differs in detailed structure and origin, e.g. the wings of insects and the wings of birds.

aneurysm: An abnormal sac-like dilation of the wall of a weakened artery.

antagonistic: Applied to things, such as muscles or drugs, which have opposite effects.

antagonistic muscles: Two muscles with opposite actions (e.g. the biceps brachii, which flexes the upper arm, and the triceps brachii, which extends the upper arm).

antibiotic: A substance such as penicillin that selectively kills or prevents the growth of microorganisms and is usually produced by another microorganism.

antibiotic resistance: A genetically determined ability of some microorganisms to tolerate high concentrations of a normally lethal agent.

antibody: A complex protein produced by an animal in response to the presence in its body of a foreign substance (an antigen) which may be harmful.

Antibody Directed Enzyme Prodrug Therapy (ADEPT): A way to target a drug to a specific tissue; used in cancer treatment.

anticodon: The triplet of bases at one end of a tRNA molecule that is complementary to a particular codon on an mRNA molecule.

antigen: A substance (often a protein) capable of binding with specific receptors on white blood cells called T-cells, and which usually triggers the production of further specific antibodies.

antigenic drift: A gradual change in the antigens carried by some viruses caused by small genetic changes.

antigenic shift: A substantial change in the antigens carried by certain viruses, such as the influenza group, caused by the exchange of genetic material between two strains of virus which often manifests itself as a new type of virus.

antigenic variation: The ability of some microorganisms to change their antigens from one generation to the next.

antioxidant: A compound that prevents or slows down oxidation of materials such as food. Some antioxidants, such as the vitamins A, C, and E, may provide some protection against cancer because they neutralize the effects of some chemicals that are potentially harmful to DNA.

antiport: Membrane protein transport system in which the transport of a substance depends on the simultaneous transport of another substance in the opposite direction.

antiretroviral drug: A drug that is designed to suppress the activity or harmful effects of a retrovirus such as the human immunodeficiency virus.

aorta: The largest artery in mammals; it leaves the heart from the left ventricle and carries oxygenated blood.

aortic sinus: A normal, expanded part of the aorta just above aortic valve, and just above the heart. There are three aortic sinuses, two of which give rise to coronary arteries.

apoplast route: Pathway of water from the soil to xylem vessels in roots via cell walls.

apoptosis: Programmed cell death.

appendix: Worm-shaped extension to the large intestine of mammals. In herbivores, it contains bacteria that help break down vegetable material. In humans, it appears to have no vitally important function.

artefact: In microscopy, an apparent structure obtained due to the preparation of the specimen.

arteries: Tubular vessels that carry blood from the heart to another part of the body.

arterioles: Small arteries, some of which have sphincter muscles in their walls that enable them to control the blood flow through an organ.

arteriosclerosis: A pathological condition involving thickening, hardening, and loss of elasticity of arteries.

artificial classification: The grouping of organisms according to characteristics such as colour, size, habitat, and locomotion, that do not necessarily reflect the evolutionary relationships of the organisms.

artificial fertilizer: An artificial, synthetic substance, typically containing mineral salts of nitrogen, phosphorus, and potassium that is added to soils as a source of nutrients for plant growth.

aryl hydrocarbon receptor (AhR): A gene regulatory protein which, when it forms a complex with the toxin dioxin, activates some types of oestrogen receptors.

Asilomar Conference: A conference convened at Asilomar Beach in 1975 to discuss, identify, and evaluate potential biohazards of recombinant technology.

aspect: The direction in which something (e.g. a wall or valley slope) is facing.

assimilated energy: In a consumer, energy that is absorbed and subsequently used to produce new tissues, or is lost in respiration and urine.

assimilation: The uptake and incorporation of substances, such as digested products, into the body of an organism.

association neurone (interneurone, relay neurone): A neurone in the central nervous system that links a sensory neurone to a motor neurone.

asthma: A respiratory disorder characterized by recurrent attacks of difficult breathing, particularly on exhalation.

atheroma: An accumulation of fatty material in the inner lining of an artery.

atherosclerosis: Progressive narrowing and loss of elasticity of arteries due to the accumulation of fatty material in the inner linings.

atmosphere: Environmental system consisting of the gaseous envelope surrounding the Earth.

ATP (adenosine triphosphate): A nucleotide consisting of a base, ribose sugar, and three phosphate groups; it is the only source of chemical energy that can be used directly in biological activities such as locomotion.

ATP cycle: A cycle of reactions in which ATP is hydrolysed to ADP with the release of free energy and then the ATP is reformed by a condensation reaction in which a phosphate group is added to ADP.

ATP synthase: An enzyme that catalyses the formation of ATP from ADP and a phosphate group.

ATPase: Adenosine triphosphatase; an enzyme that catalyses the conversion of ADP into ATP and the reverse reaction, ATP into ADP.

atrioventricular node: Specialized fibres in the heart which transmit the wave of contraction from the atria to the ventricles.

atrioventricular septum: Tissue that divides the atria and ventricles in the heart.

atrium (pl. atria): One of the two upper chambers of the heart; also known as the auricle. The right atrium receives blood from veins other than those from the lungs; the left atrium receives blood from the pulmonary veins.

attenuation: The loss of the harmful effects of a microorganism that is usually pathogenic.

autologous haematopoietic stem cell transplantation: A therapeutic procedure that extracts and filters out adult stem cells from the blood, which are then re-injected back into the patient's bloodstream to treat a disorder.

autonomic nervous system: Part of the peripheral nervous system that controls what are normally involuntary activities such as heart rate, breathing rate, body core temperature, and blood pressure. It is sometimes called the involuntary nervous system.

autoradiography: Process by which an image of a radioactively labelled specimen (e.g. large molecules, cell components, or body organs) is recorded on a photographic film by resting the specimen on the film in the dark.

autosomes: In an organism with two sets of chromosomes, pairs of chromosomes that look alike and carry the same genes.

autotroph (primary producer): An organism such as a green plant that can manufacture its own food.

auxin: A group of related plant growth factor that consists mainly of indoleacetic acid. They are involved in cellular elongation and differentiation.

average: In statistics, a term often used as a synonym of the mean, but it may also refer to other measures of central tendency such as mode and median.

axon: A cytoplasmic extension of a neurone that carries impulses away from the cell body.

axon hillock: The conical region where an axon joins a cell body.

axon terminal (synaptic bulb synaptic knob): A swelling at the tip of an axon containing mitochondria and vesicles with neurotransmitters.

axoplasm: The name given to the cytoplasm within an axon.

AZT (azidothymidine): An antiretroviral drug that acts by inhibiting reverse transcription of the viral RNA into DNA.

B cells: White blood cells that develop in bone marrow.

bacilli: Rod-shaped bacteria.

bacteriophage: A virus that attacks bacteria.

Bainbridge reflex: Response to a stretching of the walls of the right atrium which causes the heart to beat faster.

balanced diet: A diet comprising the optimal amounts of food components (carbohydrates, fats, proteins, minerals, vitamins, and fibre) in the best proportion to promote good health.

ballistic impregnation: A non-biological method of transferring a gene from one organism to another. A DNA gun fires tungsten or gold coated with DNA into the target cells.

baroreceptor: A sensory receptor that is sensitive to changes in pressure.

base: A substance that reacts with an acid to form a salt and water only.

behaviour: The observable responses of an organism to its environment.

bell-shaped curve: The shape of a curve drawn from data that has a normal distribution such that the mean, mode, and median are all equal.

Benedict's test: A food test for reducing sugars such as glucose.

benign tumour: A growth of cells that is not malignant and does not spread to other areas.

beta blocker: A drug that slows the heart rate by blocking the action of adrenaline on the heart.

beta glucose: The form of glucose molecule that combines to form cellulose in plants.

beta-pleated sheet: In proteins, a type of secondary structure of a polypeptide chain, or part of a chain, that takes the form of a pleated sheet.

bicuspid valve: A valve with two flaps that prevents backflow from the left ventricle to the left atrium.

bilayer: A layer two molecules thick; often applied to the double layer of phospholipid molecules in cell membranes.

bile: A thick, brown-green, alkaline secretion from the liver that helps emulsify fats (causes large globules of fat to break down into small droplets).

bimodal distribution: A pattern of distribution (e.g. of phenotypes in a population) in which the frequency of values in a sample have two distinct peaks.

binomial nomenclature: A system of giving each different type of organism two names: a generic name and a specific name.

biodiversity: The variety of life.

biodiversity action plan: A government initiative to help conserve wildlife.

biogeochemical cycles: The circulation of chemical elements through the Earth's environmental systems by biological, geological, and chemical processes.

biological adaptation: Anything that improves the chances of an organism to survive.

biological conservation: The active management of habitats to maintain biodiversity.

biological control: Use of a biological agent (e.g. a viral disease, predator, or the release of sterilized males) to limit the size of a pest population.

biological nomenclature: Naming organisms.

biological oxygen demand (BOD): A measure of organic pollution in water; it is the amount of oxygen (usually measured in milligrams per cubic decimetre) lost from a known volume of water kept in darkness at 20°C for five days.

biome: The largest subunit of the biosphere.

biosphere: The part of the Earth in which organisms live and are active.

biotic factor: An environmental factor that results from the activities of living organisms.

biotic phase: In relation to nutrient cycling, the living parts of a cycle through which nutrients pass.

biotic potential: The maximum rate at which a particular population can grow.

birth cohort: Individuals of a population that are born at about the same time.

birth rate: Number of births per number of adults in a population.

biuret test: A food test used to test for the presence of proteins.

bivalent: Applied to two homologous chromosomes when they come together during prophase I of meiosis.

blind spot: The area of the retina where the optic nerve leaves they eye and in which there are no rods or cones.

blood clotting: The process by which blood coagulates to form a solid plug (a clot).

blood plasma: Clear, straw-coloured fluid matrix of blood; it is about 90% water and 10% other inorganic and organic compounds, including fibrinogen, a protein involved in blood clotting.

blood pressure: The force exerted by blood on the walls of a blood vessel.

blunt ends: The ends of a fragment of double-stranded DNA that have been cut at sites which are directly opposite each other.

BOD (biological oxygen demand): The amount of oxygen (usually measured in milligrams per cubic decimetre) lost from a known volume of water kept in darkness at 20°C for five days; a measure of organic pollution in water.

body core temperature: The temperature in the part of the body that contains the vital organs (the brain, heart, lungs, and kidneys).

Bohr effect (Bohr shift): A shift to the right of the oxygen dissociation curve of haemoglobin due to an increase in the partial pressure of carbon dioxide.

bolus: A rounded mass of chewed food softened by saliva and made suitable for swallowing.

boom-and-bust: Applied to a population that has a J-shaped population growth curve.

broad-spectrum pesticide: A pesticide that affects a wide range of pests.

bronchioles: Small air-tubes leading from the bronchi into the alveoli of the lungs; the diameter of bronchioles can vary.

bronchitis: Inflammation of the bronchi (the large air tubes leading into the lungs) resulting in restricted air flow to the lungs.

bronchus: One of two large air tubes leading from the trachea to bronchioles in the lungs.

brush border: Applied to the collection of microvilli (minute finger-like extensions) on the cell surface membrane of, for example, epithelial cells in the small intestine.

buccal cavity: The cavity immediately behind the mouth in the alimentary canal.

buffer: A solution that tends to resist changes in pH when an acid or alkali is added to it.

bundle of His: Specialized fibres in the heart which transmit the wave of contraction from the atria to ventricles.

C3 plants: Plants in which the first stable compound formed during carbon dioxide fixation is a three-carbon compound, glycerate-3-phosphate.

caecum: A blind-ending sac in the alimentary canal which, in mammals, occurs at the junction between the small intestine and large intestine.

calcium ion: An important regulatory ion which (among other functions) plays a key role muscle contractions.

calcium ion channel: A channel protein in a cell membrane through which calcium ions can pass.

callus: In a plant tissue culture, a mass of undifferentiated cells.

Calvin cycle: The metabolic pathway in the light-independent stage of photosynthesis in which carbon dioxide is reduced to carbohydrate using the hydrogen from NADPH and ATP.

cAMP (cyclic AMP): A small mononucleotide molecule that acts as a second messenger within cytoplasm by bringing about intracellular responses to some non-steroid hormones such as adrenaline.

cancer: A group of diseases characterized by the uncontrolled proliferation of cells, causing a solid malignant tumour or other abnormal condition.

capillaries: The narrowest blood vessel in vertebrates, located between arteries and veins.

capillarity: The spontaneous creeping movement, due to surface tension effects, of a watery liquid in very fine tubes or channels.

carbohydrates: A group of organic compounds composed of carbon, hydrogen, and oxygen all sharing the chemical formula $C_x(H_2O)_y$.

carbon cycle: The circulation of carbon through Earth's environmental systems by biological, geological, and chemical processes.

carbon dioxide fixation: The assimilation of carbon dioxide into an organism as a usable product. In plants, carbon dioxide fixation occurs during the light-independent reactions when carbon dioxide is reduced to form a carbohydrate.

carbon monoxide: Inorganic carbon compound that combines with haemoglobin to form carboxyhaemoglobin, thereby preventing oxygenation of blood.

carbonic acid: An acid formed by the combination of carbon dioxide with water.

carbonic anhydrase: An enzyme in red blood cells that catalyses the formation of hydrogencarbonate ions from carbon dioxide and water.

carboxyhaemoglobin: Compound formed by the combination of carbon monoxide with haemoglobin.

carcinogen: A cancer-causing agent.

cardiac cycle: The cycle of events during one complete heart beat.

cardiac muscle: Heart muscle.

cardiac output: The volume of blood pumped out of the left ventricle of the heart in one minute.

cardiovascular disease: A disease of the heart and its blood vessels.

carotid artery: Main artery in the neck supplying oxygenated blood to the head.

carotid body: An area of glandular tissue, close to the carotid artery, that contains chemoreceptors involved in the control of oxygen content and acidity of the blood.

carotid sinus: A swelling in the wall of the carotid artery; it contains nerve endings that are sensitive to changes in blood pressure.

carrier molecule: A molecule of protein that binds with and transports other molecules and ions across a cell membrane by either active transport or facilitated diffusion.

carrier protein: A protein molecule that binds with and transports other molecules and ions across a cell membrane by either active transport or facilitated diffusion.

carrying capacity: The maximum number of organisms a particular environment can support during the harshest part of the year without damage to that environment.

cascade effect: Applied to a biochemical chain reaction in which a small chemical signal starts the reaction and becomes amplified in each step of the chain so that it results in a very large response.

case-control matching: A procedure used by medical scientists to measure risk factors. It involves comparing the medical histories and lifestyles of two groups of people similar in all respects other than that one group has a particular disease and the other has not.

case-control study: A procedure used by medical scientists to measure risk factors. It involves comparing the medical histories and lifestyles of two groups of people, one without a particular disease and one with a particular disease.

Casparian strip: A waterproof layer created by a corky thickening of the radial and end walls of cells in the endodermis of plant roots that is believed to influence the route that water takes as it moves through a plant.

catabolism: Type of metabolism in which complex chemicals are broken down into simpler ones.

catalyst: A substance that accelerates the rate of a chemical reaction but which can be recovered unchanged at the end of the reaction.

causal relationship: A relationship between two events or phenomena such that the occurrence or presence of one is always preceded, accompanied, or followed by the other.

cDNA (complementary DNA): A DNA molecule synthesized *in vitro* from mRNA using reverse transcriptase.

cell: The basic structural unit of living matter.

cell adhesion molecules: Glycoproteins on the surface of cell membranes that promote adhesion between animal cells.

cell biologist: A biologist who specializes in the study of cells.

cell body: The part of a neurone that contains the nucleus and most organelles.

cell cycle: The ordered series of events that take place in a cell that is actively dividing.

cell fractionation: The separation of the organelles from a cell by disruption of the cell followed by ultracentrifugation.

cell line: In tissue culture, a line of cells that can be cultured indefinitely.

cell surface membrane: The selectively permeable membrane consisting mainly of phospholipids and proteins, that covers the surface of a cell.

cell wall: A relatively rigid wall surrounding the cell surface membrane of plant cells, fungi, and some bacteria. In plants it is made of cellulose which in some cells (e.g. xylem vessels) is strengthened and made waterproof by being impregnated with lignin.

cellular respiration: The breakdown of organic molecules to release energy in the form of ATP.

cellulase: Enzyme that catalyses the hydrolytic digestion of cellulose to glucose.

cellulose: An insoluble polysaccharide made of straight chains of beta-glucose monomers.

central nervous system: In vertebrates, the part of the nervous system comprising the brain and spinal cord.

centrioles: Organelles occurring in animal cells which are involved in the organization of the spindle apparatus during meiosis and mitosis.

centromere: A constricted part of a chromosome which contains no genes; it is the region in which sister chromatids are joined.

cerebrovascular accident: A stroke; an interruption of the supply of blood to the brain causing some cells to die.

cervical cancer: A malignant growth of cells in the cervix, the region lying between the uterus and vagina of female mammals.

chain-termination method: A method of sequencing the nucleotide bases in DNA, in which the last nucleotide at one end of each the DNA is modified and labelled with a fluorescent dye.

channel protein: A membrane protein that functions as a pore which allows the diffusion of certain water-soluble solutes across the membrane.

CHD (coronary heart disease): A disease of the heart that involves a narrowing of the coronary arteries.

chemical contraception: The use of drugs to prevent the fertilization of an egg by a sperm cell.

chemiosmotic channel: A membrane structure that acts as a functional pore through which protons can pass.

chemiosmotic theory: A theory explaining how ATP is generated in mitochondria by using energy released from the movement of protons through the stalked particles of mitochondria.

chemoautotroph: An organism that synthesizes for itself organic compounds from carbon dioxide and other inorganic molecules using energy from the oxidation of inorganic substances rather than from light.

chemoreceptor: A cell particularly sensitive to a particular chemical.

chemotherapy: A medical treatment, e.g. of cancer, that involves the use of chemicals.

chi-squared test: A statistical routine used to test the significance of any difference between the results observed in an experiment or sample and the theoretically expected results.

chitin: A structural polysaccharide which is the main component of insect and other arthropod cuticles. Like cellulose, it is made from beta glucose monomers but unlike cellulose it also has a nitrogenous component.

chlorophyll: A group of green pigments found in the chloroplasts of most plants and on special membranes in cyanobacteria (blue-green algae). Chlorophyll a absorbs light and uses the energy to drive the reactions of photosynthesis.

chloroplast: A double-membraned organelle in plant cells that is the site of photosynthesis.

cholera: A diarrhoeal disease caused by the bacterium *Vibrio cholerae*.

cholesterol: A lipid found in cell membranes; it helps to strengthen the membranes. In humans, excess cholesterol may be carried by the blood and deposited in arterial walls, narrowing arteries and increasing blood pressure.

cholinergic synapse: A synapse into which the neurotransmitter acetylcholine is secreted.

chromatid: One of two copies of a chromosome which has replicated during interphase and which appears as a thread-like structure during prophase.

chromatin: A granular substance present in the nucleus during interphase which consists of a network of chromosomal material (DNA and proteins) with a little RNA.

chromatography: A technique used to analyse or separate the components of a mixture of gases, liquids, or dissolved substances (e.g. plant pigments). The technique depends on the different solubility of molecules in a moving solvent (mobile phase), and absorption on, or solubility in, an inert substrate such as paper or chalk (stationary phase). Components of the mixture are carried to different parts of the substrate (or pass through the substrate at different rates) and thus separate out. They can be identified by comparison with the movement pattern of known substances.

chromosomes: The structures that contain DNA and carry genetic information. Each chromosome has many genes arranged in a linear sequence. Chromosomes become visible as rods or thread-like structures during mitosis and meiosis.

chyme: Partially digested substances that pass from the stomach into the duodenum.

cilia: Fine, hair-like projections of the cell surface membrane which move the fluid surrounding them by producing a beating or rowing action.

ciliated epithelium: Epithelial cells lining a passage, such as that of the nose or trachea, composed of cells with cilia.

cis-fatty acids: Unsaturated fatty acids in which the two hydrogen atoms adjacent to a double bond are on the same side of the molecule. This imposes a kink in the shape of the molecule that makes it difficult for cis-fatty acid molecules to pack tightly together, consequently most are liquid at room temperature.

cisternae (sing. cisternum): Fluid-filled compartments, usually flattened, formed within the cytoplasm of a cell by membranes of the endoplasmic reticulum and Golgi body.

cistron: A portion of DNA that contains the code for a single polypeptide chain; it is sometimes called a functional gene.

classification: In biology, the arrangement of organisms into groups.

cleavage site: The specific location at which DNA is cut by a restriction endonuclease.

climatic climax: According to some ecologists, a single type of community that forms the final stage of ecological succession for a particular climatic region.

climatic factor: A factor determined by the climate, the long-term atmospheric conditions that characterize a particular locality.

climax community: According to the climax theory, the final stage of ecological succession in which the community is relatively stable and in equilibrium with its environment.

climax theory: The theory that ecological succession progresses in a series of stages from a relatively simple, unstable community to a more complex community that is in equilibrium with its environment.

clonal selection: Process that takes place in the immune system when an antigen introduced into the body activates a specific lymphocyte so that the lymphocyte proliferates to form a clone of cells specific for the antigen.

clone: A group of genetically identical cells or organisms.

cloning: Using a somatic cell from a multicellular organism to produce one or more identical organisms.

clumped distribution: A distribution pattern in which individuals are clustered together.

codominance: Applied to two alleles which, when they occur together in a diploid organism, are both expressed in separate and distinguishable ways.

codon: One of 64 base triplets of DNA or mRNA that codes for a specific amino acid or which acts as a signal to start or stop the synthesis of a polypeptide.

coenzyme: A non-protein, organic substance that is essential for the normal functioning of an enzyme.

cofactor: A non-protein group that is essential for the normal activity of an enzyme.

cohesion: The force of attraction between molecules of the same substance, often by hydrogen bonding (as in water molecules).

cohesion-tension theory: A theory explaining the mechanism of transpiration whereby the movement of water involves forces of attraction between water molecules that results in tension in the column of water that extends from the leaves to the roots.

cold blooded: Term commonly applied to animals whose body temperature is generally the same as that of the environment.

cold receptors: Skin receptors that are sensitive to a decrease in skin temperature.

coleoptiles: A conical protective sheath that surrounds the plumule (the embryonic shoot) of a germinating monocotyledon seed (a seed such as that of grass or a cereal that has only one seed leaf).

collenchyma: A plant tissue that provides mechanical support; it is composed of elongated living cells in which the cell wall is unevenly thickened with cellulose deposited at the corners.

colon: The first part of the large intestine; its main function is reabsorption of water.

colonic bacteria: Bacteria that live in the colon.

community: A group of populations of different interacting species living in the same area at the same time. Members of a community are linked by feeding relationships and are, in varying degrees, interdependent.

comparative anatomy: The study of the similarities and differences between the structures of organisms, often used to determine evolutionary relationships.

comparator: In a homeostatic control mechanism, the component that compares the actual output with the set point.

compensation point: The point at which the rate of photosynthesis in a plant is in exact balance with the rate of respiration so that the is no net loss or gain of carbon dioxide or oxygen.

competition: Interaction between organisms striving for the same environmental resource (e.g. food, space, or a mate).

competitive exclusion principle: The principle that species with the same resource requirements cannot live together permanently in the same habitat because one is bound to be a stronger competitor than the other.

competitive inhibitor: A substance that reduces the rate of an enzyme-catalysed reaction by occupying the active site in place of the normal substrate whose structure is similar.

complement proteins: Proteins that play a role in the immune system by helping to remove foreign particles and damaged cells.

complementary base pairing: The joining together of two bases by virtue of their having molecular shapes that enable them to fit together. In DNA, for example, adenine pairs with thymine, and guanine with cytosine.

complementary DNA (cDNA): A DNA molecule synthesized *in vitro* from mRNA using reverse transcriptase.

condensation: A chemical reaction in which two molecules become covalently bonded to each other through the removal of water.

conduction: Transfer of heat by contact from a warmer object to a cooler object.

conjecture: An uncertain conclusion based on incomplete evidence; sometimes used synonymously with hypothesis.

conjugated protein: A molecule that consists of a protein combined with a distinct non-protein group.

conjugation: A type of sexual reproduction in which two cells are joined together by a cytoplasmic tube through which DNA can be transferred.

connective tissue: Animal tissue in which the material in between the cells (the intercellular matrix) forms a major part.

conservation biology: The scientific study of how biological diversity and wildlife habitats can be safeguarded.

consumer: Any organism that feeds n another organism to gain energy and nutrients.

contact pesticide: A pesticide that kills a pest by making contact with the external surface.

contagious distribution: A distribution pattern in which individuals are clustered together.

continuous variation: Applied to a characteristic such as height in humans in which differences between individuals are slight and grade into each other.

control sequence: A particular fragment of DNA that is incorporated into a plasmid to enable a bacterium to express a gene inserted into the plasmid from a donor.

controlled breeding: Reproduction of organisms determined by human selection of the organisms allowed to reproduce.

convection: Transfer of heat from one place to another by the movement of fluid molecules across the heated surface.

convergence: In the retina, the connecting together via synapses of a large number of rods to one neurone.

core nucleotide sequence: A repeating nucleotide base sequence in a hypervariable region of DNA.

cork: Dead, waterproof plant tissue formed from the layer of cells immediately inside the epidermis.

coronary heart disease (CHD): A disease of the heart that involves a narrowing of the coronary arteries.

corpus luteum: The 'yellow body'; a temporary endocrine gland formed from an ovarian follicle after ovulation.

correlation: An association between two variables such that when one changes the other changes also. A correlation may be negative (in which a fall in the magnitude of one variable is associated with a rise in another) or positive (in which the rise in magnitude of one variable is associated with a rise in the magnitude of another).

cortex: The outermost part or layer found in a variety of plant and animal structures and organs.

counter-current mechanism: A process in which two fluids flow in opposite directions along adjacent channels thus maximizing the exchange of dissolved salts, gases, or heat. For example, in the gills of bony fish, water is pumped through gill filaments in the opposite direction to the flow of blood.

courtship: Behaviour, often highly ritualized, that precedes copulation in some animals.

covalent bond: A strong chemical bond in which two atoms share electrons.

crassulacean acid metabolism: A type of metabolism in plants in which carbon dioxide is taken in at night, stored in organic acids, and used in photosynthesis during the day.

cristae (sing. crista): Shelf-like infoldings of the inner membrane of mitochondria.

crop rotation: System of agriculture in which different types of crop are grown in a field in regular sequence over a number of seasons.

crossing over: A process occurring during prophase I of mitosis in which homologous chromosomes come together and exchange genetic material.

curare: A drug which blocks the action of acetylcholine and prevents the contraction of skeletal muscle.

245

cuticle: 1 A non-living waterproof layer on the outer surface of some plant cells. 2 A protective outer layer that covers the body of invertebrates; in some insects, the cuticle is waxy and reduces water loss.

cyclic AMP (cAMP): A small mononucleotide molecule that acts as a second messenger within cytoplasm by bringing about intracellular responses to some non-steroid hormones such as adrenaline.

cyclic photophosphorylation: A series of chemical reactions that take place during the light-dependent stage of photosynthesis. It involves electrons escaping from chlorophyll, being passed along a series of electron carriers, and then returning to chlorophyll. During the process, ATP is generated.

cystic fibrosis: An inheritable disorder characterized by the production of excess mucus in the airways.

cytochrome: An iron-containing protein involved in cellular respiration.

cytochrome oxidase: An enzyme that catalyses the final stage of aerobic respiration when oxygen combines with hydrogen ions to form water.

cytokines: Protein factors secreted by white blood cells in the mammalian immune system which regulate neighbouring cells.

cytokinesis: Cell division; the division of the cytoplasm to form two separate daughter cells.

cytokinin: A plant growth factor that often acts with indoleacetic acid to promote growth.

cytoplasm: The contents of a cell, excluding the nucleus, bounded by the cell surface membrane.

cytosis: Pertaining to movement of materials into cells (endocytosis) or out of cells (exocytosis).

cytosol: The watery part of cytoplasm which contains numerous small molecules, such as sugars, in solution and large molecules, such as proteins, in suspension.

dark adaptation: An increase in the sensitivity of the eye to light when a person remains in darkness or in light of low intensity.

dark band (A band): Part of a muscle sarcomere that comprises several thick myosin filaments held between thin actin filaments.

daughter chromosomes: Two chromosomes resulting from the replication of a chromosome.

death rate: Number of deaths per number of adults in a population.

deciduous trees: Trees which shed all their leaves at the end of the growing season.

decomposer: An organism (e.g. a fungus or bacterium) that obtains nutrients by feeding on dead organisms and breaking them down into simpler substances, ultimately releasing inorganic materials into the environment.

decomposition: In nutrient cycles, the breakdown of dead organisms by decomposers.

degenerate: Applied to a code that has more information than is essential; e.g. the genetic code is degenerate because it has 64 codons for 20 amino acids.

degree of freedom: A statistical term that relates to the number of free variables in a system.

dehydrated: Applied to material that has had its water removed.

deletion: A form of gene mutation in which one or more of the nucleotide bases has been lost.

deme: A local population that is largely isolated genetically from other populations of the same species, and which has clearly distinguishable characteristics.

demographic transition: A change in a human population from having high birth rates and high death rates to having low birth rates and low death rates.

demography: The statistical study of the size and structure of populations.

denaturation: The loss of the three-dimensional shape (tertiary structure) of a protein molecule.

dendrite: A fine cytoplasmic extension of a neurone that receives signals from other neurones.

dendron: A cytoplasmic extension of a neurone that carries impulses towards the cell body.

denitrification: The conversion of nitrates and nitrites to nitrogen gas.

denitrifying bacteria: Bacteria that carry out denitrification, converting nitrates and nitrites to nitrogen gas.

density gradient centrifugation: Type of cell fractionation in which cell organelles are ultracentrifuged and separated according to their densities.

density-dependent factors: Environmental factors, such as competition for limited resources (e.g. food and space), that affect a higher proportion of a population as the population density increases.

density-independent factors: Environmental factors, such as freezing temperatures, that affect the same proportion of a population no matter what its population density.

depolarization: A decrease in the potential difference across a cell membrane. Depolarization occurs during an action potential when the potential difference across the membrane of a muscle or nerve cell changes from negative to positive due to an influx of sodium ions.

detector: In a homeostatic control mechanism, the component that is sensitive to the level of output.

detritivores: Organisms that have detritus, decomposing particles of organic matter, as the major part of their diet.

development: The continuous changes in size, shape, form, and degree of complexity that accompany the growth of a multicellular organism from a single cell to an adult.

diabetes mellitus: A metabolic disorder caused by lack of insulin or a loss of responsiveness to insulin that results in an excessively high blood glucose concentration.

diaphragm: A sheet of muscular tissue, present only in mammals, which separates the thoracic (chest) cavity from the abdominal cavity.

diarrhoea: Frequent evacuation of the bowels and/or the production of large quantities of soft watery faeces.

diarrhoeal disease: A disease, such as cholera, which causes diarrhoea.

diastole: Resting phase of the cardiac cycle when all parts of the heart are relaxed.

dictyosome: Alternative name for the Golgi apparatus.

differential centrifugation: Separation of cell organelles by ultracentrifuging a homogenate at different speeds and for different durations. Each speed and duration causes specific types of organelles to form a sediment; the higher the speed and longer the duration, the smaller the organelle.

differentiation: The process in a multicellular organism by which unspecialized cells become modified and specialized to carry out specific functions. Differentiation involves changes in both the structure and physiology of cells.

diffusion: The net movement of molecules or ions from a region of high concentration to a region of low concentration.

diffusion shell: A layer of stationary air adjacent to a stomata through which water vapour diffuses.

digestion: The breakdown of large complex organic compounds into smaller, simpler materials.

dioxin: A toxic chemical by-product of the manufacture of some herbicides.

dipeptide: The product formed by two amino acids being connected by a peptide bond.

diploid: Applied to organisms with two sets of chromosomes.

directional selection: A form of natural selection which selects against phenotypes of one or other of the extremes of the distribution and which therefore tends to result in a shift of the mean to either the right or left.

discontinuous variation: Applied to variations in characteristics which fall into discrete groups that do not overlap.

disruptive selection: A form of natural selection which selects against intermediate phenotypes and favours those of the extreme.

divergent evolution: Evolutionary changes that tend to produce differences between two organisms.

DNA (deoxyribonucleic acid): A complex nucleic acid composed of two polynucleotide chains wound around each other to form a double helix; it is responsible for inheritance in most living organisms.

DNA fingerprinting (DNA profiling; genetic fingerprinting): The process of obtaining an image of the pattern of DNA fragments from a DNA source that is unique to that source.

DNA hybridization: The formation of double-stranded DNA from two single strands obtained from different sources; the extent of double-helix formation is used to determine the similarity between the two DNA sources.

DNA ligase: An enzyme that catalyses the joining together of fragments of DNA.

DNA polymerase: An enzyme that catalyses the synthesis of new DNA by the addition of nucleotides to an existing chain using DNA or RNA as a template.

DNA probe: A molecular tool for locating a specific gene. It consists of a single strand of DNA that contains a nucleotide base sequence complementary to the known sequence of bases in a target gene.

DNA profiling: The process of obtaining an image of the pattern of DNA fragments from a DNA source that is unique to that source.

DNA sequencing: Determination of the sequence of nucleotide bases in DNA.

DNA-DNA hybridization: A technique for determining the similarity of DNA from two different sources. It involves forming double-stranded hybrid DNA and determining the extent of complementary base-pairing.

dominant: Applied to an allele of a gene that is fully expressed in both the heterozygous and homozygous condition.

double circulatory system: A circulatory system in which blood passes through the heart twice during each complete circuit of the body.

drug: Any substance that alters the natural internal chemical environment of an organism and affects its normal body functions.

dual system: Name given to a homeostatic control system that has two separate mechanisms for controlling deviations: one for controlling deviations above the set point, another for controlling deviations below the set point.

ductless gland (endocrine gland): A gland which lacks ducts and which secretes a hormone directly into the bloodstream.

duplication: A type of gene mutation in which there are more than the normal number of genes.

E. coli (Escherichia coli): A rod-shaped bacterium that is a common resident of the large intestine of mammals and found in faeces; usually harmless, some strains can cause severe food poisoning.

ECG (electrocardiogram): A graphical recording of the electrical changes occurring as the heart beats.

ecological niche (niche): The function or role of an organism within a community.

ecological stability: The ability of an ecosystem to resist changes and/or to return to its original state after being changed.

ecological succession (succession): The sequence of change within an ecosystem from initial colonization to a relatively stable climax community.

ecologist: A biologist who specializes in the study of the relationship between organisms and their natural environment.

ecology (environmental biology): The scientific study of organisms in relation to all aspects of their living and non-living environments.

economic damage threshold: Applied to the size of a pest population that is just at the point of causing economic damage.

economic injury level: Applied to the size of a pest population that is causing significant economic harm.

ecosystem: A discrete, relatively stable and self-contained system comprising a community of organisms and their abiotic and biotic environments.

ecotourism: A form of tourism which focuses on the natural environment. It typically involves travelling to areas where biodiversity is high and the scenery spectacular.

ectotherm: An organism whose source of body heat is largely external; ectotherms that cannot control their body temperatures are commonly referred to as 'cold blooded', but many ectotherms (e.g. snakes) can regulate their body temperature by behavioural mechanisms.

edaphic factor: A factor related to the biological, physical, and chemical composition of a soil.

effector: A structure such as a gland, organ, or muscle that brings about a particular action or effect.

egested: Applied to material that has passed through the gut unchanged and has been eliminated from the body.

electrocardiogram (ECG): A graphical recording of the electrical changes occurring as the heart beats.

electron carrier: A molecule that readily accepts an electron and passes it on to another electron carrier with a higher affinity for the electron.

electron transport chain: A series of electron carriers along which an electron is passed from one redox reaction to the next.

electroporation: A non-biological method of transferring a gene from one organism to another. A burst of electricity creates temporary pores in the target cells, allowing donor DNA to enter.

embolus: An obstruction, such as fat, a blood clot, or an air bubble, in a blood vessel.

embryonic stem cell: An undifferentiated cell in an embryo that has the potential to gives rise to any type of specialized cells.

emigration: In population statistics, the number of individuals moving out of a population from another area.

empirical science: A method of study which involves systematic observation or experimentation rather than speculation or mere theorizing. Observations are made, experiments performed, and facts gathered primarily for their own sake without regard to theories.

emulsion test: A test of lipids in which the test substance is added to absolute alcohol. After being left a sufficient time for fats to dissolve in the alcohol, distilled water is added and the mixture shaken. If fats are present, the mixture appears cloudy because of the formation of fine droplets of fat (emulsion) in the water.

endemic: Applied to a disease present in a particular region all the time.

endergonic: Applied to a chemical reaction that requires an input of energy.

endergonic reaction: A chemical reaction that requires an input of free energy.

endocrine gland (ductless gland): A gland that lacks ducts and which secretes a hormone directly into the bloodstream.

endocytosis: An active, energy-consuming process that results in the transport into the cytoplasm of a cell of substances too large to pass through the cell surface membrane by simple diffusion, facilitated diffusion, or active transport through membrane proteins.

endodermis: Innermost layer of cells in the cortex of roots of plants; it forms a cylinder one cell thick around the stele.

endopeptidase: An enzyme that helps to break down proteins by catalysing the hydrolysis of internal peptide bonds within a polypeptide chain.

endoplasmic reticulum: A system of parallel membranes enclosing fluid-filled channels within the cytoplasm of eukaryotic cells.

endosymbiosis: A mutually beneficial relationship in which one of the partners lives inside the cell of another.

endothelium: A single layer of cells that lines internal body surfaces of fluid-filled vessels, such as blood vessels.

endotherm: An animal which controls its body temperature by internal physiological means, independently of the environment.

endothermic homoiotherms: Animals that can regulate their body temperature by internal, physiological mechanisms.

end-plate potential: A graded potential that occurs across the membrane of a muscle fibre at a neuromuscular junction in which the change in the potential difference across the membrane is proportional to the intensity of the stimulus.

end-product inhibition: The slowing down of a chemical reaction by the product formed at the end of the reaction.

energy: The capacity to do work; a product of force and distance.

energy budget: The balance of energy input and output in a biological system (e.g. an organism, trophic level, or ecosystem).

energy flow diagram: A diagram showing the movement of energy through an ecosystem.

entropy: A measure of disorder or randomness within a system.

environment: The surroundings of an organism including abiotic and biotic factors.

environmental biology (ecology): The scientific study of organisms in relation to all aspects of their living and non-living environments.

environmental resistance: The factors that limit the size of a population.

environmental resistances (selection pressures): Environmental factors that limit the growth of populations.

enzyme: A protein that acts as a biological catalyst for a reversible chemical reaction.

enzyme specificity: Refers to an enzyme's ability to catalyse only one or a limited number of chemical reactions.

enzyme-substrate complex: The combination of an enzyme and its substrate attached to its active site.

epidemic: Applied to a disease that unexpectedly spreads rapidly and affects a large number of people at the same time.

epidemiology: The study of the occurrence and spread of diseases.

epidermal cells (epidermis): The outermost layer of cells on the surface of an organism.

epinephrine (adrenaline): A hormone secreted by the adrenal medulla (the central portion of the adrenal gland) that prepares the body for action, e.g. by increasing the heart rate. It also causes the breakdown of glycogen in the liver, boosting blood glucose concentrations.

epithelial tissue: A tissue consisting of a sheet or tube of one or a few layers of closely packed cells lining the surfaces of organs and the walls of body cavities.

erythrocytes: Red blood cells.

ethane: A plant growth substance released as a gas from ripening fruit, nodes of stems, ageing leaves, and flowers which is involved in seed dormancy, fruit ripening, and leaf fall.

ethics: A code of behaviour considered by a particular group of people as being correct.

eukaryotes: Organisms that have a well defined nucleus and double-membraned organelles.

eukaryotic: Related to eukaryotes.

eutrophication: The enrichment of water by the addition of nutrients.

evaporative cooling: The loss of heat from an object when water on the surface of the object changes from a liquid to a gas.

evolution: The formation of new species from pre-existing ones; the process by which the earliest forms of life have been transformed into the diversity of species that live on Earth today.

exchange pools: In relation to nutrient cycles, those parts of a cycle where an element is held for a relatively short period of time.

excitatory presynaptic cell: A neurone the end of which secretes a neurotransmitter that decreases the membrane potential of a target cell, making it more excitable.

excited state: The condition of an electron when at its highest energy level.

exergonic reaction: A chemical reaction that results in the release of free energy.

exocytosis: The ejection of material from a cell by emptying the contents of a membrane-lined vesicle or vacuole at the surface of a cell by fusion with the cell surface membrane.

exons: Coding regions of genes in eukaryotes.

explant: In plant tissue culture, the part of a plant (e.g. a stem cutting) that is to be cultured.

exponential growth phase (log phase): Phase of population growth during

ex-situ conservation: Wildlife conservation outside the natural habitats of the organisms being conserved (e.g. in zoos and wildlife parks).

extracellular material: Material outside a cell.

extrinsic protein: A membrane protein embedded in the outer layer of a cell membrane.

facilitated diffusion: The movement of ions or molecules from a high concentration to a low concentration across a cell membrane by carrier molecules; it does not require the expenditure of energy.

FAD (flavine adenine dinucleotide): An electron carrier and hydrogen acceptor in the Krebs cycle.

faeces: Bodily waste eliminated from the alimentary canal; faeces contain a mixture of excretory products and egested material including bile, undigested food, bacteria, and mucus.

fast-twitch fibre: A type of muscle fibre characterized by a relatively fast contraction time, high anaerobic capacity, and a rich supply of blood causing the fibre to appear red. It fatigues quickly.

fat: A lipid that is solid at room temperature.

fatigue: Exhaustion of a muscle fibre, e.g. due to depletion of the neurotransmitter in neuromuscular junction or to the excess of lactate and hydrogen ions.

fermentation: A process that takes place during glycolysis and anaerobic respiration in which NADH is oxidized without the generation of ATP.

fertility drug: A drug that stimulates the development of ovarian follicles and ovulation, increasing the chances of an egg becoming fertilized by a sperm cell.

fertilizer: A substance that is added to soils as a source of nutrients for plant growth.

fetal haemoglobin: Type of haemoglobin present in the blood of a fetus. It has an oxygen dissociation curve to the left of maternal haemoglobin.

FEV_{10} (forced expiratory volume$_{10}$): The volume of air forced from the lungs during a ten second period of maximum exhalation.

fibrous protein: A relatively insoluble protein consisting of chains of polypeptides that are mainly linear and strengthened by cross-linkages (e.g. keratin)

fibrous root: A root that consists of a mass of fibrous structures.

Fick's law: A law stating that the rate of diffusion in a given direction across an exchange surface is directly proportional to the area of the exchange surface and to the concentration difference across the surface, and inversely proportional to the length of the diffusion pathway.

fitness: In evolution, the ability of an organism to pass on its alleles to subsequent generations.

five' cap (5' cap): A modified guanine nucleotide attached to one end of mRNA (the 5' end); it helps to protect the mRNA from hydrolysis and signals the point of attachment when mRNA reaches a ribosome.

fixed action patterns: A complex form of stereotyped, unlearned behaviour characterized by relatively fixed patterns of coordinated movement.

flagellum (pl. flagella): A thread-like extension of the cell surface specialized for locomotion or for movements of fluids close to the cell. In eukaryotes it contains microtubules while these are absent in prokaryotes.

flavine adenine dinucleotide (FAD): An electron carrier and hydrogen acceptor in the Krebs cycle.

fluid-mosaic model: A model which suggests that cell membranes are made of a bilayer of phosopholipid in which a mosaic of proteins are inserted.

follicle: Any small sac-like structure or cavity; in the ovary, it consists of a fluid-filled ball of cells containing an oocyte.

follicle-stimulating hormone (FSH): Hormone secreted by the anterior pituitary gland which stimulates the growth of ovarian follicles in females and the production of sperm in males.

follicular phase: Phase of menstrual cycle in which a follicle containing an oocyte develops.

food chain: The sequence of organisms, usually starting with a producer and ending with a top consumer, in which each organism is the food of the next organism in the chain.

food web: A diagrammatic representation of feeding relationships within a community; it consists of an interconnected group of food chains.

founder effect: Genetic drift resulting from establishing a new population with relatively few individuals from a parent population; generally, there is a decrease in the genetic diversity of the new population.

fovea: A small depression in the retina opposite the lens that consists only of cones.

frame quadrat: An area marked out on the ground with a frame (typically, metal, plastic, or wooden) of any convenient shape within which a population is sampled.

frame shift: A DNA mutation occurring when the number of nucleotides inserted or deleted is not a multiple of three.

Frank-Starling effect: The effect of blood in the heart at the end of diastole; the greater the volume of blood, the more the ventricular walls are stretched and the greater the force of contraction and stroke volume.

free radicals: A chemical group that has unshared electrons that react readily with other chemicals; free radicals are implicated as a cause of damage to DNA and the development of cancers.

freeze-fracturing: A method of preparing a specimen for electron microscopy by freezing the specimen in liquid nitrogen and then splitting it, usually along the middle of a bilipid layer, to expose internal structures.

frequency of occurrence: How often a species occurs, expressed as a proportion of the total number of sample areas which contains the species (e.g. if 10 areas are sampled and 3 contain a particular species, it has a frequency of occurrence of 3 out of 10).

FSH (follicle-stimulating hormone): Hormone secreted by the anterior pituitary gland which stimulates the growth of ovarian follicles in females and the production of sperm in males.

fundamental niche: The area in which a species is physiologically capable of living.

fungicide: A pesticide that kills fungi.

G_0 phase: Phase of the cell cycle in which cells are at rest and not undergoing mitosis.

gamete: A haploid cell such as an egg or a sperm.

gated: Applied to a protein or ion channel in a cell membrane that opens or closes in response to a particular stimulus (e.g. a change in voltage across the membrane).

gel electrophoresis: A technique for analysing and separating particles (e.g. proteins or nucleic acid fragments) on the basis of their size and electrical charge, by their differential migration through a gel in an electrical field.

gene (cistron): A discrete unit of inheritance that consists of a particular sequence of nucleotide bases in DNA (or RNA in certain viruses) and which holds the information for the synthesis of one polypeptide.

gene cloning: The making of multiple, identical copies of a fragment of DNA that includes one or more genes.

gene expression: The transformation of the information encoded in a gene to produce a functional polypeptide.

gene locus: The position of a gene on a chromosome.

gene mutation: A sudden random event that results in a new allele.

gene pool: The sum of all the alleles in a population.

gene replacement: A form of gene therapy in which an abnormal, mutated allele is replaced by a normal one.

gene supplementation: A form of gene therapy in which one or more copies of a normal allele are added to a cell without removing any of the pre-existing abnormal, mutated alleles; the added alleles are dominant and are added to mask the effects of the recessive, mutant alleles.

gene test: A test carried out to identify mutant alleles associated with a disease.

gene therapy: The treatment of a disease by altering the genes of a patient with the disease. It involves correcting a mutant allele by repairing it, swapping it for a normal allele, or by regulating its activity so that it no longer produces harmful effects.

generator potential: A graded potential across the plasma membrane of a receptor cell in which the change in the potential difference across the membrane is proportional to the intensity of the stimulus.

genetic bottleneck: A phenomenon characterized by a reduction in genetic diversity which occurs when the size of a population is significantly reduced, e.g. by a catastrophe.

genetic code: The 64 different triplets of nucleotide bases by which genetic information needed to make polypeptides is carried in DNA.

genetic diversity: Variations in a species due to differences in the alleles carried by individuals.

genetic drift: Random changes in the frequency of alleles within a population; genetic drift tends to become more significant as a population becomes smaller.

genetic fingerprinting (DNA fingerprinting): The process of obtaining an image of the pattern of DNA fragments from a DNA source that is unique to that source.

genetic recombination: The process in sexually reproducing organisms by which offspring acquire different characteristics to their parents as a result of the exchange of genetic information that occurs when homologous chromosomes cross over during prophase I of meiosis.

geneticists: Biologists who specialize in the study of the genetic factors that determine characteristics acquired by inheritance.

genetics: The scientific study of inheritance.

genome: The complete complement of genes that occur in a species.

genotype: The genetic composition of an organisms.

geographical isolation: The separation of populations by geographical barriers such as a mountain range, an ocean, or a river.

geotropism: A movement of a part of a plant towards or away from gravity.

germ-line gene therapy: The treatment of a genetic disorder by altering the genetic composition of sperm or egg cell, or of a germinative cell that gives rise to the gametes.

GI (glycaemic index): Ranking of carbohydrate-containing foods based on their overall effect on blood glucose levels. Slowly absorbed foods have a low GI while foods absorbed into the bloodstream quickly have a higher GI.

gibberellins: A plant growth factor that acts as a growth promoter by increasing cell division and cell elongation in stems and leaves.

gill: Plate-like or filamentous outgrowth of tissue well supplied with blood which acts as a respiratory organ in some aquatic organisms, such as fish.

gill lamellae (sing. gill lamella): Thin plate-like structures that increase the surface area of gills. There are two types: primary and secondary. The secondary lamellae are smaller and grow out of the primary lamellae.

gill plate: Alternative name for secondary gill lamella.

GL (glycaemic load): A measure of the effect of foods on blood glucose concentrations calculated as the product of the percentage of carbohydrate in a portion of the food and its glycaemic index.

global stability: The tendency of a community to return to its original state after a large disturbance.

global warming: The increase in the average temperature of the Earth's near surface air and oceans.

globular protein: A roughly spherical protein that has at least one highly folded polypeptide chain; globular proteins are generally more soluble in water than are fibrous proteins.

glucagon: A polypeptide hormone secreted by alpha cells in the islets of Langerhans of the pancreas, which tends to increase blood glucose concentrations by increasing the conversion of glycogen to glucose.

gluconeogenesis: The conversion of non-carbohydrate substances such as amino acids and glycerol into glucose.

glucose: A hexose (six-carbon) monosaccharide sugar that in nature usually occurs in one of two forms: alpha glucose and beta glucose.

glucose tolerance test: A diagnostic test for diabetes mellitus in which the patient swallows a sugar solution and the doctor measures the blood glucose concentration at intervals.

glycaemic index (GI): Ranking of carbohydrate-containing foods based on their overall effect on blood glucose levels. Slowly absorbed foods have a low GI while foods absorbed into the bloodstream quickly have a higher GI.

glycaemic load (GL): A measure of the effect of foods on blood glucose concentrations calculated as the product of the percentage of carbohydrate in a portion of the food and its glycaemic index.

glycerol: An alcohol derived from a three-carbon sugar. It is a component of fats and oils.

glycogen: A polysaccharide made of alpha glucose monomers. It is the main storage carbohydrate in animals.

glycogenesis: The formation of glycogen from glucose by condensation reactions.

glycogenolysis: The breakdown of glycogen to glucose by hydrolysis.

glycolipid: A complex lipid molecule that has a carbohydrate component.

glycolysis: The first stage of respiration which takes place in the cytoplasm with or without oxygen. During glycolysis, glucose is broken down to pyruvate, NAD is reduced, and ATP is generated.

glycoprotein: Any molecule containing a sugar combined with a protein.

glycosidic bond: The covalent bond that joins two glucose monomers that have combined as a result of a condensation reaction.

glycosuria: The presence of glucose in the urine.

GnRH (gonadotrophin-releasing hormone): A hormone released by the hypothalamus that triggers the secretion of follicle-stimulating hormone from the anterior pituitary gland.

goblet cells: Wine glass-shaped, mucus secreting cells that occur in the epithelium of e.g. the small intestine.

Golgi apparatus: Stacks of flattened, membrane bound sacs (cisternae) present in eukaryotic cells, involved in processing, storing, and packaging materials.

Golgi body: One of the stacks of flattened membrane bound sacs that make up the Golgi apparatus.

gonadotrophin-releasing hormone (GnRH): A hormone released by the hypothalamus that triggers the secretion of follicle-stimulating hormone from the anterior pituitary gland.

graded potential: A change in the potential difference across a membrane that is proportional to the intensity of the stimulus.

gradient distribution: A distribution pattern in which the distribution of individuals varies smoothly over a sampling area.

grana (sing. granum): A stack of chlorophyll-containing membrane-lined sacs in chloroplasts.

gravid: Applied to a female, e.g. a fish, that is carrying eggs.

greenhouse effect: The heating effect of certain gases that trap long-wave radiation in the Earth's atmosphere, causing the atmosphere to retain heat.

greenhouse gas: A gas (e.g. methane, carbon dioxide, and chlorofluorocarbons) that trap long-wave radiation.

gross productivity: In primary producers, the total amount of organic matter produced or energy assimilated. It includes the organic material (or energy) used in respiration as well as the organic matter or energy stored as new plant tissue.

ground state: The condition of a an electron when at its lowest energy level.

guard cells: Specialized cells in the epidermis of plants that form the boundary of stomata.

guide strand: One strand of the double-stranded siRNA which combines with mRNA by complementary base pairing and interferes with gene expression.

gut: The alimentary canal or part of it.

guttation: The release of droplets of water from a plant due to the upward force of root pressure.

H zone: The area in the middle of a muscle sarcomere where actin and myosin filaments do not overlap; it contains thick myosin filaments only.

Haber process: An energy-demanding industrial process that synthesizes ammonia from hydrogen using high temperatures, high pressures, and an inorganic catalyst.

habitat: The particular locality in which an organism lives.

haemoglobin: A large, conjugated globular protein molecule consisting of four polypeptide chains each attached to an iron-containing group.

haemophilia: An inherited condition in which blood does not clot normally.

halophyte: Plant specially adapted to tolerate conditions in which soils have a high salt content.

haploid: Applied to organisms with one set of chromosomes.

Hardy–Weinberg equation: An equation used to calculate the frequency of genotypes in a population.

Hb^S: The symbol used in genetics to represent the allele responsible for sickle cell anaemia.

HCG (human chorionic gonadotrophin):
In female mammals, a hormone produced by the placenta during pregnancy which stops the degeneration of the corpus luteum; it is involved in maintaining pregnancy and (in humans) suspending menstruation.

health: A state of complete physical, mental, and social well-being.

healthy lifestyle: A way of living, particularly with regard to diet and exercise, that promotes health.

heart: In mammals, a four-chambered muscular organ that pumps blood around the body.

heart rate: The number of ventricular contractions per minute.

heat gain centre: Part of the thermoregulatory centre in the hypothalamus that is activated by a decrease in blood temperature.

heat loss centre: Part of the thermoregulatory centre in the hypothalamus that is activated by an increase in blood temperature.

heat receptors: Skin receptors that are sensitive to an increase in skin temperature.

heavy drinking: Drinking alcohol significantly above the recommended limits.

helicases: Enzymes that unwind the double helix of DNA during replication.

helper T cells: A class of white blood cells which interact with B cells in the presence of a specific antigen to stimulate the proliferation of the B cells into plasma cells that secrete antibodies against the antigen.

hepatic artery: The artery that carries oxygenated blood to the liver.

hepatic portal vein: The vein that carries deoxygenated blood from the small intestine to the liver.

herbicide: A pesticide that kills plants, especially broad-leaved weeds.

herbivore: An animal that feeds on plant material.

herd immunity: A general immunity against a contagious disease given to a population when a large number of people are vaccinated at the same time.

heroin: A narcotic drug that binds to opiate receptors and blocks the action of nerve impulses that cause pain.

heterosomes: In mammals, the sex chromosomes (X and Y).

heterozygous: In a diploid organism, having two different alleles for a particular gene.

hexoses: Six-carbon sugars such as glucose.

histamine: A substance released by mast cells in response to an irritant that contributes to allergic responses, e.g. dilation of blood vessels, in the eyes, nose, and skin.

histones: Small protein molecules that bind to eukaryotic DNA and play a key role in chromatin structure.

HIV (human immunodeficiency virus): An RNA retrovirus responsible for AIDS (acquired immunodeficiency syndrome).

homeostasis: The regulation and maintenance of relatively constant conditions within an organism or other system.

homeotherm (homoiotherm): An organism that maintains a relatively constant body temperature despite fluctuations in environmental temperature.

homogenous distribution: A distribution pattern in which individuals are clustered together.

homoiotherm (homeotherm): An organism that maintains a relatively constant body temperature despite fluctuations in environmental temperature.

homologous chromosomes: Pairs of chromosomes in a diploid organism which normally appear similar and carry the same genes, although the alleles on each chromosome may differ.

homologous structures: Anatomical structures in different species that have a similar evolutionary origin and development in the body (e.g. the pentadactyl limbs of mammals).

homozygosity: The presence in diploid organisms of pairs of identical alleles for one or more genes.

homozygous: In a diploid organism, having two identical alleles for a particular gene.

hormone: A chemical that is produced in one part of the body and transported to another part where it produces specific effects.

human chorionic gonadotrophin (HCG):
In female mammals, a hormone produced by the placenta during pregnancy which stops the degeneration of the corpus luteum; it is involved in maintaining pregnancy and (in humans) suspending menstruation.

Human Genome Project: The international scientific project that identified the whole of the human genome and sequenced the approximately 3 billion nucleotide base pairs that make up human DNA.

human immunodeficiency virus (HIV): An RNA retrovirus responsible for AIDS.

hybrid: Any offspring of a cross between two genetically dissimilar individuals.

hybrid vigour: A vitality and state of good health often observed when offspring are produced from a cross between two different pure-breeding individuals.

hydrogen bonds: Individually weak electrostatic bonds that occur between a slightly positively charged hydrogen atoms and another atom with a slight negative charge.

hydrolysis: A chemical reaction that results in the breakdown of a large molecule by the addition of water.

hydrolytic enzyme: An enzyme that catalyses a hydrolysis; digestive enzymes are hydrolytic enzymes.

hydrophobic: Applied to something that has an aversion to water.

hydrophyte: A flowering plant adapted to living in freshwater or very wet conditions.

hydrosphere: Environmental system consisting of all the aquatic environments.

hydrotropism: A movement of a part of a plant towards or away from water.

hydroxyl group: The –OH group.

hyperglycaemia: A condition characterized by an abnormally high blood glucose concentration.

hyperpolarization: A change in potential difference across a plasma membrane so that it becomes more negative than the normal resting potential.

hypertension: High blood pressure.

hypertonic: Applied to a solution that (usually) has a lower water potential than a cell immersed in it, and which causes the cell to shrink as water is lost from it by osmosis.

hypervariable region: A non-coding region of DNA that is made of a variable number of repeated sequences of nucleotide bases. The number and length of the repeated sequences varies between individuals.

hypoglycaemia: A condition characterized by an abnormally low blood glucose concentration.

hypothalamus: Part of the brain that is involved in many autonomic functions (e.g. thirst, sleep, and hunger) and which contains centres for thermoregulation, ionic regulation, and osmoregulation.

hypothesis: A scientific explanation of an observation or phenomenon. A hypothesis is usually based on facts already known or research carried out, and is expressed in such a way that it can be tested or appraised as a generalization about a phenomenon.

hypotonic: Applied to a solution that (usually) has a higher water potential than a cell immersed in it, and which causes the cell to expand as water enters it by osmosis.

I band (light band): A zone in a muscle sarcomere bisected by the Z line. It contains thin actin filaments.

IAA (indoleacetic acid): A major plant growth factor, synthesized in the tips of shoots, that promotes growth by increasing the rate of cell elongation and which is involved in phototropisms.

ileum: The final part of the small intestine in mammals.

immigration: In population statistics, the number of individuals moving into a population from another area.

immobilized enzyme: An enzyme (or cells containing a specific enzyme) that is encapsulated within an agar bead or attached to an inert solid material.

immune system: A complex body system consisting of interacting hormones, cells, and other adaptive mechanisms that defends the body against diseases caused by microorganisms or abnormal malignant cell growths.

immunoglobulin: One of a group of highly variable proteins made by white blood cells which has antibody activity.

immunological memory: A function of memory cells produced in the blood in response to an initial exposure to an antigen; the long-lived memory cells enable the immune system to mount a larger and more rapid secondary response to that specific antigen.

inbreeding: Mating between two closely related individuals, or self-fertilization.

incubation period (incubation time): In relation to an enzyme-catalysed reaction, the length of time over which a reaction takes place.

independent assortment: The fact that genes on non-homologous chromosomes are inherited independently of each other so that gametes may contain any combination of parental alleles.

indicator community: A community containing specific indicator species that provides information about the environmental conditions prevailing at a particular location.

indicator species: A species the presence of which provides information about environmental conditions because it requires a particular condition or set of conditions in which to live.

indoleacetic acid (IAA): A major plant growth factor, synthesized in the tips of shoots, that promotes growth by increasing the rate of cell elongation and which is involved in phototropisms.

induced fit theory: A theory that suggests that in the formation of an enzyme-substrate complex the substrate brings about a change in shape of the active site of the enzyme as it binds to it.

induced pluripotent stem cell (IPSC): A cell that has been converted into a totipotent cell by the insertion of special genes.

industrial melanism: An increase in the frequency of dark forms of an organism associated with the soot and smoke produced by high levels of industrial pollution.

infectious disease: A disease capable of being transmitted from one organism to another.

inherited mutation: A change in the genetic make-up that is present in gametes and passed on from parent to offspring.

inhibitory presynaptic cell: A neurone the end of which secretes a neurotransmitter that increases the membrane potential of a target cell, making it less excitable.

initial rate of reaction: The rate at the start of a chemical reaction.

innate behaviour: Inborn behaviour that is not learned.

inorganic catalyst: A non-carbon based catalyst, such as a metal.

insecticide: A pesticide that kills insects.

insertion: A form of gene mutation in which one or more extra nucleotides is added to a section of DNA.

in-situ conservation: Wildlife conservation in the natural habitats of the organisms being conserved.

instinctive: Applied to behaviour that is inherited (i.e. under genetic control) and which is not modified by learning.

insulin: A polypeptide hormone secreted by beta cells in the islets of Langerhans of the pancreas, which tends to decrease blood glucose concentrations by causing glucose to be converted to glycogen.

integrated pest management: System of pest control that, after considering fully the biological and environmental context, uses the most appropriate mix of chemical and biological control methods.

intercropping: System of plant cultivation in which two different crops are planted in the same field.

International Human Genome Project: An international project to determine the complete sequence of the 3 billion DNA subunits (bases), identify all human genes, and make them accessible for further biological study.

interneurone (relay neurone, association neurone): A neurone in the central nervous system that links a sensory neurone to a motor neurone.

interphase: A growth stage in the cell cycle during which the chromosomes are in a more or less dispersed state, and during which DNA and organelles are replicated.

interspecific competition: Competition between individuals belonging to different species.

interspecific interactions: Interactions between individuals belonging to different species.

interspecific variation: Variation between organisms belonging to different species.

interstitial space: The area or spaces between cells.

intraspecific competition: Competition between members of the same species.

intraspecific interactions: Interactions between individuals belonging to the same species.

intraspecific variation: Variation between organisms belonging to the same species.

intravenous hydration: Rehydration with a saline solution administered by injection into a vein.

intrinsic protein: A membrane protein embedded in the inner layer of a cell membrane.

intron: A non-coding sequence of nucleotide bases in a eukaryotic gene.

inversion: A gene mutation in which there is a reversal of the nucleotide sequence in a section of DNA.

involuntary nervous system: Name sometimes given to the autonomic nervous system because many of its actions are not usually under conscious control.

IPSC (induced pluripotent stem cell): A cell that has been converted into a totipotent cell by the insertion of special genes.

islets of Langerhans: Endocrine tissue within the pancreas that contains alpha cells which secrete glucagon and beta cells which secrete insulin.

isolating mechanism: A factor that prevents breeding between two populations and which may eventually lead to speciation.

isomerism: The phenomenon of two or more compounds having the same chemical formula but different structural formulae.

isotonic: Applied to a solution that has the same water potential as a cell and which does not cause any change in volume of the cell when it is immersed in it.

karyotype: The total chromosome complement of a species or individual.

kb (kilobase pairs): A unit of measurement for the size of a DNA fragment equal to 1000 base pairs.

killer T cells: A class of large granular white blood cells that recognize and kill a variety of cancerous cells and virus-infected cells.

kilobase pair (kb): A unit of measurement for the size of a DNA fragment equal to 1000 base pairs.

kinesis (pl. kineses): A random movement of an organism in which the rate of movement is related to the intensity of the stimulus, but not to its direction.

kinetic energy: The energy of motion; energy actually in the process of doing work.

klinokinesis: A behavioural response to a stimulus such that an organism changes its rate of turning with changes in stimulus intensity.

knee jerk reflex: An involuntary, unlearned kick of the lower leg following a tap on the tendon just below the patella (kneecap).

Krebs cycle: A series of aerobic reactions occurring in mitochondria in which carbon dioxide is produced, hydrogen is removed from carbohydrate molecules, and ATP is generated.

lac operon: A gene switch in the genome of *Escherichia coli* and some other bacteria that ensures the bacterium produces enzymes which catalyse the breakdown of lactose only when lactose is present.

lactase: An enzyme that catalyses the breakdown by hydrolysis of lactose to galactose and glucose.

lactate: A salt of lactic acid formed in animal cells as a result of anaerobic respiration.

lactate fermentation: A process that takes place during glycolysis and anaerobic respiration in animal cells, in which pyruvate is converted into lactate using hydrogen released from NADH. Lactate fermentation enables NAD to be regenerated.

lacteals: One of the small lymph vessels in the centre of a villus in the small intestine that absorb fats.

lactose: A disaccharide (also known as milk sugar) formed by a condensation reaction that combines glucose with galactose.

lactose intolerance: An intolerance to milk or milk products because of an inability to digest lactose.

lag phase: Initial part of an S-shaped population growth curve in which the number of individuals in a population rises gently.

large intestine: Part of the alimentary canal of mammals, concerned mainly with reabsorption of water and formation of faeces.

larynx: Structure located between the pharynx and trachea that contains a pair of elastic membranes (vocal cords) which can vibrate and make sounds.

law of segregation: A law formulated by Gregor Mendel (in modern terms) stating that the characteristics of a diploid organism are determined by alleles occurring in pairs. Of a pair of such alleles, only one can be carried in a single gamete.

leaching: The removal of nutrients from soil by the downward movement of water, by which the nutrients are carried into the drainage system of rivers and lakes.

leaf abscission: Leaf fall; the process by which leaves are shed from shrubs and trees.

leguminous plants: Plants belonging to the pea and bean family; they have root nodules containing nitrogen-fixing bacteria.

lepthaemoglobin: A type of haemoglobin that occurs in the root nodules of leguminous plants.

leucocytes: White blood cells.

LH (luteinizing hormone): A hormone secreted by the anterior pituitary gland which stimulates testosterone production in male mammals. In mature female mammals it stimulates the oocyte development, ovulation, and the development of the corpus luteum.

lifestyle: A way of living, such as active or sedentary, heavy alcohol drinking or tee-totaller, smoker or non-smoker, heavy eater or light eater.

ligament: Tissue connecting one bone to another.

light adaptation: A decrease in the sensitivity of the eye to light when a person remains in a brightly lit environment.

light band (I band): Part of a muscle sarcomere that comprises several thick myosin filaments held between thin actin filaments.

light compensation point: The light intensity at which the rate of photosynthesis in a plant is in exact balance with the rate of respiration so that there is no net loss or gain of carbon dioxide or oxygen.

light-dependent reactions: Stage of photosynthesis that can only take place in the presence of light. In plants, it occurs in the grana of chloroplasts. It involves energy being absorbed by chlorophyll and used to manufacture ATP and NADPH.

light-independent reactions: Stage of photosynthesis that can take place in the dark provided there is sufficient ATP and NADPH. In plants, it takes place in the stroma of chloroplasts and results in the reduction of carbon dioxide to make carbohydrates during the Calvin cycle.

lignin: An impermeable complex organic molecule that is incorporated into some cellulose cell walls to strengthen them.

limiting factor: In a biological process controlled by several factors, the factor that is at the least favourable value and which thus determines the rate of the process.

Lincoln index: An index for estimating the population size using data collected from the mark-release-recapture technique. It is summarized by the equation $N = S_1S_2/R$, where N is the estimate of population size, S_1 is the number of individuals captured and marked in the first sample, S_2 is the total number of organisms in the second sample, and R is the number of recaptures.

linear code: A feature of a code such as the genetic code which is always read in the same direction from the same starting point to a finishing point.

lipids: A group of very variable organic molecules that are insoluble in water but are soluble in organic solvents such as alcohol, ether, chloroform, and benzene.

liposome transfer: A non-biological method of transferring a gene from one organism to another.

lithosphere: Environmental system consisting of all the rocks and soils that form the outer rigid layer of the Earth's crust.

loading p(O_2): The partial pressure of oxygen at which haemoglobin becomes 95% saturated with oxygen.

local chemical mediators: Chemicals such as histamines or prostaglandins that only affect cells around the area in which they are released.

local stability: The tendency of a community of organisms within an ecosystem to return to its original state after a small disturbance.

lock-and-key theory: A theory of enzyme action which proposes that the shape of a substrate fits that of the active site exactly before an enzyme-substrate complex is formed.

locus: The position of a gene within a DNA molecule, or the specific position a particular gene occupies on a chromosome.

log phase (exponential growth phase): Phase of population growth in a sigmoid growth curve.

lugworms: Marine bristle worms (polychaetes) that live in burrows in muddy sand.

lungs: The internal organs of gaseous exchange in amphibians, reptiles, birds, and mammals. In mammals, the lungs are a pair of elastic sacs in the chest cavity, linked to the air by a system of tubes.

luteal phase: Phase of the oestrus cycle or menstrual cycle during which an ovarian follicle forms the corpus luteum (yellow body).

luteinizing hormone (LH): A hormone secreted by the anterior pituitary gland which stimulates testosterone production in male mammals. In mature female mammals it stimulates the oocyte development, ovulation, and the development of the corpus luteum.

lymph: A milky or colourless fluid which drains from the tissue fluid into the lymphatic system.

lymph nodes: Clumps of tissue in the lymphatic system that act as filters and are sites of white blood cell formation.

lymphatic system: In vertebrates, a system of blind-ending tubules that drains excess tissue fluid (lymph) and transports it to two veins in the upper chest cavity.

lymphocyte: A white blood cell which has a large nucleus and clear cytoplasm. Those that complete their development in bone marrow are called B cells; those that complete their development in the thymus are called T cells.

lysins: Substances that promote the breakdown of cells.

lysis: The breakdown of cells.

lysosome: An organelle bound by a single membrane and which contains hydrolytic enzymes.

lysozymes: Enzymes occurring in tears that have an antibacterial action.

macrophage: A large white blood cell that moves through tissue engulfing bacteria.

magnification: The process of enlarging the appearance of something without necessarily revealing more detail.

malignant tumour: A cancerous growth of cells that can spread sufficiently to impair the function of one or more organs.

maltase: An enzyme that hydrolyses the breakdown of maltose to glucose.

mark-release-recapture technique: A technique for estimating the population density of mobile animals within a designated area. A sample of the population is captured, marked, and released. After leaving the individuals to mix with the population, a second sample of individuals is captured and the proportion of marked individuals (recaptures) is used to estimate the population size, e.g. by using the Lincoln index.

mass flow: The transport of substances in a fluid which is moving from one location to another along a hydrostatic pressure gradient.

matrix: 1 In connective tissue, the non-living substance between cells. 2 In mitochondria the inner fluid region enclosed within the inner membrane; it contains enzymes, DNA, and ribosomes.

Mb (megabase pairs): A unit of measurement for the size of a DNA fragment equal to one million base pairs.

mean: A measure of the average value of data comprising a sample. It is calculated as the sum of the measurements divided by the total number of measurements (sample size).

mechanical digestion: Physical breakdown of large chunks of food into smaller pieces, e.g. by chewing with teeth.

mechanical energy: The energy a body has by virtue of its motion, state of deformation, or position.

mechanoreceptor: A receptor sensitive to mechanical stimuli such as changes in pressure.

median: The central value in a sample.

medicine: A chemical that can be applied internally or externally to demonstrate, relieve, or cure disease or the symptoms of disease.

megabase pair (Mb): A unit of measurement for the size of a DNA fragment equal to one million base pairs.

meiosis: Reduction division; nuclear division that results in four daughter cells being formed with half the number of the chromosomes of the parent cell.

meiosis I: First phase of meiosis in which homologous chromosomes are separated.

meiosis II: Second phase of meiosis in which sister chromatids are separated.

melanic: Applied to an organism that is a dark colour due to the production of melanin in the cells covering its body surface.

melanin: A dark brown or black pigment.

melanism: A condition in which an organism appears to be a dark colour due to the production of melanin in the cells covering its body surface.

menopause: The period of life during which ovulation and menstruation ceases in women.

menstrual cycle: In sexually mature, non-pregnant human females, the monthly cycle of changes associated with ovulation, the thickening of the uterus in preparation to receive a fertilized ovum and, in the absence of fertilization, menstruation.

menstruation: The discharge of blood and the uterine lining during menstruation.

meristem (meristematic tissue): Tissue in an actively growing plant in which nuclei are dividing by mitosis and new permanent plant tissue is forming.

mesophyll: The middle layer of cells in a leaf, lying between the two epidermal layers.

mesophyte: Plant adapted to soils with average water conditions, not too dry nor too wet.

messenger RNA (mRNA): Ribonucleic acid that consists of a single strand of nucleotides formed by transcription from DNA and which carries the genetic code for the synthesis of a polypeptide chain.

metabolic poison: A toxin, such as cyanide, that disrupts metabolism.

metabolic rate: The rate of metabolism.

metabolism: The sum of all the chemical reactions that take place in an organism.

metastasis: The movement from their primary site of formation of cancer cells to other sites in which they colonize and grow.

microevolution: The gradual change in allele frequency within a population that results from natural selection.

microfibrils: Microscopic fibres, e.g. of cellulose or keratin.

microhabitat: The precise locality in which an organism lives.

micro-injection: A non-biological method of transferring a gene from one organism to another. A very fine pipette is used to inject DNA into a host cell nucleus.

microtubule: A very fine, hollow cylindrical tube in cytoplasm made of the protein tubulin.

microtubule organizing centres (MTOCs): In eukaryotic cells, structures that are involved in the formation of microtubules in the spindle apparatus.

microvillus (pl. microvilli): One of a number of minute finger-like extensions of the cell surface membrane of some animal cells ((e.g., in the epithelium of the small intestine).

middle lamella: The layer between two adjacent plant cells; it is made largely of pectin.

minimal medium: In relation to the growth of microorganism and fungi, the minimal type of nutrients required for reproduction and growth.

mitochondrion (pl. mitochondria): Double-membraned organelle in eukaryotic cells that is the main site of aerobic cellular respiration and the production of ATP.

mitosis: Nuclear division in which one parent cell normally produces two identical daughter cells.

mitotic cell cycle: Changes that take place in cells that are actively dividing by mitosis.

mode: The most frequently occurring value among the data in a sample.

molecular biologists: Biologists who specialize in the study of biological molecules.

monoclonal antibody: An antibody produced by clones of cells formed by the fusion of a particular antibody-producing white blood cell and a cancer cell.

monoculture: Cultivation of a single plant species or a single crop variety.

monohybrid inheritance: Inheritance involving a single pair of contrasting factors.

monomer: One of the chemical subunits which serve as building blocks for large multi-unit organic molecules (polymers).

mononucleotide: A chemical such as ATP that consists of a single nucleotide.

monosaccharides: Simple sugars; a group of soluble, sweet-tasting carbohydrates that share the chemical formula $(CH_2O)_n$.

monosynaptic reflex arc: A reflex arc in which a sensory neurone makes a direct synaptic connection to a motor neurone with no interneurone.

monounsaturated fatty acid: A fatty acid molecule such as oleic acid (a major constituent of olive oil) that has a single double bond.

morphine: A depressant drug that binds to opiate receptors and blocks the action of nerve impulses that cause pain.

mortality: Death rate.

motile: Able to move from one place to another independently.

motor end-plate (neuromuscular junction): A special plate-like synapse between a motor neurone and a muscle fibre.

motor neurone: A neurone that transmits nerve impulses from the central nervous system to effector organs.

motor unit: A motor neurone and all of its associated muscle fibres which contract as a single unit.

mRNA (messenger RNA): Ribonucleic acid that consists of a single strand of nucleotides formed by transcription from DNA and which carries the genetic code for the synthesis of a polypeptide chain.

MRSA: An acronym for methicillin-resistant *Staphylococcus aureus*, used to describe those examples of this organism that are resistant to commonly used antibiotics.

mucigel: Gelatinous material on the surface of roots in soil.

mucoproteins: Complex proteins found in mucus.

mucosa: The membrane lining the gut wall that includes mucus-secreting cells.

multicellular: An organism made of many cells.

multifactorial disease: A disease that has many causes.

multiple alleles: A condition in which there are more than two alternative forms of a particular gene in a species.

multiple sclerosis: A neurological disease that results from the gradual deterioration the myelin sheath around neurones in the peripheral nervous system.

muscle fibre recruitment: The sequential enlistment of different numbers and types of muscle fibre during the contraction of a whole muscle. First slow-twitch muscle fibres are brought into action, and then fast-twitch muscle fibres.

muscle tissue: Contractile animal tissue involved in locomotion and the movement of internal body parts.

mutagen: An environmental agent that increases the rate of spontaneous mutations.

mutant: An organism or cell that differs from its parent because it contains genetic material that has been altered by one or more mutations.

mutated: Applied to a cell or organism that has undergone a mutation.

mutation: A sudden random change in the genetic material of a cell that may be inherited if it occurs in cells that produce gametes or cells that produce offspring asexually.

myelin sheath: An insulating sheath of fatty material that forms around some vertebrate neurones enabling nerve impulses to be transmitted quickly.

myofibrils: Contractile, small thread-like structures that make up a muscle fibre.

myogenic: Applied to contractions of heart muscle cells which arise spontaneously from within the muscle cells and without any need for nervous stimulation.

myoglobin: A protein consisting of one polypeptide chain linked to an iron-containing prosthetic group which can combine irreversibly with oxygen; it acts as a store of oxygen for use when oxygen released from haemoglobin is exhausted.

myosin: A contractile protein that forms the thick filaments in sarcomeres of muscle fibres.

myosin head: The globular end of a myosin molecule that forms a cross-bridge with actin.

myxomatosis: A highly infectious viral disease of rabbits characterized by swelling of the mucous membranes and the formation of skin tumours.

NAD (nicotinamide adenine dinucleotide): An organic molecule, formed from two nucleotides, which acts as a coenzyme. NAD acts as an electron carrier and hydrogen acceptor.

NADP (nicotinamide adenine dinucleotide phosphate): A hydrogen acceptor and electron carrier that plays an important role in photosynthesis.

natality: Birth rate.

natural classification: The grouping of organisms according to characteristics that reflect the evolutionary relationships of the organisms.

natural fertilizer: A naturally occurring substance such as farmyard manure, crop residues, and compost, that is added to soils as a source of nutrients for plant growth; sometimes called organic fertilizer.

natural killer cells (killer T cells): A class of large granular white blood cells that recognize and kill a variety of cancerous cells and virus-infected cells.

natural selection: The action of natural processes (e.g. predation, disease, competition for food, and temperature changes) which result in less well adapted members of a species producing fewer offspring than better adapted members. The mechanism that, according to Darwin, is the driving force behind evolution.

negative feedback: Control of the performance of a system such that any deviation in the output of the system switches on a correcting mechanism that brings the output back to normal.

negative phototaxis: The movement of a whole organism away from light.

negative tropism: A movement (usually resulting from growth) of part of a plant away from a directional stimulus such as light or gravity.

neo-Darwinism: A modern theory of evolution based on that of Charles Darwin who argued that new species evolve from ancestral types by the action of natural selection on variations within populations, but which also incorporates modern discoveries about genes, genetic change, and molecular biology.

nerve cell (neurone): A cell specialized for the transmission of nerve impulses; it contains a cell body and cytoplasmic extensions that may be up to a metre long.

nerve fibre: A long extension of a neurone that transmits nerve impulses rapidly from one part of the body to another.

nerve impulse: A wave of depolarization that passes along a neurone and is caused by changes in the permeability of the plasma membrane that can be detected as an action potential.

nerve tissue: Animal tissue consisting of nerve cells and supporting cells; the nerve cells carry information as nerve impulses quickly from one part of the body to another.

nervous system: Body system that conveys information rapidly from one part of the body as nerve impulses transmitted along interconnecting neurones.

net productivity: In primary producers, gross productivity minus respiratory loss.

neuromuscular junction (motor end-plate): A special plate-like synapse between a motor neurone and a muscle fibre.

253

neurone (nerve cell): A cell specialized for the transmission of nerve impulses; it contains a cell body and cytoplasmic extensions that may be up to a metre long.

neurotransmitter: A chemical messenger that is released from a synaptic bulb in the end of a neurone, passes across a synaptic cleft, and affects the activity of a target cell.

niche (ecological niche): The function or role of an organism within a community.

nicotinamide adenine dinucleotide (NAD): An organic molecule, formed from two nucleotides, which acts as a coenzyme

nicotinamide adenine dinucleotide phosphate (NADP): A hydrogen acceptor and electron carrier that plays an important role in photosynthesis.

nicotine: A drug found in tobacco that mimics the action of the neurotransmitter acetylcholine.

ninety-five percent (95%) confidence limit: For data obtained by sampling a population, the range of values around the sample mean in which there is a 95% probability that it contains the true mean.

Nissl granules: Granular structures in nerve cells that include ribosomes.

nitrifying bacteria: Bacteria that carry out nitrification, the conversion of ammonium salts into nitrites and nitrates.

nitrogen cycle: The circulation of nitrogen through the Earth's environmental systems by biological, geological, and chemical processes.

nitrogen fixation: The conversion of atmospheric nitrogen into nitrogenous chemicals that can be absorbed into plants.

nitrogenase: An enzyme possessed by nitrogen-fixing bacteria that catalyses the reduction of nitrogen to ammonia.

nitrogen-fixing bacteria: Bacteria that carry out nitrogen fixation, the conversion of atmospheric nitrogen into nitrogenous chemicals (e.g. ammonia) that can be absorbed into plants.

nodes of Ranvier: Small gaps in the myelin sheath encapsulating a neurone.

non-overlapping code: A code in which each piece of the code is used only once, as in the genetic code in which each base is part of only one triplet and is therefore involved in specifying only one amino acid.

non-competitive inhibitor: A chemical that slows down the rate of an enzyme-catalysed reaction without competing with the substrate for a place on the active site; a non-competitive inhibitor does not resemble the shape of the substrate.

non-cyclic photophosphorylation: A series of chemical reactions that take place during the light-dependent stage of photosynthesis. It involves electrons escaping from chlorophyll, being passed along a series of electron carriers, and then being taken up by NADP. During the process, ATP is generated.

non-decremental: Applied to the wave of depolarization that passes along a neurone or muscle cell that does not decrease in amplitude from its point of origin to its point of destination.

non-infectious disease: A disease that is not transmitted from one organism to another.

non-overlapping: Applied to the genetic code in which each base is part of only one triplet.

non-polar substance: A substance consisting of molecules with covalent bonds in which electrons are shared equally between atoms of similar electrical charge.

non-specific immune system: Part of the immune system that does not distinguish one infective agent from another.

norm (set point or reference point): In a homeostatic control mechanism, the level at which the output of a product or factor is supposed to be kept constant.

normal distribution: In statistics, a continuous distribution of randomly collected data in which the mean, median, and mode are equal. Graphically, the normal distribution takes the form of a symmetrical, bell-shaped curve.

nucleic acid: A complex, long, thread-like macromolecule made of a chain of nucleotides. The two main types are DNA and RNA.

nucleoid: The DNA-rich region in a prokaryote.

nucleolus (pl. nucleoli): A small dense round body within non-dividing eukaryotic cells. It contains DNA and ribosomal RNA and is involved in the synthesis of ribosomes.

nucleotide: An organic compound consisting of a base, sugar, and phosphate. Single nucleotides include adenosine triphosphate. Nucleotides can combine to form the polynucleotides DNA and RNA.

nucleotide base: A base that occurs in a nucleotide. In DNA, these are adenine, thymine, guanine, and cytosine; in RNA, they are adenine, uracil, guanine, and cytosine.

nucleus: A conspicuous organelle in eukaryotic cells that contains DNA and is surrounded by a nuclear envelope. During nuclear divisions, the DNA in a nucleus condenses to form chromosomes.

null hypothesis: A hypothesis used in statistical analyses which postulates that there will be no significant difference between two or more sets of results.

nutrient cycle: The circulation in the biotic and abiotic environment of nutrients used by living organisms.

obesity: A condition resulting from the storage of excessive amounts of fat, particularly under the skin and around abdominal organs.

oestrogen: A hormone produced mainly by the ovary of female mammals. It is also produced by the adrenal cortex and, in pregnant females, the placenta. It plays a primary role in the development of secondary female sexual characteristics and the growth and function of reproductive organs (including the repair of the uterine wall after menstruation). In males, oestrogens are produced in relatively small amounts by the adrenal cortex and testes.

oestrus: In a female mammal, the period when ovarian follicles develop and ovulation takes place, and when she is sexually most receptive and most fertile.

oestrus cycle: The reproductive cycle of non-primate mammals in which oestrus is the most prominent event.

oils: Lipids that are liquid at room temperature (20°C).

oncogene: An abnormal gene that is involved in triggering cancers.

oocyte: An egg cell.

operator gene: A sequence of nucleotides at the start of an operon to which an active repressor molecule can act. It forms part of the operon.

operculum: Hard bony flap covering the gills of bony fish.

operon: A cluster of genes common in prokaryotes that act as a functional unit, interacting and regulating the production of specific polypeptides.

opsonins: Antibodies that bind to antigens, facilitating them for ingestion by phagocytes.

optimum: Applied to the level of an environmental factor that brings about the most suitable rate of a biological process, e.g. the highest rate of an enzyme-catalysed reaction.

oral rehydration solution (oral rehydration treatment): A solution made up of water, salts, and glucose, designed to replace fluids lost from the body, e.g. as a result of a diarrhoeal disease.

organ: A discrete multicellular structure composed of different tissues which is adapted to carry out one or more functions (e.g. the heart or lungs in animals).

organ system: Two or more different organs working together to provide a common function.

organelle: A discrete structure with a specific function found within a cell.

organophosphate: A constituent of some insecticides and nerve gases that inhibits the action of acetylcholinesterase, tending to cause muscles to contract continuously.

orthokinesis: A behavioural response to a stimulus such that an organism changes its speed of movement with changes in stimulus intensity.

osmosis: The net movement of water molecules from a region of high water potential (e.g. a dilute solution) to a region of low water potential (e.g. a concentrated solution) through a selectively permeable membrane.

outbreeding: Mating between two unrelated individuals, e.g. a cross between two different pure breeds.

output: In a homeostatic control mechanism, the product or factor that is being controlled.

ovulation: The release of an egg cell from the ovary.

ovulatory phase: Phase of the oestrus cycle or menstrual cycle in which ovulation, the release of an egg cell, occurs.

ovulatory surge: The sudden increase in luteinizing hormone that occurs just before ovulation in a female mammal.

oxidation: Any reaction involving the addition of oxygen, the removal of hydrogen, or the loss of an electron.

oxidative decarboxylation: A process that occurs during aerobic respiration when carbon atoms derived from acetyl coenzyme A are oxidized to carbon dioxide.

oxidative phosphorylation: The process by which ATP is generated during aerobic respiration using energy released from redox reactions in electron transport chains.

oxygen dissociation curve: A curve on a graph showing the percentage saturation of a substance (e.g. haemoglobin) with oxygen against the partial pressure of the surrounding oxygen.

oxygenation: The addition of oxygen to a molecule or solution.

pacemaker: A region of specialized muscle cells in the sinoatrial node (a region located in the right atrium) of the mammalian heart which initiates and maintains the heartbeat.

Pacinian corpuscle: A receptor sensitive to changes in pressure and touch.

pain receptor: A sensory receptor that responds selectively to potentially damaging stimuli.

palisade cells: Cells that pack closely together to form a photosynthesizing layer in the mesophyll of leaves.

pandemic: Applied to a disease that affects people over a wide area, such as a continent or even the whole world.

papers: Peer-reviewed articles published in scientific journals.

papillary muscles: Muscles that attach tendinous cords to the wall of the heart.

parasympathetic nervous system: Part of the autonomic nervous system that helps to create internal body conditions found during rest and sleep, e.g. by decreasing the heart rate.

parenchyma cells: Relatively unspecialized roughly spherical plant cells that have a thin cellulose cell wall; they form the basic packing tissue of stems and roots.

partial pressure: The pressure exerted by each component of a mixture of gases. For example, if the total pressure of gases in the atmosphere is 760 mm Hg and oxygen makes up 21% of the atmosphere, the partial pressure of oxygen would be $0.21 \times 760 = 160$ mm Hg.

passive immunity: A temporary form of immunity provided by antibodies from an outside source; e.g. the passive immunity given a fetus by the antibodies provided from the mother's blood.

patchy distribution: A distribution pattern in which individuals are clustered together.

pathogen: A potentially disease-causing microorganism.

pathogenic: Pertaining to a pathogen.

pathologist: A medical scientist who specializes in the study of diseased tissues, organs, and organ systems.

PCR (polymerase chain reaction): An *in vitro* method of cloning DNA by incubating the starter DNA with special primers, DNA polymerase, and nucleotides.

peak expiratory flow: A lung function test which is a measure of how quickly a person can exhale through a peak flow meter. It is usually measured at the same time as the forced vital capacity (the volume of air exhaled with force after inhaling as deeply as possible).

peak flow meter: A device for measuring peak expiratory flow.

pentadactyl limb: A limb, characteristic of all tetrapod vertebrates, that terminates in five digits, or one that has evolved from an ancestral form possessing five digits.

pentose: A five-carbon monosaccharide sugar.

peptide bond: Chemical covalent bond linking two amino acid monomers in a dipeptide or polypeptide.

percentage cover: The proportion of an area covered by the growth form of a plant or a sedentary animal.

percentage frequency: How often a species occurs, expressed as a percentage of the total number of sample areas that contain the species (e.g. if 3 out of 10 sample areas contain a particular species, it will have a percentage frequency of 30%).

pericycle: A layer of vascular tissue between the endodermis and conducting tissue (xylem and phloem).

peripheral nervous system: All the nerves and associated tissue outside the brain and spinal cord.

peristalsis: Wave-like contractions of the muscular walls of the gut that push the gut contents along.

permease: A general name for a protein that transports substances across a cell membrane.

peroxisome: A small cell organelle bound by a single membrane and containing detoxifying enzymes including catalases that break down hydrogen peroxide.

pest: Any organism that is undesirable because it is harmful to humans economically, medically, or aesthetically.

pesticide: Any substance that kills a pest, especially a toxic chemical.

pH: A numerical scale ranging from 0 to 14 that indicates the acidity (pH 0 to 6.9), neutrality (pH 7.0), or alkalinity (pH 7.1 to 14) of a solution; $pH = \log10 (1/H^+)$ where H^+ is the hydrogen ion (proton) concentration.

phagocyte: A white blood cell that can engulf foreign particles and some microorganisms by phagocytosis.

phagocytosis: 'Cell eating'; the process of engulfing large solid particles (e.g. bacteria) into a cell by endocytosis.

pharynx: The passage leading from the buccal cavity to the oesophagus in the vertebrate alimentary canal.

phenotype: The observable characteristics of an individual organism resulting from interactions between inherited factors and environmental factors.

pheromone: A chemical substance released into the external environment and used in communication between organisms of the same species.

phloem: Plant tissue that transports nutrients and other vital chemicals from one part of a vascular plant to another.

phosphate PO_4^-: a salt or ester of phosphoric acid; an important component of DNA, RNA, and ATP.

phosphodiester bond: The covalent bond (-O-P-) between a sugar group and a phosphate group, e.g. that links nucleotides in the polynucleotide chains of DNA and RNA.

phospholipid: A lipid combined with a phosphate group. The most common consist of glycerol attached to two fatty acid chains and a phosphate group.

phosphorylase: An enzyme that helps the breakdown of glycogen to glucose by catalysing the conversion of glycogen to a phosphate of glucose.

photolysis: The breakdown of water into hydrogen ions (protons) and hydroxyl ions by the action of light energy.

photometer: A device for measuring water uptake into a plant and (indirectly) water loss by transpiration from a plant.

photophosphorylation: In photosynthesis, the addition of a phosphate group to ADP to make ATP, using light energy from the Sun.

photorespiration: Process in which products of photosynthesis are oxidized with the release of carbon dioxide but with no ATP being produced.

photosynthesis: The manufacture of complex organic chemicals using light energy from the Sun. In green plants, it takes place in chloroplasts and results in the conversion of carbon dioxide and water into glucose.

phototropism: A movement of a part of a plant towards or away from light.

phylogenetic relationships: Evolutionary relationships; two organisms that have a close phylogenetic relationship have a similar evolutionary history and common ancestors.

physiologist: A biologist who specializes in the functioning of healthy cells, tissues, organs, organ systems, and organisms.

piliferous layer: Vertical section of roots in which hair cells occur.

pin frame (point quadrat): A device for sampling stationary organisms (usually plants) at specific points. Typically, it consists of a pin-frame – legs of adjustable height supporting a cross-bar with holes down which pins, usually 10, are inserted.

pinocytosis: 'Cell drinking'; the process by which cells can take in drops of liquid by an infolding of the cell surface membrane to form small vesicles (endocytosis).

pioneer species: The first colonizers of a newly formed area in which no life had previously existed.

pituitary gland: An endocrine gland attached to the base of the brain that controls the activities of many other endocrine glands.

placenta: In pregnant female mammals, a temporary, spongy, double-layered structure formed from the tissues of the embryo and mother, by which the embryo is attached to the uterus wall.

plagioclimax: A relatively stable climax community resulting from ecological succession that has been arrested, deflected, either directly or indirectly, as a result of human activity.

plant growth factor: One of a group of plant chemicals that plays an important part in the coordination of plant growth and development.

plasma: The fluid part of blood, including proteins (especially fibrinogen) involved in the clotting process.

plasma membrane: The cell surface membrane; the membrane surrounding a cell that acts as a selective barrier regulating the cell's composition.

plasmids: Small rings of DNA in a bacterial (prokaryotic) cell; plasmids replicate independently of the large circular DNA in the nucleoid and can be transferred from one bacterial cell to another.

plasmodesmata: Threads of cytoplasm that pass through the cell walls of adjacent plant cells and which form a channel from one plant cell to another.

pluripotent: Applied to a cell that has the ability to divide and produce a number of different specialized cells.

pneumoconiosis: A lung disease characterized by damaged alveoli and caused by accumulation of dust, especially from coal, asbestos, or silica.

poikilotherm: An organism whose body temperature follows that of the external environment.

point mutation: A gene mutation caused by a change of a single nucleotide base.

point quadrat (pin frame): A device for sampling stationary organisms (usually plants) at specific points. Typically, it consists of a pin-frame – legs of adjustable height supporting a cross-bar with holes down which pins, usually 10, are inserted.

polar substance: A substance consisting of molecules with an unequal distribution of electrical charge so that one part has a slight negative charge and another part has a slight positive charge.

poly(A) tail: The 3′ end of mRNA which is modified by having 50–250 adenine nucleotides incorporated into it.

polymerase chain reaction (PCR): An *in vitro* method of cloning DNA by incubating the starter DNA with special primers, DNA polymerase, and nucleotides.

polymerization: The formation of polymers by condensation reactions.

polymer: A large organic molecule made of subunits (monomers).

polynucleotide: A chain of nucleotides; DNA and RNA are polynucleotides.

polypeptide chain: A chain of ten or more amino acids joined together by peptide bonds.

polyribosome (polysome): An aggregate of 5–50 ribosomes on the same mRNA, simultaneously translating the genetic information on the mRNA into polypeptides.

polysaccharide: A chain of monosoaccharides (simple sugars) joined together by glycosidic links.

population: 1. A group of organisms, all of the same species, that occupies a particular area. 2. In statistics, all the possible individuals or items from which a sample is taken.

population bottleneck: A situation (e.g. an environmental catastrophe) that causes a reduction in the size of a population such that it is no longer genetically representative of the original population.

population density: The number per unit area of individuals in a population.

population genetics: The study of heredity and the genetic composition of populations.

population growth: The change in the number of individuals in a population.

population growth curve: A graphical representation of population growth that may take many forms including an S-shaped curve and a J-shaped curve.

population growth rate: The change in number of individuals in a population per unit time.

population pyramid: A graph that shows the distribution of a population by age and sex.

population size: The number of individuals in a population.

porphyrin: A class of pigments containing a metal. Chlorophyll is a magnesium-containing porphyrin while haemoglobin is an iron-containing porphyrin.

positive feedback: A situation that occurs when a small change in the output of a system causes a further change in the same direction (e.g. when a small increase in body temperature causes a further increase).

positive tropism: A movement (usually resulting from growth) of part of a plant towards a directional stimulus such as light or gravity.

potential energy: The capacity to do work by virtue of a body's position.

photometer: A device for measuring water uptake into a plant and (indirectly) water loss by transpiration from a plant.

predator: An organism that kills another organism in order to eat it.

predator–prey relationship: A form of interspecific interaction involving a predator and its prey.

pre-mRNA: The mRNA initially transcribed from DNA in eukaryotes which has introns and exons.

prey: An organism that is killed and eaten by a predator.

primary consumer: An organism that feeds on a primary producer.

primary lactase deficiency: Lactase deficiency which develops as a person ages. It is usually inherited and is due to a fault in the gene that carries the instructions for making lactase.

primary producer (autotroph): An organism such as a green plant that can manufacture its own food.

primary productivity: The production of new organic matter or assimilation of energy by primary producers (autotrophs, such as plants).

primary receptor: A receptor consisting of a single neurone, one end of which detects changes in a particular stimulus.

primary structure: In relation to proteins, the sequence of amino acids in a polypeptide chain.

primary succession: Ecological succession that takes place in newly formed areas where no life previously existed.

primer: In the polymerase chain reaction, a short nucleotide sequence which is bound to the regions of the DNA to be copied to ensure that DNA replication is initiated at the required points.

principle of limiting factors: A principle which states that when a biological process such as photosynthesis depends on more than one essential factor being favourable, its rate at any given moment is limited by the factor at the least favourable value.

products: Substances formed as a result of a chemical reaction.

progesterone: A hormone produced by the corpus luteum and (in pregnancy) the placenta of female mammals. High concentrations inhibit the secretion of follicle stimulating hormone. A sharp decrease, along with a decrease in oestrogen, triggers menstruation. Production of progesterone prevents ovulation during pregnancy and stimulates the development of the placenta and mammary glands.

prokaryotes: Organisms characterized by not having a well defined nucleus or double-membraned organelles.

promoter: The base sequence in DNA that initiates transcription.

prophase I: A stage in the first division of meiosis in which homologous chromosomes form bivalents, cross over, and exchange genetic material.

proprioceptor: A sensory receptor in muscles, tendons, and joints which conveys information about the physical state and position of skeletal muscles and joints.

prospective study: A tool used by medical scientists to measure risk factors associated with a disease; it involves recruiting large numbers of healthy people in a similar age group, obtaining their detailed medical histories and screening them over a long period to identify those who develop a disease and those who do not.

prostaglandins: Modified fatty acids derived from plasma membranes that act as local regulators in the mammalian immune system.

prosthetic group: A non-protein group attached tightly to a protein molecule and cannot be removed easily, e,g. the iron-containing haem group in haemoglobins.

proteases: Enzymes that catalyse the hydrolysis of proteins.

protein carrier: A protein molecule in cell membranes that transports chemicals from one side to another.

proteins: Nitrogen-containing organic molecules made of one or more polypeptide chains.

proto-oncogene: A normal gene that codes for proteins which stimulate cell growth and cell division. Mutations can convert a proto-oncogene into an oncogene.

provirus: Viral DNA that has become integrated into a host's genome and which can be passed on to the host's daughter cells.

pulmonary circulation: The circulation of blood to, through, and from the lungs.

pulmonary vein: Vein carrying oxygenated blood from the lungs to the left atrium of the heart.

pulse: The rhythmic expansion and recoil of the arteries resulting from the wave of pressure produced by contractions of the left ventricle of the heart.

punctuated code: A code that include 'punctuation'; e.g. the genetic code that contains 'start' and 'stop' codons.

pyramid of biomass: A diagram showing the quantitative relationships between the biomass (usually total dry mass) of organisms at each trophic level within an ecosystem. Each trophic level is represented by a rectangle, the length of which is proportional to the biomass of organisms at that level.

pyramid of energy: A diagram showing the quantitative relationships between the energy stored within organisms at each trophic level within an ecosystem. Each trophic level is represented by a rectangle, the length of which is proportional to the energy entering each trophic level over a given time.

pyramid of numbers: A diagram showing the quantitative relationships between the number of organisms at each trophic level within an ecosystem. Each trophic level is represented by a rectangle, the length of which is proportional to the number of organisms at that level.

pyrogens: Chemical substances, such as some bacterial toxins, that raise the set point in a mammal's thermoregulatory system and induce a fever.

pyruvate: A salt of pyruvic acid, a three-carbon compound produced during glycolysis.

Q_{10}**:** The temperature coefficient; a measure of the effect of a 10°C rise in temperature on the rate of a chemical reaction.

Q-PCR (quantitative polymerase chain reaction): A technique which analyses the products of the polymerase chain reaction as they are produced by a technique. This is achieved by using special fluorescent dyes which react with specific base sequences in the amplified product and which can be measured by the Q-PCR instrument. In terms of base sequences, the instrument can tell a researcher both what is present and how much is present, allowing samples from different sources to be compared quantitatively.

quadrat: An area marked out on the ground within which a population is sampled.

quantitative data: Data relating to objective, numerical measurements.

quantitative polymerase chain reaction (Q-PCR): A technique which analyses the products of the polymerase chain reaction as they are produced by a technique. This is achieved by using special fluorescent dyes which react with specific base sequences in the amplified product and which can be measured by the Q-PCR instrument. In terms of base sequences, the instrument can tell a researcher both what is present and how much is present, allowing samples from different sources to be compared quantitatively.

quaternary structure: The fourth level of protein structure which refers to the overall shape of a protein molecule that consists of more than one polypeptide chain.

R plasmids: Small circular pieces of DNA in bacteria that contain genes for antibiotic resistance.

radiation: Transfer of heat as rays, or electromagnetic radiation, especially infrared waves.

radiotherapy: Use of radiation (e.g. X-rays) to treat cancers.

random genetic drift: Random changes in the frequency of alleles within a population; genetic drift tends to be more significant as a population become smaller.

random sampling: Selecting individuals from a population in a manner that ensures that every organism within the population has an equal chance of being selected.

rDNA technology (recombinant DNA technology): A form of genetic engineering in which pieces of DNA from different sources are combined artificially.

realized niche: The area in which a species actually lives.

receptor activation: The binding of a neurotransmitter to specific protein receptor molecules on the postsynaptic membrane (the membrane of the target cell).

receptor cell: A specialized cell sensitive to a particular stimulus.

recessive: Applied to an allele that in diploid organisms is only fully expressed in the homozygous condition; it is masked in the heterozygous condition.

reciprocal cross: In genetic breeding experiments, a pair of crosses in which the phenotypes of the parents are reversed.

reciprocal inhibition: The inhibition of the stretch reflex in antagonistic pairs of muscles; when one of the muscle pair contracts its sends nerve impulses to its opposing muscle causing it to relax.

recognition sequence: A specific sequence of nucleotide bases on DNA which is recognized by a particular restriction endonuclease.

recombinant DNA: DNA that has been made from DNA fragments from different sources.

recombinant DNA technology (rDNA technology): A form of genetic engineering in which pieces of DNA from different sources are combined artificially.

redox reaction: A chemical reaction that involves a simultaneous oxidation and reduction.

reduction: Any reaction involving the removal of oxygen, the addition of hydrogen, or the gain of an electron.

reference point (set point norm): In a homeostatic control mechanism, the level at which the output of a product or factor is supposed to be kept constant

reflex: A rapid, involuntary, unlearned response to a specific stimulus.

reflex action: A rapid, involuntary response to a specific stimulus.

reflex arc: The nerve pathway that brings about a reflex action. A typical reflex arc consists of a receptor, sensory neurone, interneurone, motor neurone, and effector.

refractory period: Brief period after excitation when a neurone or muscle cell becomes at first completely inexcitable (the absolute refractory period) and then less excitable (the relative refractory period).

refutation: The process of proving something wrong.

regulator gene: A gene that regulates the activity of another gene or genes, e.g. by producing a repressor molecule that acts on an operator gene next to a structural gene.

relative refractory period: Brief period after excitation when a neurone or muscle cell becomes less inexcitable.

relay neurone (interneurone, association neurone): A neurone in the central nervous system that links a sensory neurone to a motor neurone.

renal artery: Artery carrying oxygenated blood to the kidneys.

renal vein: Vein carrying deoxygenated blood from the kidneys.

repolarization: The process by which a neurone or muscle cell regains its resting potential after it has discharged an action potential.

representative sampling: Selecting individuals from a population in an objective unbiased manner which ensures that every member of the population has an equal chance of being selected, and which reflects the true characteristics of the population.

repressor molecule: A molecule that inhibits the expression of a gene.

reproductive isolation: The prevention of breeding between individuals due to differences in breeding behaviour, the timing of breeding cycles, or anatomy.

reservoirs: In relation to nutrient cycles, those parts of a cycle where an element is held in large quantities for prolonged periods of time.

residence time: In relation to nutrient cycles, the length of time an element is retained in a reservoir or exchange pool.

resilience: A type of ecological stability which refers to the ability of an ecosystem to return to its original state after being changed.

resistance: A type of ecological stability which refers to the ability of an ecosystem to resist a change following a disturbance.

resolution: In microscopy, the minimum distance that two points can be separated and still be distinguished as two separate points.

resolving power: A measure of a microscope's ability to reveal detailed structure: the minimum distance that two points can be separated and still be distinguished as two separate points.

respiration: The breakdown of organic molecules to release energy in the form of ATP.

resting potential: The potential difference across the plasma membrane of a neurone that is not conducting a nerve impulse, or a muscle cell that is not contracting.

restriction endonuclease: An enzyme that recognizes specific nucleotide base sequences in DNA and cuts up the DNA at where these recognition sequences occur.

restriction mapping: The process of using different specific restriction endonucleases to cut DNA into a series of fragments of different length, which can be separated and their size determined by gel electrophoresis. The fragments can be arranged to show a 'restriction map' of the different regions of DNA, with the distance between each site defined in terms of the number of base pairs.

restriction site: A specific nucleotide sequence on a DNA strand that is recognized by a particular restriction endonuclease as the site at which the DNA strand is to be cut.

retrovirus: An RNA virus which reproduces by synthesizing DNA from its RNA by reverse transcription, and then inserting the DNA into a chromosome of the host.

reverse transcriptase: An enzyme that uses RNA as a template for the synthesis of DNA.

reverse transcription: The synthesis of DNA using RNA as a template.

rhodopsin: A rose-purple photosensitive pigment in rod cells of the retina.

ribose: A five-carbon sugar that, for example, forms part of ribonucleic acid.

ribosome: A very small organelle, 10–20 nm in diameter, involved in protein synthesis.

risk assessment procedure: The process of estimating how harmful a specific activity might be.

risk factor: A factor that increases the probability of a particular disease occurring.

risk management: The use of safe practices so that risks associated with specific activities can be reduced to acceptable levels.

risk: The likelihood that carrying out a specific task might be harmful.

RNA interference: The interference of gene expression at the translation stage.

RNA polymerase: An enzyme that catalyses the transcription of DNA into mRNA.

RNA splicing: The cutting out and removal of introns from pre-mRNA by spliceosomes and the subsequent joining together of the exposed ends of adjacent exons to form the functional mRNA.

root cap: A thin layer of cells at the tip of the root of a plant that protects the growth region behind it.

root hair cells (root hairs): Fine, thin-walled cells with tube-shaped extensions going it into the soil.

root meristem: The region of growth, just behind the root tip, in which cells divide by mitosis.

root pressure: A pressure in roots that pushes water up the xylem to shoots.

roots: Structures that anchor plants in the soil and which are responsible for the absorption of water and mineral salts.

running mean: When taking a series of quadrat samples to estimate population density, an arithmetic mean calculated after each new quadrat is sampled.

saliva: A watery fluid produced by glands in the buccal cavity.

salivary amylase: An enzyme in the saliva of humans that digests starch by hydrolysis.

Salmonella: A group of bacteria that cause food poisoning.

salmonellosis: An intestinal disease, one of the most common forms of food poisoning, caused by *Salmonella* bacteria.

saltatory conduction: The means by which a nerve impulse is transmitted rapidly along myelinated neurones. The wave of depolarization appears to leap from one node of Ranvier to the next.

sap: Watery liquid contained within the vascular tissues of plants and vacuoles of cells.

saprobiotic nutrition: Nutrition carried out by organisms which absorb organic materials in solution from their environment, with the organic material originating from dead organisms.

sarcolemma: The plasma membrane of a muscle cell.

sarcomere: The functional unit of a muscle. It contains filaments made of the contractile proteins actin and myosin. A single sarcomere extends from one Z line to the next.

sarcoplasm: The name given to the cytoplasm in a muscle fibre.

sarcoplasmic reticulum: The name given to the endoplasmic reticulum in a muscle fibre.

saturated fatty acids: Fatty acids that have only single covalent bonds between carbon atoms in the hydrocarbon chain.

Schwann cell: A cell that forms the myelin sheath encapsulating vertebrate neurones in the peripheral nervous system.

scientific method: A systematic way of studying natural phenomena characterized by objectivity and gathering evidence for and against a particular viewpoint. There is no one scientific method, but a typical scientific method involves making observations, formulating a hypothesis, and testing the hypothesis by carrying out experiments from which conclusions are made.

sclerenchyma: Structural plant tissue comprising two types of cell: elongated fibres and roughly spherical sclereids (stone cells). In both, the cell wall is thickened with lignin.

second messenger: A small soluble non-protein molecule such as cyclic AMP that relays a signal to the cell's interior in response to a signal received by a signal receptor protein on the plasma membrane.

secondary consumer: An organism that feeds on a primary consumer.

secondary lactase deficiency: Lactase deficiency which occurs when a person cannot produce enough lactase because of damage to the small intestine through injury or a disease such as coeliac disease, inflammatory bowel disease, or Crohn's disease.

secondary metabolites: Chemicals produced by an organism but which are not essential for the growth or reproduction of the organism that produces them (e.g. antibiotics produced by microorganisms).

secondary productivity: The rate at which energy or organic matter is built up into the tissues of a consumer.

secondary receptor: A receptor that consists of a modified epithelial cell linked to a sensory neurone.

secondary structure: The second level of protein structure which refers to the shape of a polypeptide chain or part of a chain.

secondary succession: Ecological succession that takes place in areas such as a ploughed field where life is already present but which has been disturbed in some way.

sedentary animal: An animal, such as a sea anemone, which is able to move from one place to another, but with very slow or restricted movement.

selection pressures (environmental resistances): Environmental factors that limit the growth of populations.

selective advantage: Members of a species are said to have a selective advantage if they have one or more characteristics that make them better adapted to an environment than other members of the species.

selective permeability: A property of cell membranes through which some materials can pass but not others.

semiconservative replication: Type of DNA replication in which the double strand of DNA unwinds to produce new DNA with one strand from the parent molecule and one new strand.

semilunar valves: Half-moon-shaped valves in the pulmonary artery and aorta that prevent the backflow of blood to the heart.

sense organ: A structure in a multicellular organism that contains receptors that are particularly sensitive to specific stimuli.

sensitivity: Ability to detect changes in the environment.

sensory adaptation: A decline in the generator potential when a receptor is stimulated continuously with a stimulus of constant intensity. It results in a decrease in frequency of the nerve impulses transmitted by the sensory neurone.

sensory neurone: A neurone that carries information from sensory receptors to the central nervous system.

septum: Tissue separating two cavities or masses of tissue, like the one separating the left and right sides of the heart.

seral stage (sere): A stage in ecological succession which is characterized by a particular community of organisms.

sere (seral stage): A stage in ecological succession which is characterized by a particular community of organisms.

serum: Blood plasma with the protein fibrinogen removed so that it does not clot when stored.

sessile animal: An animal such as a barnacle that is fixed to its substrate and unable to move itself from one place to another.

set point (norm, reference point): In a homeostatic control mechanism, the level at which the output of a product or factor is supposed to be kept constant.

Sewall Wright effect (genetic drift): Random changes in the frequency of alleles within a population; genetic drift tends to be more significant as a population becomes smaller.

sex chromosomes: In humans, the X and Y chromosomes that determine the sex of an individual.

sex linked: Applied to a characteristic such as haemophilia that has the tendency to occur more frequently in one sex than the other because the gene responsible for it is carried on a sex chromosome.

sexual reproduction: Reproduction involving the fusion of haploid gametes to produce a diploid cell (the zygote).

shade plant: A plant adapted to photosynthesizing in low light intensities.

sickle cell anaemia: An inherited blood disorder caused by a mutation that results in the change of one amino acid in the beta chains of haemoglobin. The disease is characterized by red blood cells that are sickle shaped, less efficient at carrying oxygen, and are removed more rapidly than normally from the circulation.

sign stimulus: A stimulus that induces a species-specific behaviour.

silent mutation: A mutation that does not cause a change in gene expression and does not therefore affect the phenotype (e.g. a change from one codon to another which codes for the same amino acid).

simple proteins: Proteins that consist of amino acid monomers only.

single-factor disease: A disease caused by one factor.

sinoatrial node: The main pacemaker region in the right atrium of the heart.

siRNA (small interfering RNA): Double-stranded RNA that plays a key role in inhibiting gene expression at the translation stage. One strand called the guide strand combines with mRNA by complementary base pairing and interferes with gene expression.

sister chromatids: Two copies of a chromosome, visible during mitosis and meiosis, joined together at the centromere.

skin cancer: Malignant growth of cells in the skin.

sliding filament theory: A theory that explains how muscles contract by actin and myosin filaments being interconnected by cross-bridges, and the two types of filament sliding past each other by means of a ratchet-like mechanism.

slow-twitch fibre: A type of muscle fibre characterized by a relatively slow contraction time, high aerobic capacity, and a rich supply of blood causing the fibre to appear red. It fatigues slowly.

small interfering RNA (siRNA): Double-stranded RNA that plays a key role in inhibiting gene expression at the translation stage.

sodium-potassium pump: A metabolic energy-consuming pump involved in the active transport of sodium and potassium across cell membranes.

solute: The substance dissolved in a solvent.

somatic nervous system: That part of the peripheral nervous system which includes the sensory neurones that transmit nerve impulses to the central nervous system, and motor neurones which send impulses to skeletal muscle.

somatic-line gene therapy: The treatment of a genetic disorder by altering the genetic composition of body cells only; the genetic composition of sperm or egg cells, or of germinative cells that give rise to the gametes, is not altered.

Southern blotting: A technique in which DNA fragments separated by gel electrophoresis are transferred onto a thin nylon membrane over which several sheets of blotting paper or filter paper are laid.

spatial summation: A phenomenon occurring on a postsynaptic membrane in which the membrane potential results from the sum of all the excitatory and inhibitory effects of the different neurotransmitters released at any one time from the presynaptic membranes to which it is connected.

Spearman's rank correlation: A statistical test that uses a ranking system to assess the degree of correlation existing between two sets of data. The two sets of data are placed in rank order next to each other so that they can be compared statistically.

speciation: The process by which two or more species evolve from one original species.

species: A group of organisms sharing similar features which are reproductively isolated from other species and which can interbreed to produce fertile offspring.

species diversity: The variety of life in an area. In ecology, it is usually expressed as an index that takes into account both the number of individuals and the number of species.

species diversity index: A measure of the variety of life in an area that takes into account both the number of individuals in each species and the number of species.

species richness: A measure of the variety of life based on the number of species in an area.

specific immune system: Part of the immune system responsible for defending against particular types of foreign agents.

sphincters: Circular muscles that close or contract an opening.

sphygmomanometer: A device for measuring blood pressure.

spiracles: 1 In insects, openings in the cuticle that allow air to enter trachea. 2 In fish, openings through which water is drawn for gaseous exchange in the gills.

spleen: An important organ in the immune system that stores red blood cells and destroys old cells.

spliceosomes: Intracellular structures formed of RNA and protein molecules which carry out RNA splicing.

spores: Small, usually single-celled, reproductive bodies from which new organisms arise when conditions are suitable.

squamous epithelium: Flattened epithelial cells.

stabilizing selection: A form of natural selection which tends to select against phenotypes of both extremes of the distribution and which therefore tends to select for the average phenotype.

stalked particles: Structures on the inner membrane of mitochondria through which protons can move.

standard error: The standard deviation divided by the square root of the sample size how reliable; a measure of how reliable is a sample as an estimate of the true mean of the whole population.

standing crop: The amount of dry mass of a particular group of organisms (e.g. a species within a community) measured at any one time.

starch: A polysaccharide consisting of alpha glucose monomers found in plants.

stem cell: An undifferentiated cell in an embryo or adult that gives rise to one or more specialized cells.

stems: Main body axis of plants, most of which are erect and above ground and which bear leaves.

stereotyped response: A behavioural response to a particular stimulus that is always similar.

sticky ends: The ends of a fragment of double-stranded DNA that have been cut at sites which are offset from each other.

stimulus: A detectable change in the external or internal environment.

stomata (sing. stoma): Small pores in leaves (and sometimes stems) of terrestrial plants through which gas and water vapour are exchanged.

stomatal aperture: The opening between two adjacent guard cells that forms a stoma.

stratified distribution: A distribution pattern in which individuals are uniformly distributed in different sub-areas of a habitat.

stratified random sampling: A procedure for sampling populations in biology fieldwork in which the area to be studied is divided into sub-areas within which random sampling is carried out in each.

stress: A psychological condition occurring when a person perceives an imbalance between the demands being made of him or her and the ability to meet those demands.

stretch reflex: A reflex contraction of a muscle in response to its being stretched.

stroke: An interruption in the supply of blood to the brain.

stroke volume: The volume of blood ejected from the left ventricle during one heartbeat.

stroma: The gel-like matrix of chloroplasts which contains enzymes required for photosynthesis and DNA.

structural gene: A gene that encodes for a polypeptide or enzyme required for an organism's structure or metabolism.

structure: In anatomy, the gross external parts of biological material that are visible to the naked eye and optical microscope.

strychnine: A drug that acts as a stimulant in small doses but which in high doses can cause severe muscle spasms which can be fatal.

suberin: Waxy waterproof material found, for example, in cork.

substitution: A type of gene mutation in which one type of nucleotide base is replaced by another.

substomatal air space: The air space just behind stomata in the mesophyll of a leaf.

substrate: In relation to enzyme-catalysed reactions, the reactant being catalysed.

substrate level phosphorylation: A reaction in which the formation of ATP by the addition of a phosphate group to ADP is coupled with the exergonic breakdown of a high-energy substrate molecule.

succession (ecological succession): The sequence of change within an ecosystem from initial colonization to a relatively stable climax community.

sucrose: A disaccharide made of glucose and fructose; a non-reducing sugar.

sugar: A sweet-tasting, crystalline, soluble monosaccharide or disaccharide.

sugar-phosphate backbone: The interconnecting sequence of sugar-phosphate groups that make up the main axis of the polynucleotides DNA and RNA.

sun plant: A plant adapted to photosynthesizing in high light intensities.

supernatant: The clear fluid lying above a precipitate after a suspension has been centrifuged or ultracentrifuged.

supernormal stimulus: A stimulus that triggers a greater than normal response.

suppressor T cell: A white blood cell involved in immune responses.

survival curve: For a given population, the line on a graph showing the percentage of individuals born at about the same time that are still living at different ages.

survival of the fittest: A phrase associated with Darwin's theory of evolution in which individuals best adapted to an environment have the greatest chances of surviving long enough to produce viable offspring while those less well adapted will tend not to survive.

sympathetic nerve: A nerve of the sympathetic nervous system.

sympathetic nervous system: Part of the autonomic nervous system that helps to prepare the body for physical activity, e.g. create internal body conditions found during rest and sleep.

symplast route: Relates to the direction water takes when it goes through the cytoplasm of adjacent root cells.

synapse: The junction between one neurone and another.

synaptic bulb (axon terminal, synaptic knob): A swelling at the tip of an axon containing mitochondria and vesicles with neurotransmitters.

synaptic cleft: A narrow gap separating the membrane of an axon terminal from the membrane of a target cell.

synaptic knob (synaptic bulb, axon terminal): A swelling at the tip of an axon containing mitochondria and vesicles with neurotransmitters.

synaptic vesicle: A sac-like organelle in the synaptic bulb of a neurone in which neurotransmitter is stored.

synports: Transporters involved in a form of cotransport in which two substances are transported together in the same direction across cell membranes.

systematic sampling: A procedure for taking population samples in biological fieldwork in which the first sampling point is chosen at random but all others are taken at fixed points from this.

systemic circulation: In mammals, blood circulation to all parts of the body other than the lungs.

systemic pesticide: A pesticide that is taken into a plant, e.g. through its leaves or roots, and is translocated throughout its body.

systemic veins: The veins of the body, excluding the pulmonary vein.

systole: The phase of the cardiac cycle in which the heart contracts and pumps blood from one part to another.

T cells: White blood cells that mature in the thymus.

T tubules: Membranous tubules formed by extensions of the sarcolemma that allow an electrical impulse and substances carried in the extracellular fluid (e.g. glucose, oxygen, and ions) to reach the deep regions of a muscle fibre.

taproot: The long, straight main root of a vascular plant.

target cell: The end-cell upon which a hormone or a neurone has its effect.

taxis (pl. taxes): A movement of a whole organism towards or away from a stimulus.

taxon (pl. taxa): In taxonomy, a grouping of organisms that share certain features. There is a hierarchical series of taxa; in most classification systems the largest is Kingdom and the smallest species.

TDI (tolerable daily intake): The amount of a potentially toxic substance that can be ingested daily over a lifetime without any appreciable health risk to the most vulnerable in a population.

telomeres: Structures at the ends of chromosomes that have a protective function.

temperature coefficient (Q_{10}): A measure of the effect of a 10°C rise in temperature on the rate of a chemical reaction.

template: The molecular pattern of a structure that serves as a pattern for the production of another compound.

temporal summation: The effect of repeated stimulation of one presynaptic neurone on the membrane potential of a postsynaptic membrane. Each stimulus causes the release of neurotransmitters from the presynaptic membrane which produces a graded potential in the postsynaptic membrane. The graded potentials evoked by a rapid succession of two or more stimuli are added together until a threshold is reached and an action potential occurs in the postsynaptic membrane.

tendinous cords: Tendons connecting heart valves to papillary muscles.

tendon: Tissue connecting a muscle to a bone.

tension: A stretching or pulling force.

terminator: The base sequence in DNA that terminates transcription.

territory: In animals, an area defended for feeding or reproduction.

tertiary consumer: An organism that feeds on a secondary consumer.

tertiary structure: The third level of structure of proteins which refers to the overall shape of a whole polypeptide chain.

tertracyclne: An antibiotic which binds to bacterial ribosomes, interfering with translation.

tetanus: A smooth sustained contraction of a muscle motor unit resulting from the fusion of a number of twitches.

theoretical science: Science based on logical reasoning from first principles and not dependent on observation and experimentation.

thermodynamics: The study of the relationship between usable energy (free energy), heat, and mechanical work.

thermoregulation: The control of body temperature by either physiological or behavioural means.

thin filament: One of two types of protein filament in a muscle sarcomere; it consists of actin, troponin, and tropomyosin.

thrombus: A blood clot that remains at its point of formation.

thylakoid: A flattened, fluid-filled sac within a chloroplast. The sac is lined with a membrane that is impregnated with chlorophyll.

thymus: A gland in the neck in which certain white blood cells (T cells) mature.

tissue: A group of similar cells and the intercellular substances associated with them that work together to carry out one or more functions in an organism.

tissue culture: The culture outside of an organism (e.g. in test tubes or on Petri dishes) of plant or animal tissue.

tissue fluid: The fluid bathing cells inside an organism.

tolerable daily intake (TDI): The amount of a potentially toxic substance that can be ingested daily over a lifetime without any appreciable health risk to the most vulnerable in a population.

topographic factor: A factor related to the surface features of a substrate (e.g. the land surface or sea bed).

totipotency: The ability of a cell to generate any of the specialized cells of a mature organism.

totipotent: Applied to a cell that has the ability to generate any of the specialized cells of a mature organism.

trachea: 1 In mammals, the tubular structure that leads from the buccal cavity and pharynx into the bronchi and which allows the flow of air to and from the lungs. 2 In insects, the air tubes that lead from the external surface to tracheoles allowing the flow of air to and from respiring tissue.

tracheal system: The system of air tubes in an insect.

tracheoles: The small blind-ended air tubes that lead from tracheae to respiring tissue in insects.

transcription: The process by which the genetic information in DNA directs the synthesis of mRNA.

transcriptional factor: A regulatory protein that binds to DNA and stimulates transcription of a specific gene.

transducer: A structure that changes energy from one form to another.

transduction: The process of converting one form of energy to another (e.g. the transformation by a retinal cell of light energy to the electrochemical energy of a nerve impulse).

transect sampling: In fieldwork, sampling using a line transect, belt transect, continuous transect, or interrupted transect.

trans-fatty acids: Unsaturated fatty acids in which the two hydrogen atoms adjacent to a double bond are on opposite sides of the molecule. This allows trans-fatty acid molecules to pack tightly together, consequently many are solids at room temperature.

transfer RNA (tRNA): One of the single-stranded cloverleaf-shaped RNA molecules that carries a specific amino acid to a ribosome for polypeptide synthesis in the translation of the genetic information in mRNA.

translation: The process by which information the genetic information in mRNA directs the synthesis of a polypeptide chain.

transpiration stream: The upward-moving column of water in a plant from root cells next to the soil to cells just beneath stomata.

triacylglycerols (triglycerides): Fats and oils that are made of glycerol attached to three fatty acids by ester bonds.

tricuspid valve: The valve between the right atrium and right ventricle that prevents backflow of blood to the atrium.

triglycerides: Fats and oils that are made of glycerol attached to three fatty acids by ester bonds.

triplet: A sequence of three nucleotide bases that make up a codon in the genetic code.

triplet code: A code that is based on three items, such as the genetic code which is based on non-overlapping sequences of three bases.

tRNA (transfer RNA): One of the single-stranded cloverleaf-shaped RNA molecules that carries a specific amino acid to a ribosome for polypeptide synthesis in the translation of the genetic information in mRNA.

trophic efficiency: The percentage of energy that flows from one trophic level to the next.

trophic level: A feeding level in an ecosystem consisting of organisms of a similar nutritional type. Producers form the first trophic level, primary consumers the second, and so on.

tropism: A movement (usually resulting from growth) of part of plant towards or away from a directional stimulus.

tropomyosin: A tube-shaped protein associated with thin filaments in a muscle fibre that, in the absence of calcium ions, prevents myosin heads from attaching onto actin.

troponin: A protein associated with thin filaments in a muscle fibre which, in the presence of calcium ions, causes tropomyosin to change shape and allow myosin heads to attach to actin.

tubercle: A nodule or small rounded swelling.

tuberculosis: A disease of characterized by nodules, especially in the lungs; caused by the bacterium *Myobacterium tuberculosis*.

tumour: A growth caused by an abnormal proliferation of cells.

tumour suppressor gene: A gene that inhibits cell division.

turgid: Applied to cells that are fully expanded with water.

turgor pressure: In plants, the pressure of water in the vacuole and cytoplasm pressing against the cell wall.

twin method: The use of identical twins to study the effects of inheritance and environmental factors an individual trait, e.g. susceptibility to disease.

twitch: A single brief contraction of a muscle motor unit in response to a stimulus.

type I diabetes: A form of diabetes mellitus that usually occurs suddenly in childhood; an autoimmune disease in which the immune system attacks the beta cells, destroying a person's ability to secrete insulin.

type II diabetes: A form of diabetes mellitus that usually occurs in adulthood. It may be caused by a gradual loss in the responsiveness of cells to insulin or to an insulin deficiency.

ultracentrifugation: A technique for separating cell organelles in suspension by spinning the suspension in a tube at various speeds. Each speed causes the sedimentation of particular organelles according to their density.

ultrafiltration: The process by which small molecules and ions are separated as they pass through the basement membrane of capillaries to produce tissue fluid.

ultrastructure: The fine, detailed structure of a material, such as a cell, as revealed by the resolving power of electron microscopes at high magnification.

uniform distribution: A distribution pattern in which individuals are evenly spaced apart.

unit membrane model: A model of cell membrane structure which suggested that all membranes have the same basic structure of a bilipid layer covered on the inside and outside by a protein layer.

universal code: Applied to the genetic code as it is more or less the same for all organisms.

unlearned behaviour: Behaviour that is inherited and is not learned.

unloading p(O_2): The partial pressure of oxygen at which haemoglobin becomes half saturate (50% saturation).

unmyelinated neurone: A neurone that lacks a myelin sheath.

unsaturated fatty acids: Fatty acids that have one or more double covalent bonds between carbon atoms in the hydrocarbon chain.

utilitarian: In relation to nature conservation, refers to conserving biological resources for their usefulness or economic value.

vaccination: The administration of a vaccine to confer immunity.

vacuolar route: Relates to the direction water takes when it goes through the vacuoles of adjacent root cells.

vacuole: A fluid-filled, membrane-bound sac within a cell. A plant vacuole contains sap.

vagus nerve: A nerve of the sympathetic nervous system that leads from the medulla oblongata (the hind brain) to the heart, lungs, and part of the alimentary canal.

variance to mean ratio: A statistic used to measure population dispersion.

vascular bundle: Transport system of plant tissue containing phloem and xylem that runs through a plant from the roots and up the stem and into leaves.

vasoconstriction: A narrowing of blood vessels.

vasodilation: A widening of blood vessels.

veins: Blood vessels that carry blood to the heart.

venae cavae: The two major veins that lead directly into the right atrium of the heart.

venous return: The volume of blood that flows from the veins into the right atrium of the heart.

ventilation: Rhythmic movements by which a respiratory medium (air or water) is pumped into and out of a respiratory organ.

ventilation rate: The volume of fluid passing across a respiratory organ per minute; in mammals, the product of tidal volume and number of breaths per minute.

ventricle: One of a pair of chambers in the heart; the right ventricle pumps blood to the lungs, and the left ventricle pumps blood to the rest of the body.

venules: Small vessels that convey blood from capillaries to veins.

vertical gene transmission: Transfer of genetic information, e.g. antibiotic resistance in bacteria, from one generation to the next.

Vibrio cholera: The bacterium responsible for cholera.

villi (sing. villus): A finger-like projection from the surface of the small intestine.

visual acuity: The ability to distinguish fine details in an image.

voluntary nervous system: Part of the nervous system that is under conscious control.

warm blooded: Name commonly given to endotherms (birds and mammals) that can keep their body core temperature at a level higher than that of the environment.

water potential: A measure of the pressure exerted by water molecules, e.g. on a membrane.

X linkage: Applied to a characteristic determined by gene that is carried on the X chromosome and which therefore tends to be expressed in males more than females.

xeromorphic: Describes structural features that adapt plants to dry conditions.

xerophytes: Plants adapted to conditions of water scarcity.

xylem: Woody plant tissue consisting of a number of different cell types that provide mechanical support and transport water.

Z line: A protein band that defines the boundary between one sarcomere and the next in a muscle fibre.

zwitterions: An ion carrying both a positive and a negative charge which cancel each other out and which is therefore electrically neutral.

zygote: The diploid cell formed by the fusion of a haploid sperm cell and a haploid egg cell.

Index